BIOACTIVE FOOD PROTEINS AND PEPTIDES

APPLICATIONS IN HUMAN HEALTH

BIOACTIVE FOOD PROTEINS AND PEPTIDES

APPLICATIONS IN HUMAN HEALTH

Edited by
Navam S. Hettiarachchy

Associate Editors
Kenji Sato • Maurice R. Marshall • Arvind Kannan

CRC Press
Taylor & Francis Group
Boca Raton London New York

CRC Press is an imprint of the
Taylor & Francis Group, an **informa** business

CRC Press
Taylor & Francis Group
6000 Broken Sound Parkway NW, Suite 300
Boca Raton, FL 33487-2742

First issued in paperback 2016

© 2012 by Taylor & Francis Group, LLC
CRC Press is an imprint of Taylor & Francis Group, an Informa business

No claim to original U.S. Government works

ISBN 13: 978-1-138-19899-9 (pbk)
ISBN 13: 978-1-4200-9314-8 (hbk)

Visit the Taylor & Francis Web site at
http://www.taylorandfrancis.com

and the CRC Press Web site at
http://www.crcpress.com

Contents

Preface

This book features sections that are in a continuum with our earlier edition, *Food Proteins and Peptides: Chemistry, Functionality, Interaction, and Commercialization*, an attempt to comprehensively cover food proteins' and peptides' bioactivities, proteomics, biomarkers, therapeutic and nutraceutical uses, commercialization trends, and challenges and opportunities. The sections that are covered in this book provide a human health perspective of food-derived proteins and peptides; potentials for large-scale production with advances in technology; and challenges and opportunities for the future of health, nutritional, medical, and biosciences.

The first section addresses chemistry and bioactivity. It looks at proteins and peptides as allergens, antihypertensive agents, antimicrobials, antioxidants, and anticancer agents. Findings on the bioavailability and toxicity of food-derived peptide and intestinal functions are also included. The second section involves chapters on therapeutic peptides, including recent developments in proteomics, bioavailability, and opportunities for designing future peptide-based foods. The book is intended to provide a comprehensive review of bioactive proteins and peptides obtained from food sources. A most distinctive feature of this book is the selection of chapters contributed by eminent researchers from all over the world.

About the Editors

Dr. Navam Hettiarachchy, an IFT fellow, is a university professor in the Department of Food Science and is also associated with the Institute of Food Science and Engineering at the University of Arkansas, Fayetteville. She earned her BS in chemistry from Madras University, and MS and PhD in biochemistry at the University of Edinburgh Medical School, Edinburgh, Scotland, and the University of Hull, England, respectively. She also earned a postgraduate diploma in nutrition from the Indian Council of Medical Research, Hyderabad, India. She has served as a professor in three major universities in the United States. Prior to those appointments, she was a faculty member in the Department of Biochemistry in a major medical school, and her research was focused on nutrition in health and disease. Her current research program focuses on an integrated approach to protein chemistry and biopeptides, value-added nutraceuticals, product development and food safety. The spin-off of one of Dr. Hettiarachchy's research projects resulted in the formation of a new company, Nutraceutical Innovations LLC, which is involved in utilizing rice bran with a total system approach in generating several nutraceutical and functional food ingredients. Her research program has pioneered anticancer research in rice and soybean peptides, and demonstrated potent anticancer activities in various types of cancer cell lines.

Professor Maurice Marshall is a faculty member in the Food Science and Human Nutrition Department, University of Florida, where he has taught and researched food chemistry/biochemistry for 30 years. He received his BS in biology from Duquesne University, his MS from Rutgers University, and his PhD from The Ohio State University both in the area of food science, specializing in food chemistry/biochemistry. His research focuses on analysis of food constituents including biogenic amines, antioxidants, and pesticide residues. He is director of the Southern Region USDA-IR-4 Project to assist specialty crop growers in obtaining registered low-risk crop protection tools. His additional research focuses on food enzymes, specifically polyphenol oxidases in fruits and vegetables and proteases from stomachless marine organisms, as well as unique browning inhibitors from insects, mussels, and crustaceans. He also works on the hydrolysis of waste seafood proteins for bioactive peptide generation and their action as antioxidants, antidiabetic, anticancer, and high blood pressure regulation.

Professor Kenji Sato has been a faculty member for 22 years at the Division of Applied Life Sciences, Graduate School of Life and Environmental Sciences, Kyoto Prefectural University in Kyoto City, Japan, where he is a professor. He obtained his undergraduate degree and a PhD in agriculture from Kyoto University. After spending a year at Kyoto University as a post-doctoral fellow he joined the faculty at Kyoto Prefectural University. His current research interest focuses on bioactive peptides, especially the metabolic fate of these peptides. His lab conducts research on food-derived peptides to demonstrate their absorption, metabolic fate, and efficacy, and to elucidate their molecular mechanisms of action.

Dr. Arvind Kannan is currently working as a postdoctoral fellow in the Food Science Department at the University of Arkansas. He received his BS in biochemistry from Madras University, India, and a MS in medical laboratory technology from Birla Institute of Technology & Science (BITS), India. He then worked as a research associate with the Madras Diabetes Research Foundation, India, studying markers of coronary artery disease and diabetes before getting overseas experience as a research associate at the National Tsing-Hua University in Taiwan. He worked on bacterial gene and protein expression and biophysical experiments related to protein structure in the structural proteomics program of Professor Chin Yu. Dr. Kannan obtained his PhD in cell and molecular biology from the University of Arkansas in 2009. His research was on characterizing a bioactive peptide from rice bran for cancer antiproliferative activity using human cancer cell lines. He is currently furthering studies on the bioactive peptide, and is involved in projects dealing with bioactive proteins and peptides. He also serves as a scientist at Nutraceutical Innovations LLC, a start-up firm that focuses on fermentation of rice bran to generate bioactives for value-added functional and nutraceutical uses.

Contributors

Amaya Aleixandre
Department of Pharmacology
Faculty of Medicine
Universidad Complutense
Madrid, Spain

Rotimi E. Aluko
Department of Human Nutritional
  Sciences
The Richardson Centre for
  Functional Foods and
  Nutraceuticals
University of Manitoba
Winnipeg, Manitoba, Canada

Dominique Bouglé
Service de Pédiatrie CH de Bayeux
Bayeux, France

Saïd Bouhallab
INRA
Rennes, France

A. Wesley Burks
Duke University Medical Center
Durham, North Carolina

Wasaporn Chanput
Kasetsart University
Bangkok, Thailand

Ben O. de Lumen
Department of Nutritional Sciences
  and Toxicology
University of California
Berkeley, Califonia

Blanca Hernández-Ledesma
Food Research Institute (CIAL,
  CSIC-UAM)
Universidad Autónoma de Madrid
Madrid, Spain

Navam Hettiarachchy
Department of Food Science
University of Arkansas
Fayetteville, Arkansas

Yasumi Horimoto
University of Guelph
Guelph, Ontario, Canada

Chia-Chien Hsieh
Department of Nutritional Sciences
  and Toxicology
University of California
Berkeley, California

Aruna S. Jaiswal
Department of Anatomy and Cell
  Biology
UF Shands Cancer Center
University of Florida
Gainesville, Florida

Arvind Kannan
Department of Food Science
University of Arkansas
Fayetteville, Arkansas

Hiroshi Kawakami
Department of Food Science and
  Nutrition
Kyoritsu Women's University
Tokyo, Japan

Se-Kwon Kim
Department of Chemistry
Pukyong National University
Busan, South Korea

Hannu J. Korhonen
MTT Agrifood Research Finland
Jokioinen, Finland

Masashi Kuwahata
Department of Nutrition Science
Graduate School of Life and
    Environmental Sciences
Kyoto Prefectural University
Kyoto, Japan

Maurice Marshall
Food Science and Human Nutrition
    Department
University of Florida
Gainesville, Florida

Yasuki Matsumura
Division of Agronomy and
    Horticultural Science
Graduate School of Agriculture
Kyoto University
Kyoto, Japan

Marta Miguel
Department of Pharmacology
Faculty of Medicine
Universidad Complutense
Madrid, Spain

Shuryo Nakai
Professor Emeritus
University of British Columbia
Vancouver, British Columbia
Canada

Yasushi Nakamura
Division of Applied Life Sciences
Graduate School of Life and
    Environmental Sciences
Kyoto Prefectural University
Kyoto, Japan

Satya Narayan
Department of Anatomy and Cell
    Biology
UF Shands Cancer Center
University of Florida
Gainesville, Florida

Harekrushna Panda
Department of Anatomy and Cell
    Biology
UF Shands Cancer Center
University of Florida
Gainesville, Florida

Eun Young Park
Division of Applied Life Sciences
Graduate School of Life and
    Environmental Sciences
Kyoto Prefectural University,
Kyoto, Japan

Susanna Rokka
MTT Agrifood Research Finland
Jokioinen, Finland

Kenji Sato
Division of Applied Life Sciences
Graduate School of Life and
    Environmental Sciences
Kyoto Prefectural University
Kyoto, Japan

Soichi Tanabe
Hiroshima University
Hiroshima, Japan

Chibuike C. Udenigwe
Department of Human Nutritional
    Sciences
University of Manitoba
Winnipeg, Manitoba, Canada

Isuru Wijesekara
Department of Chemistry
Pukyong National University
Busan, South Korea

Jianping Wu
Department of Agricultural, Food,
    and Nutritional Science
University of Alberta
Edmonton, Alberta, Canada

Linping Wu
China Agriculture University
Beijing, China

Hironori Yamamoto
Department of Clinical Nutrition
Institute of Health Biosciences
University of Tokushima Graduate
    School
Tokushima, Japan

1

Food Proteins and Peptides as Bioactive Agents

Arvind Kannan, Navam Hettiarachchy, and Maurice Marshall

CONTENTS

1.1 Introduction

This chapter covers bioactive proteins and peptides emerging as health functional foods, and their ability to arrest disease propagation. Many naturally occurring compounds from foods such as rice, peas, vegetables, fruits, and so on have been found to possess properties that help to slow disease progression, inhibit pathophysiological mechanisms, or suppress activities of pathogenic molecules. Proteins and peptides play significant roles in such activities (Hartmann and Meisel, 2007). They are thus gaining importance as nutraceuticals that benefit aspects of health and nutrition. Moreover, certain diseases and disorders that do not have convincing treatment strategies or 100% cure can benefit from the proteins and peptides naturally present in many foods that possess antidisease characteristics. These antidisease effects can be classified into antiproliferative, antimutagenic, anti-inflammatory, anticancerous, or antioxidative properties that are manifested in many diseases including cancer, diabetes, and inflammatory disorders (Mine and Shahidi, 2006). This chapter provides a comprehensive review of the chief bioactive proteins and peptides determined as antidisease agents.

With the emergence of new detection platforms these diseases have become increasingly prevalent in the last decade. Such diseases have vicious cascades involving interplay of several molecules that not only act on their own but also trigger their partners or adjoining cells and molecules and cause damage to tissue. Many drugs have been discovered and modeled to reduce the ill effects of certain pathogenic molecules participating in these disease processes. Unfortunately they cannot be administered to the masses on a large scale as is a vaccine because they are expensive, and moreover they only delay the disease progression; they do not completely cure the disease. A cheaper alternative is to identify foods or food constituents having functional properties so that they can be consumed as a prescribed or a surrogate to expensive drugs on a large scale if proven therapeutic. Thus, evaluating compounds present in foods for biological activities such as anticancer and antimutagenic among others, can prove effective as they not only can be recommended for subjects who are already on treatment for cancer but also as a preventive measure for those genetically predisposed to diseases.

Proteins are made up of 20 different amino acids arranged diversely into several folds or conformations giving each protein its own unique structure, and hence function. Amino acids that constitute the proteins can be cleaved at specific sites on the proteins to yield peptides of different sizes. Bioactive peptides are considered fragments of proteins that, upon digestion using specific proteolytic enzymes or fermentation, impart positive functions or benefits that influence human health. They normally remain dormant until they are acted upon by specific proteases. Gastrointestinal proteolytic enzymes release the peptides, and the small fraction released and absorbed is sufficient for imparting biological functions. This chapter covers several aspects of bioactive proteins and peptides including preparation, processing effects, structurally functional attributes to impart bioactivity, evaluation of biological activities and their significance, the need for evaluating bioavailability, and finally emergence of biopeptides as novel health-enhancing foods on a commercial scale.

This chapter provides a comprehensive list of food proteins, mainly peptides that serve as modulators in human health and nutrition and also features current trends in technology, analyses, and characterization of food peptides for biological activity.

1.2 Proteins and Peptides as Nutraceuticals in Health and Disease

1.2.1 Preparation of Proteins and Peptides for Bioactivity: Physical and Chemical Processing

The methods for studying bioactive peptides are varied, although many rely on *in vitro* methods for demonstrating biological activity. Generation and identification of bioactive peptides have been performed in a number of ways. Peptides have been produced *in vitro* by hydrolysis methods using digestive enzymes or have been isolated and characterized (Shahidi and Zong, 2008). Often, peptides have been synthesized and used for *in vitro* studies. In some cases, the peptides have been identified based on comparison of sequences with those of known biological activity (e.g., opioid peptides; Pihlanto-Leppala, 2000). Other methods rely on food-processing techniques, such as using heat, pH changes, or the ability of microbial enzymes to hydrolyze proteins (e.g., during fermentation; Rutherfurd-Markwick and Moughan, 2005).

Generation of bioactive peptides during physical treatments involved in food processing stages can also cause structural and chemical changes with potential detrimental effects to the proteins and bioactive peptides. Such effects can prevent their release due to the formation of hydrolysis-resistant covalent bonds. Meisel and FitzGerald (2003) have shown the effects of dephosphorylation to be attributed to subsequent loss of bioactivity such as mineral-binding capacity.

Food processing can also damage proteins to such an extent as to render the bioactive peptides either inactive following digestion or prevent them from being released from the parent protein (Korhonen and Pihlanto, 2003). Damaged proteins are frequently digested in a different manner, as resistant peptide bonds can be generated from heat or alkali treatments. Hence, peptides that would not normally occur naturally may be generated (Rutherfurd-Markwick et al., 2005).

Technologies involving food processing such as using heat, pH changes, or the ability of microbial enzymes to hydrolyze proteins, such as during fermentation, can be utilized for generation of biopeptides. Addition of functional groups to peptides to promote functionality as a chemical modification has also been in practice ever since the first kind of casein-derived peptides were phosphorylated, for example, the caseinophosphopeptides (Mellander, 1950) for enhanced bone calcification in infants. During food-processing stages bioactive peptides have been liberated. For example, during manufacture of milk products, hydrolyzed milk proteins have been liberated. The plasmin present within milk can contribute to hydrolyze milk proteins to obtain bioactive peptides (Daalsgard, Heegaard, and Larsen, 2008). Bacterial starter cultures that contain proteases help to break down protein to peptides similar to fermentation where long oligopeptides can be released upon breakdown of proteins such as caesins (Jauregi, 2008). Cheese contains phosphopeptides as natural constituents and secondary proteolysis during cheese ripening leads to the formation of bioactive peptides and have been documented as being angiotensin converting enzyme (ACE)-inhibitory (Pihlanto-Leppala, 2001).

Fermentation of milk products is a natural way in which bioactive peptides can be generated, and different micro-organisms are likely to generate different ranges of peptides, possibly with different health effects. It was also shown with ACE-inhibiting peptides present in fermented milk products, that they had the ability to resist degradation but were able to absorb directly and inhibit ACE in the aorta (Takono, 1998). Several trials using fermented milk products have been shown to release casokinins, having the ability to lower blood pressure using lactic acid bacteria (LAB) as fermentors because they can act as extracellular proteases thus releasing different forms of caseins from milk, although not all LAB can function effectively to release antihypertensive peptides (FitzGerald et al., 2004).

Owing to losses in function, inconsistent fermentability among strains of similar bacterial families, potential damaging effects to proteins, and generation of inadvertent peptides, treatment due to physical food processing is not preferred. On the other hand, use of enzymatic hydrolysis using specific digestive enzymes is preferred over physical processing for the generation of bioactive peptides.

Selection of proteases plays critical roles in the liberation of bioactive peptide fragments. Pancreatic enzymes (preferably pepsin) and trypsin and chymotrypsin have been popularly used to obtain peptides from food proteins. ACE-inhibitory peptides have most commonly been produced using trypsin. Tryptic, chymotryptic, and peptic hydrolysis of casein protein has resulted in many short peptides possessing biological activities such as immunomodulatory, ACE-inhibitory, and antioxidative properties (Migliore-Samour and Jolles, 1988; Suetsuna, Ukeda, and Ochi, 2000). One of the peptides (Ile-Ile-Ala-Glu-Lys) isolated from β-lactoglobulin by tryptic hydrolysis by Nagaoka et al. (2001) was found to have hypocholesterolemic action that suppressed absorption of cholesterol in Caco-2 cells *in vitro*. Another peptide (Trp-Leu-Ala-His-Lys) derived from β-lactoglobulin has been shown to possess antihypertensive activity resulting from ACE-inhibition (Pihlanto-Leppala et al., 2000). Figure 1.1 depicts an arrangement of peptides obtained by treatment of pepsin and cyanogen bromide.

A number of peptides derived from casein have been found to be immunostimulatory. Not only does each of the caseins produce different bioactive peptides, but they bear effects depending on the enzymatic process used to generate them (Gill et al., 2000). For example, pancreatin and trypsin digests of α-casein and β-casein were shown to inhibit proliferation of murine spleen cells and rabbit Peyer's patch cells (Otani and Hata, 1995), whereas digests prepared using pepsin and chymotrypsin had no effect. Soybean protein isolate hydrolyzed consecutively with pepsin and pancreatin followed by

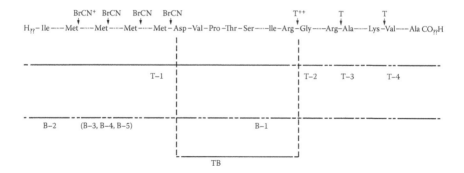

FIGURE 1.1
Peptide fragments formed by action of pepsin and cyanogen bromide. BrCN: position of cleavage of a peptide bond by cyanogen bromide. T: position of cleavage of the peptide bond by trypsin. (From Lapuk V. Kh. et al. *Chemistry of Natural Compounds*. Vol. 5. 1969. Reproduced with permission from Springer publishers.)

separation of the digest on a SP-sepharose column produced ACE-inhibitory fractions (Farzamirad and Aluko, 2008). Antihypertensive peptides were also derived from whey protein digests (Pihlanto-Leppala et al., 2000) as well as from casein (Saito, 2008).

Pepsin-liberated peptides from lactoferrin were found to disrupt the cell membrane for penetration into the cell to cause antimicrobial effects (Schibli, Hwang, and Vogel, 1999). There is evidence to suggest that lactoferricin is generated by digestion of lactoferrin in the human stomach (Kuwata et al., 1998). Hydrolysis of the whey proteins α-lactalbumin and lactoferrin has led to preparations that are capable of modulating the immune system. In a recent study conducted by Beaulieu et al. (2007), a malleable protein matrix composed of fermented whey proteins had immunomodulatory properties thereby stimulating the immune system.

Peptides with opioid activity have been generated by digesting the parent protein with digestive enzymes: pepsin, pepsin followed by trypsin, or chymotrypsin alone (Pihlanto-Leppala et al., 2001). Hamel, Kielwein, and Teschemacher (1985) demonstrated that casomorphins are produced during cheese ripening due to the proteolytic activity of certain bacteria. The β-casomorphins were the first opioid peptides identified from food proteins and, to date, they are the most studied of all the opioid peptides, with β-casomorphin-11 (Miesel and Frister, 1989) and β-casomorphin-7 being characterized as *in vivo* digestion products (Rutherfurd-Markwick and Moughan, 2005). Hartmann and Meisel (2007) have reviewed food-derived peptides with biological activity including opioid and other biofunctional peptides.

Other proteases such as alcalase, alkaline protease, and proteinases have also been used to obtain bioactive peptides mainly from cereal sources

TABLE 1.1

Bioactive Peptides Obtained from Various Food Proteins by Enzymatic Hydrolysis

Protein	Protease	Peptide	Activity
Casein	Trypsin	Phe-Phe-Val-Ala-Pro	ACE-inhibition
	Trypsin-Chymotrypsin	Val-Glu-Pro-Ile-Pro-Tyr-Gly-Leu-Phe	Immunomodulation
	Pepsin	Tyr-Phe-Tyr-Pro-Glu-Leu	Antioxidative
β-Lactoglobulin	Trypsin	Ile-Pro-Ala-Val-Pke-Lys-Trp-Leu-Ala-His-Lys-Ala-Leu-Pro-Met-His-Ile-Arg	Bactericidal ACE-inhibition
Rice albumin	Trypsin	Gly-Tyr-Pro-Met-Tyr-Pro-Leu-Pro-Arg	Ileum contraction Immunostimulation
Soybean	Proteinase S	Leu-Leu-Pro-His-His	Antioxidative
	Alcalase	Low molecular weight peptides	Antihypertensive
Wheat germ	Alkaline protease	Ile-Val-Tyr	ACE-inhibition Antihypertensive
Genetically modified soy protein	Trypsin and Chymotrypsin	Arg-Pro-Leu-Lys-Pro-Trp	Antihypertensive

including soybean and wheat germ. Such proteases have been used widely to generate peptide hydrolysates on a commercial basis.

In order to release bioactive peptides, digestive enzymes can be secreted by the digestive system or produced by resident microflora in the gut. Experimental findings have clearly shown that intestinal microflora are able to modify the immunomodulatory effects of foodborne bioactive peptides (Sutas, Hurme, and Isolauri 1996), and one way in which probiotic bacteria may be effective is by enhancing the formation of efficacious bioactive peptides. Table 1.1 lists bioactive peptides obtained from various food proteins by enzymatic hydrolysis and their bioactivities.

In addition to the generation of bioactive peptides by the digestive process, peptides can also be generated during the food manufacturing process itself (e.g., partially hydrolyzed milk products for hypoallergenic infant formulas). In such cases, the bioactive components are ingested as part of the food. Cheese ripening is another process that results in the generation of various peptides, including ACE inhibitors. A number of bacterial species used in the cheese manufacturing process are used for generating bioactive peptides (Rutherfurd-Markwick and Moughan, 2005).

1.2.2 Evaluation of Bioactivities

The need to evaluate bioactivities of food-derived proteins and peptides have proven to be natural alternatives to expensive and laborious treatment protocols mainly advocated for chronic conditions. Recent developments of bioactive proteins and peptides for the promotion of human health and prevention of chronic diseases have emphasized the need for evaluating more and more peptides for disease-fighting properties. Although a great deal of work related to production and evaluation of bioactive peptides against ACE activity or blood pressure lowering activity has been done, other chronic illnesses such as cancer and heart disease are receiving priority as well. Food-derived peptides have been shown to have an impact on a range of physiological functions, including influencing intestinal transit, modifying nutrient absorption and excretion, immunomodulatory effects, and antihypertensive activity (Mine and Shahidi, 2006).

On the other hand, advances in nuclear medicine, receptor-mediated delivery, peptide conjugates, and using peptides as carriers for tumor treatment are on the rise. The specific binding of peptides to their receptors can be used to meet the key requirement in tumor targeting. Because of their small size, peptides exhibit faster blood clearance and higher target-to-background ratios compared to macromolecular compounds (Langer and Beck-Sickinger, 2001). Coupling cytotoxic drugs to macromolecular carriers has been shown to be a promising approach for efficient drug targeting. In the past few years, peptides were introduced as carriers. Different conjugates, composed of a peptide carrier and a cytotoxic moiety, have been investigated so far. Anticancer drugs were coupled to analogues of luteinizing hormone-releasing hormone, bombesin, somatostatin, and neuropeptide Y (Langer and Beck-Sickinger, 2001). The authors suggest that suitable candidates maintained their binding affinity and could preserve the cytotoxic activity *in vitro* and *in vivo*, resulting in a peptide-mediated selective chemotherapy.

Typically, evaluation of growth inhibitory patterns, cell survival, and proliferative indices are included in the battery of screening tests for compounds predicted to have some bioactivity against cancerous states or cancer cell proliferation. For evaluation of other bioactivities, usually after inducing the cells to undergo disease-states followed by inhibition of growth by test peptides and quantifying cell death by means of color, cell-titer based assays are done. Dye exclusion assays are indicators for cell survival whereas metabolic (cytotoxic) assays are more specific and prognostic in nature (Elgie et al., 1996). Usually a combination of quantifiable (cell survival) and mechanistic (cytotoxicity) assays are recommended to arrive at a possible mode of action for the test substances. On the other hand, a few assays have been developed solely for the purpose of oncology research. One such is the clonogenic assay which also relies on the effectiveness of an agent on cancer cell survival and proliferation indices and differs in that clones of cells that are capable of forming tumors are used rather than non-tumorigenic cancer cells. It finds application especially in research laboratories to determine effects of drugs or radiation on proliferating tumor cells (Franken et al., 2006).

The ACE-inhibitory peptides' screening test uses the ability of peptides to inhibit the activity of ACE enzyme, and quantify ACE activity spectrophotometrically (Vermeirssen, Van Camp, and Verstraete, 2002). Antimutagenic activity is done using a strain of bacteria as in AMES testing, and antimicrobial activity is determined by quantitatively estimating the number of culture colonies that survive upon peptide exposure. Other *in vivo* tests rely on measuring markers in blood upon administering the test peptide in specific dose levels. This helps to profile the expression of certain proteins that are upregulated or downregulated and serve as markers for disease propagation or arrest.

Flow cytometry has been introduced in the last decade as a powerful tool to examine the nature of action the test peptides follow when inhibiting or arresting the growth of proliferating cells in real-time (Carey, McCoy, and Keren, 2007). Fluorescence-activated cell sorting is a specialized type of flow cytometry. It provides a method for sorting a heterogeneous mixture of biological cells into two or more containers, one cell at a time, based upon the specific light scattering and fluorescent characteristics of each cell. This aspect can be used to sort cells based on the changes they undergo upon treatment. For example, the population of cells entering into either the G1 or S phase of the cell cycle especially when examining cancer cell cycle control upon test peptides' treatment can be well examined.

1.2.2.1 Anticancerous and Antitumorigenic

Chemotherapy by far has been the choice of treatment for metastasizing cancers, with advancement related to reduce possible cytotoxic side effects. Monoclonal antibody therapy became popular and the only mode to target the group of cancerous cells or tissue but was not an effective agent, particularly under metastasis. Targeted delivery then became the focus of eliminating the cancerous cell (Carlson et al., 2007). Figure 1.2 shows how multivalent targeted interactions at the cell surface recognize cellular types and establish signal transduction systems enough to target the cell and mediate complement-mediated lysis.

To establish efficient and reliable therapeutic delivery into cancer cells, a number of delivery agents and concepts have been investigated in recent years. Among many improvements in targeted and controlled delivery of therapeutics, cancer cell-targeting peptides have emerged as the most valuable approach. Peptides can be incorporated into multicomponent gene-delivery complexes for cell-specific targeting. In contrast to larger molecules such as monoclonal antibodies, peptides have excellent tumor penetration, making them ideal carriers of therapeutics to the primary tumor site. Peptides and hence peptidomimetics are excellent alternative targeting agents for human cancers, and they may alleviate some of the problems with methods related to antibody targeting (Haubner and Decristoforo, 2009).

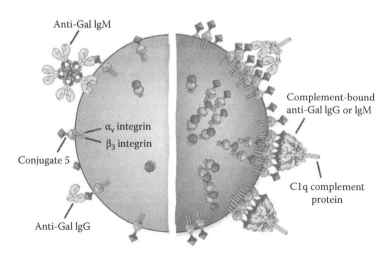

FIGURE 1.2
Targeted cancer cell therapy. Bifunctional molecules (red and blue) target membrane receptors and recruit antibodies (green and purple) that kill cells with high levels of receptors. Cancer cells (right) have more of the receptors than normal cells (left). (From Carlson, C.B., Mowery, P., Owen, R.M., Dykhuizen, E.C., and Kiessling, L.L. 2007. *ACS Chem Biol.* 2: 119–27. Copyright 2007, American Chemical Society. With permission.)

Depending upon the properties associated with each peptide or group of peptides, the nature of functional activity can be determined. Certain peptides may have a high affinity for specialized receptors on certain cancer cells, whereas other peptides even though bioactive may not cause the same effect. Hence, it is important to determine the characteristics of each group of peptides or peptide to be able to relate to their individual biological activities. Evaluation of biological activity thus becomes a critical aspect, and most often calls for concentrating the bioactivity of the substance in a form readily deliverable, and free from toxic side effects including the ability to resist the gastrointestinal environment. We first focus on peptides and protein hydrolysates able to target and slow down specific cancer types.

Proteins, peptides, and amino acids have been implicated in preventing the development of different types of cancer. Bowman–Birk protease inhibitor (BBI), a water-soluble protein isolated from legumes, has shown anticarcinogenic activity *in vitro* and in animal models and is now intensively studied as a cancer chemopreventive agent in clinical trials (Armstrong et al., 2000). Another protease inhibitor, soybean Kunitz trypsin inhibitor, was reported to suppress ovarian cancer cell invasion by blocking urokinase upregulation (Kobayashi et al., 2004). Bovine lactoferrin and lactoferricin from bovine milk were able to inhibit lung metastasis and angiogenesis in mice transplanted with murine melanoma, lymphoma, or colon carcinoma 26 cells. Specifically lactoferricin (Lfcin-B), a peptide derived from a bovine milk protein, lactoferrin (LF-B), was found to induce apoptosis in THP-1 human monocytic leukemic cells. The mechanism of action was related to the pathway medi-ated by production of the intracellular reactive oxygen species (ROS) and, activation of certain metal-dependent endonucleases (Yoo et al., 1997). The anticancer activities of these proteins may, at least partially, be attributed to encrypted bioactive peptides.

Amino acids may play an indirect role in impeding cancer induction. Many protein diets including soy protein contain nonessential amino acids that promote glucagon production. This in turn downregulates the activity of insulin and produces IGF-1 (insulin-like growth factor-1) antagonists. In addition, the presence of low levels of essential amino acids in many diets (vegan diets) may decrease hepatic IGF-1 synthesis. This causes a negative effect in cancer formation. Hence these diets when free of excess fat may prove protective against prostate, breast, and also colon cancers that are linked to insulin resistance (McCarty et al., 1999).

The RGD (Arg–Gly–Asp) peptide sequence is a common site for proteins in blood and extracellular matrix to bind to and be able to exert biological functions. It has been shown to bind to tumor cells and interfere with cancer cell cycle progression. Peptides designed to possess this particular sequence when tested for their abilities to arrest cancerous cell growth exhibited posi-tive characters. They inhibited tumor growth formation in a dose-dependent manner and also induced apoptosis and G1 phase cell cycle arrest in breast cancer cell lines (Yang et al., 2006).

Numerous peptides in different sizes from various sources have been indi-cated to render an anticancer effect in *in vivo* studies. Table 1.2 lists the pro-teins/peptides that have thus far been implicated in anticancer effects.

Lunasin, a novel chemopreventive peptide from soybeans, has been found to suppress chemical carcinogen and viral oncogene-induced transformation of mammalian cells and inhibit skin carcinogens in mice (Lam, Galvez, and de Lumen, 2003). Lunasin is a 43-amino-acid peptide containing 9 aspartic

TABLE 1.2

Proteins/Peptides with Anticancer Effects

Type of Cancer	Anticancer Agent	Source	Reference
Prostate	Proteins	Soy	Bylund et al., 2000
Breast	Protein	Soy	Badger et al., 2005
	Peptides with RGD sequence		Yang et al., 2006
	Calcaelin protein	Mushroom *Calvatia caelata*	Ng, Lam, and Wang, 2003
Lung	α-lactalbumin	Milk	Hakansson et al., 1995
	Casein and other proteins	Animal	Mahaffey et al., 1987
Leukemia	Lfcin-β peptide	Lactoferrin-β protein	Yoo et al., 1997

acid residues at the C-terminus, a tripeptide arginine–glycine–aspartic acid cell adhesion motif, and a predicted helix whose structure is similar to a conserved region of chromatin-binding proteins (Galvez et al., 2001). Detailed information is presented and reviewed in Chapter 13, "Lunasin: A Novel Seed Peptide with Cancer-Preventive Properties."

Other bioactive peptides from soybean with cancer-preventive effects have also been reported. Azuma et al. (2000) and Kanamoto et al. (2001) demonstrated that a high MW fraction (HMF) of proteinase-treated soybean protein isolate suppressed colon and liver tumorgenesis in experimental animals. Aglycopeptide isolated from soybean hydrolysate containing mainly D, E, P, G, and L has been shown to be cytotoxic against P388D1 mouse lymphoma cells (Kim et al., 1999). Kim et al. (2000) describe an anticancer peptide from the hydrophobic peptide fraction of thermoase-treated soy protein hydrolysate. The peptide was identified as a nonapeptide with a MW of 1157 Da and a sequence of X-MLPSYSPY. Soy proteins have been studied as a possible cause of reduction in prostate cancer disease progression. With soy proteins it is only a possibility that they may inhibit prostate cancer inasmuch as they affect tumor cell apoptosis and not inhibition of cancer cell proliferation Bylund et al., 2000). Recently peptide fraction isolated from rice bran by enzymatic hydrolysis and membrane fractionation was shown to inhibit growth of human colon, liver, as well as breast cancer cells *in vitro* (Kannan et al., 2008; Kannan, Hettiarachchy, and Narayan, 2009). The authors identified that a GI-resistant <5 kDa peptide fraction was significantly potent in reducing cancer cell proliferation compared to higher molecular weight fractions and non-GI resistant fractions.

Other plant sources of anticancer peptides reported include those of buckwheat and ginseng. A peptide isolated from buckwheat seeds with an MW of 4 kDa showed antiproliferative activity against hepatoma, leukemia, and breast cancer cells (Leung and Ng, 2007). A hydrophobic fraction of ginseng peptide isolate demonstrated high antitumor activity against mouse lymphoma cell line P388D1 (Kim et al., 2003). The peptides present in the fraction were identified as either dipeptides or tripeptides composed entirely of tryptophan. A novel protein, calcaelin, isolated from the mushroom *Calvatia caelata*, was found to reduce the cell viability of breast cancer cells (Ng, Lam, and Wang, 2003).

Dairy milk proteins and their peptide derivatives play a role in cancer prevention (Parodi, 2007). In a study, milk treatment onto transformed cell lines resulted in an apoptotic picture. The protein responsible was identified as a multimeric form of α-lactalbumin. This effect was first observed in human lung cancer cell line (A549). Cell viability was almost completely reduced (98%), and the cells exhibited characteristic features of apoptosis. Other cell lines affected by the same protein were a variety of epithelial cell lines (Caco-2, HT-29, NCI) kidney cell lines (Vero, GMK), and mouse cell lines (WEHI 164, B9; Hakansson et al., 1995). Animal studies conducted to determine the effects of feeding protein-rich diets as against electrolyte or dextrose/amino-acid diets observed that progression of pulmonary metastatic disease was reduced, and so did the tumor weight when fed with protein-rich diets (casein) (Mahaffey et al., 1987).

Kim et al. (1995) studied the anticarcinogenicity of hydrophobic peptide fractions isolated from cheese slurries and found that the purified peptide fractions exhibited high cytotoxic activity against tumor cell lines SNU-C2A, SNU-1, and P388D1. Egg protein hydrolysates were found to exert an antiproliferative effect on mouse lymphoma cells (Yi et al., 2003). Glycopeptides from pronase-treated ovomucin showed antitumor effects in a double-grafted tumor system in mice, which has been associated with their antiangiogenic activity (Oguro et al., 2001).

In addition, anticancer peptides are present in marine animals and microorganisms. Picot et al. (2006) assessed the antiproliferative activity of 18 fish protein hydrolysates *in vitro* and identified hydrolysates of 3 blue whiting, 3 cod, 3 plaice, and 1 salmon as significant growth inhibitors against two human breast cancer cell lines. Shark-cartilage-derived peptides have shown antiangiogenic activity, as evaluated in a variety of *in vitro* and *in vivo* models (Bukowski, 2003).

1.2.2.2 Antihypertensive

A wide range of ACE-inhibitory peptides has been identified. The majority of these peptides are low molecular weight peptides derived mostly from different caseins, lactalbumins, and lactoglobulins. Several whey protein hydrolysates obtained from peptic, tryptic, and proteinase-k digests have proven hypotensive when fed to spontaneously hypertensive rats (SHR; Costa et al., 2005), as were peptides derived from the whey fraction of skimmed milk fermented by Lactobacillus helveticus. Ovalbumin hydrolyzed with chymotrypsin yielded a peptide that imparted vasoactive effects contributed by nitric oxide released from endothelial cells (Matoba et al., 1999). Likewise there are several peptides identified as possessing hypotensive action. Meisel et al. (2006) and Guang and Phillips (2009) have provided a comprehensive list of ACE-inhibitory peptides derived from plant foods. Cereal storage proteins can also be potential sources of ACE-inhibitory peptides (Loponen, 2004) and ACE-inhibitory peptides have been identified in soybean, mung bean, sunflower, rice, corn, wheat, buckwheat, broccoli, mushroom, garlic, and spinach. Although the active peptides have not been sequenced, peanut (Quist, Phillips, and Saalia, 2009), chickpea (Pedroche et al., 2002), and potato (Pihlanto et al., 2008) protein hydrolysates also display strong ACE-inhibitory activity. ACE-inhibitory activity has also been found in traditional fermented soybean products, such as natto (Okmato et al, 1995), tempeh (Gibbs et al, 2004), and douchi (Zhang et al., 2006).

The most common way to produce ACE-inhibitory peptides is through enzymatic hydrolysis of food proteins. The specificity of the proteolytic enzyme and process conditions influence the peptide composition of hydrolysates and thus their ACE-inhibitory activities (Ven et al., 2002). The combination of pepsin–pancreatin or pepsin–chymotrypsin–trypsin is usually used to simulate the gastrointestinal degradation of food proteins in humans. Pepsin treatment (as was observed) cannot effectively elicit ACE-inhibitory peptides from buckwheat protein, whereas pepsin treatment followed by chymotrypsin and trypsin leads to a significant increase in ACE-inhibitory activity. Alcalase generates more potent ACE-inhibitory hydrolysates than other studied proteases from sources such as corn gluten (Yang et al., 2007), wheat germ (Matsui et al, 1999), potato tubers (Pihlanto et al., 2008), soy (Chiang et al., 2006), and peanut (Quist et al., 2009) proteins.

The mechanism of action of ACE inhibition involves several peptides and enzymes at the renin–angiotensin system (RAS). Renin cleaves angiotensinogen and releases angiotensin I (Asp-Arg-Val-Tyr-Ile-His-Pro-Phe-His-Leu). When angiotensin I is cleaved off its last two amino acid residues His–Leu by angiotensin converting enzyme, angiotensin II is produced (Figure 1.3).

Angiotensin II is a potent vasoconstrictor that controls blood pressure and certain hormones (Meisel et al., 2006). Peptides that inhibit ACE activity are potential hypotensive agents. It must also be emphasized that exclusive ACE inhibition may not cause a hypotensive effect because many physiological factors are involved in the different pathways involved. Table 1.3 lists food peptides having antihypertensive activities.

1.2.2.3 Cholesterol-Lowering Effect and Antiobesity Activity

Several synthetic drugs and natural extracts with cholesterol-lowering effects have been explored for their potential in the prevention and treatment of hypercholesterolemia, one of the important risk factors for heart disease. Literature indicates that soybean proteins can reduce blood cholesterol levels in experimental animal models as well as in human subjects (Potter, 1995).

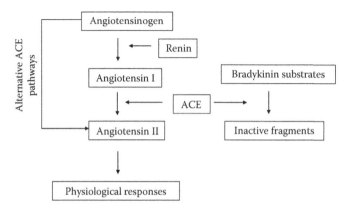

FIGURE 1.3
Renin–angiotensin system.

TABLE 1.3

Commercially Available Food Products Containing Bioactive Peptides

Product Name	Manufacturer	Type of Food	Bioactive Peptides	Health Claim
Calpis AMEEL S (Japan) or Calpico (Europe)	Calpis Co., Japan	Sour milk	VPP, IPP from â- and ê-CN	Hypotensive
Evolus	Valio, Finland	Fermented milk, calcium-enriched	VPP, IPP from â- and ê-CN	Hypotensive
BioZate	Davisco, USA	â-LG hydrolysate	Whey peptides	Hypotensive
C12 Peption	DMV, Netherlands	Ingredient	Casein-derived dodecapeptide FFVAPFPEVFGK	Anticariogenic, antimicrobial, antithrombotic
Peptide Soup	NIPPON, Japan	Soup	Bonito-derived peptides	Hypocholesterolemic
Casein DP Peptio Drink	Kanebo, Japan	Soft drink	Casein-derived dodecapeptide FFVAPFPEVFGK	Reduces stress
BioPURE-GMP	Davisco, USA	Whey protein hydrolysate	Glycomacropeptide	Helps mineral absorption
CholesteBlock	Kyowa Hakko, Japan	Drink powder	Soy peptides bound to phospholipids	
CSPHP	Ingredia, France	Milk drink, confectionery	âS1-CN (ƒ91–100)a: YLGYLEQLLR	
ProDiet F200	Arla Foods, Denmark	Ingredient	CPP	
Capolac	Suntory, Japan		CPP	
Tekkotsu Inryou	Asahi, Japan		CPP	
Kotsu Kotsu calcium CE90CPP	DMV, Netherlands		CPP (20%)	

Sirtori et al. (1977) first observed in a clinical study that the substitution of animal proteins with soy protein resulted in a 22–25% decrease in LDL cholesterol and a 20–22% decrease in total cholesterol in hypercholesterolemic patients. The hypocholesterolemic effect of soy protein was later confirmed by more animal and clinical studies (Sagara et al., 2004; Wang et al., 2004), and a soybean-rich diet has become the most potent dietary tool for treating hypercholesterolemia, although the mechanism has not yet been fully established. The U.S. Food and Drug Administration (FDA) recommended a daily intake of 25 g of soybean protein for a lower level of serum cholesterol and reduction in the risk of cardiovascular disease.

Opioid peptides have been demonstrated to play an important role in the control of food intake, which is implicated in its potential antiobesity activity. It functions by inhibiting the absorption of dietary lipids, increasing postprandial energy output, and accelerating lipid metabolism. Several food-derived and synthesized peptides have been reported to have antiobesity activity. Anorectic peptides from soybeans (e.g., LPYPR and PGP) have been shown to exert antiobesity activity through decreasing food intake and body weight (Wang and de Mejia, 2005). These authors also found that soybean protein hydrolysates decreased serum or hepatic triacylglycerol levels and body weight in rats. A synthetic peptide [epsilon]-polylysine has been demonstrated to have an antiobesity function in mice by inhibiting intestinal absorption of dietary fat (Tsujita et al., 2006).

1.2.2.4 Antioxidative Activity

Food peptides and proteins have been found to possess free radical quenching properties implicated in several oxidative insult states. Phosphopeptides derived from egg yolk protein phosvitin were shown to possess antioxidative properties in cultured intestinal epithelial cells (Katayama et al., 2006). The soy proteins have been identified as antioxidative in cardiovascular disorders. They are able to inhibit lipid (LDL) oxidation, a chief promoter of cardiovascular disease pathology (Engelman et al., 2005). Both *in vivo* and *in vitro* experiments that studied antioxidative effects of soy proteins suggest that soy proteins, soy-derived peptides, and amino acids can be considered as the antioxidative components of soy protein, exclusive of the soy isoflavones. For example, Takenaka et al. (2003) have shown a reduction in paraquat-induced oxidative stress in rats by dietary soy peptide. Also soy protein consumption has been shown to decrease serum lipid concentrations (total and LDL cholesterol, triacylglycerols) significantly. A study reported that a soy protein peptic hydrolysate reduced serum cholesterol levels more than just the intact soy protein. The hypocholesterolemic peptides from soy protein ranged between 1 and 10 Kda. But the composition of peptide remains elusive. Likewise Lovati et al. (1992) reported soy 7S globulin peptides activated LDL receptors in HepG2 cells possibly

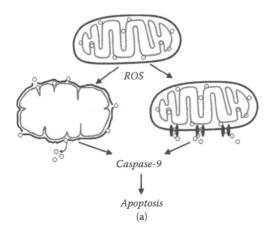

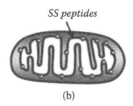

FIGURE 1.4
(a) Antioxidative peptide mechanism. General mechanism of cytochrom C mediated mito-chondrial damage. Filled circle: cytochrome C. (From Szeto H.H. 2006. *AAPS J.* 8: E277–E283. With permission from *AAPS Journal.*) (b) Internalized SS peptides within mitochondrion. SS peptide (Caco-2 cells were incubated with Dmt–D–Arg–Phe–atnDap–NH$_2$) internalized within mitochondria resulting in targeting of mitochondria against cellular death by pro-oxidants. (From Szeto, H.H. 2006. *AAPS J.* 8: E277–E283. With permission from *AAPS Journal.*)

arresting oxidative stress, as also observed in a study by Manzoni et al. (2003). β-conglycinin revealed peptide fragments that contained a His–His segment thought to be responsible for antioxidative and lipid peroxidative actions. Cell-permeable mitochondrial-targeting peptides have also been proven as antioxidants. Szeto (2006) illustrates with her research how certain peptides can be internalized within mitochondria and prevent oxi-dant-induced cellular death (Figure 1.4).

Other free-radical scavenging peptides include carnosine and anser-ine. The histidine containing anserine was thought to work in conjunction with carnosine to affect antioxidatve and peroxidative effects. A sequence of peptides obtained from tryptic hydrolysate of casein was found to contain Tyr–Phe–Tyr–Pro–Glu–Leu. The Glu–Leu portion was found to contain superoxide anion scavenging activity (Suetsuna, Ukeda, and Ochi, 2000). Likewise several casein hydrolysates were found to scavenge free radicals and exert antioxidative actions. Lactoferrin and lactoferricin from whey protein function as potent antioxidants because they are able to sequester iron. They are also capable of effecting antimicrobial activities (Satué-Gracia et al., 2000).

1.2.2.5 Antimicrobial Peptides

Peptides with antimicrobial properties have been identified in a variety of natural sources. Antimicrobial peptides display inhibitory effects against microbe-caused food deterioration and invasion of a wide range of pathogens *in vivo*, including bacteria, fungi, and viruses as well as parasites. The action mode and effectiveness of these biologically active peptides as antimicrobial agents vary depending on their structural characteristics, that is, peptide size, amino acid composition, charge, hydrophobicity, and secondary structure (Dathe et al., 1999). Moreover, they show varied selectivity and sensitivity on target micro-organisms. Antimicrobial peptides possess certain common features throughout the broad source. Most antimicrobial peptides are composed of less than 50 amino acids with approximately 50% being hydrophobic amino acids, and often fold into amphipathic 3D structures (Rydlo, Miltz, and Mor, 2006). Having an excess of basic amino acids (lysine and arginine), antimicrobial peptides usually bear a net positive charge. Being cationic and amphipathic are two important structural features that account for the antimicrobial activity of these bioactive peptides.

Bioactive peptides with antimicrobial properties such as those produced by lactic acid bacteria (LAB) in fermented foods and many food component peptides are good candidates as food additives. Advantages of antimicrobial peptides over chemical preservatives include fewer adverse effects introduced, lower intensity of heat treatment required (minimal processing), and retention of organoleptic and nutritional properties of food, such as less acidic or lower salt content (Galvez et al., 2007). In therapeutic applications, antimicrobial peptides are superior to conventional bactericidal antibiotics because they kill bacteria faster and are not affected by antibiotic resistance mechanisms that are often encountered by other antibiotics (Rydlo et al., 2006).

Antimicrobial activity of bacteriocins against pathogenic micro-organisms has been elucidated at cellular and molecular levels. The inhibition has been associated with the interaction of peptides with membrane and other cellular components. Many bacteriocins elicit their lethal effect on target micro-organisms by disintegrating the cell wall, increasing the permeability of the cell membrane, and causing efflux of small molecules such as amino acids and adenosine triphosphate (Nagao et al., 2006). The positively charged antimicrobial peptides traverse the negatively charged lipopolysaccharide-containing outer wall of Gram-negative bacteria or the acidic polysaccharides-containing outer wall of Gram-positive bacteria by competitively displacing the divalent cations that bridge and neutralize the polysaccharides (Chan and Li-Chan, 2006). At low peptide concentrations, this leads to outer cell wall disruption, whereas at higher peptide concentrations bacteria tend to be partially lysed and disintegrated. In addition to the interaction with cell membranes, bacteriocins, and other antimicrobials, peptides exert an inhibitory effect on certain micro-organisms also through interaction with intracellular anionic components such as DNA and RNA, thereby inhibiting protein synthesis and cell division (Cudic and Otvos, 2002).

Fish, cereals, and dairy products are considered the best sources of anti-microbial peptides. Lactoferricin, a peptide derived from peptic hydrolysis of the milk protein lactoferrin, acts synergistically with certain immuno-globulins to exert antimicrobial activity (Ha and Zemel, 2003). Further research into the human lactoferrin protein revealed a domain present in the N-terminal region that represented the antibacterial portion of the whole protein. The lactoferrins differ in their makeup of amino acid sequence at the N-terminal region thus comprising differential antibacterial actions (Recio and Visser, 1999).

Cereal proteins from wheat, rye, and barley yielded a class of small pep-tides that are basic in nature and are called thionins (Garcia-Olmeda et al., 1998). These thionins are rich in cysteine and lysine having about 45 amino acid residues. The thionins are known to inhibit fungal and bacterial growth *in vitro* (Duvick et al., 1992).

Nisin has long been considered a potent antimicrobial agent derived from the strain of Lactococcus lactis. Of late, nisin-coated films are used in the food industry to impart antibacterial activity, prolong shelf life, and improve food quality and safety. A study showed nisin-coated film inhibition of M. luteus ATCC 10240 (Mauriello et al., 2005). Figure 1.5 shows the structure of nisin.

Several wheat proteins have been shown to possess families of short pep-tides identified as possessing bactericidal activities. For instance, a family of 6 peptides called defensins and another family of 8 peptides rich in glycine were identified in wheat. Maize also contains peptides of 6–8 amino acids composed mostly of cysteines (Egorov et al., 2005).

1.2.2.6 Antimutagenic Peptides

Several animal studies have identified the influence on carcinogenesis of certain peptides and amino acids derived from dietary proteins. It has been shown that the predominant protein in milk, casein, and its peptides, pos-sesses antimutagenic properties. Whey protein has been shown to be superior as an antimitogenic agent thereby suppressing tumor formation specifically when testing for colon and mammalian carcinogenesis in animal models. The effects have been accompanied by increased glutathione levels in serum and tissues as well as increase in lymphocyte population phagocytosis and natural killer, T helper, and cytotoxic T cell activity.

Whey protein components have been studied of which lactoferricin has become a popular candidate for an antitumor agent. It has the ability to inhibit intestinal and other tumors. Lactoferrin acts by induction of apop-tosis, inhibition of angiogenesis, modulation of carcinogen-metabolizing

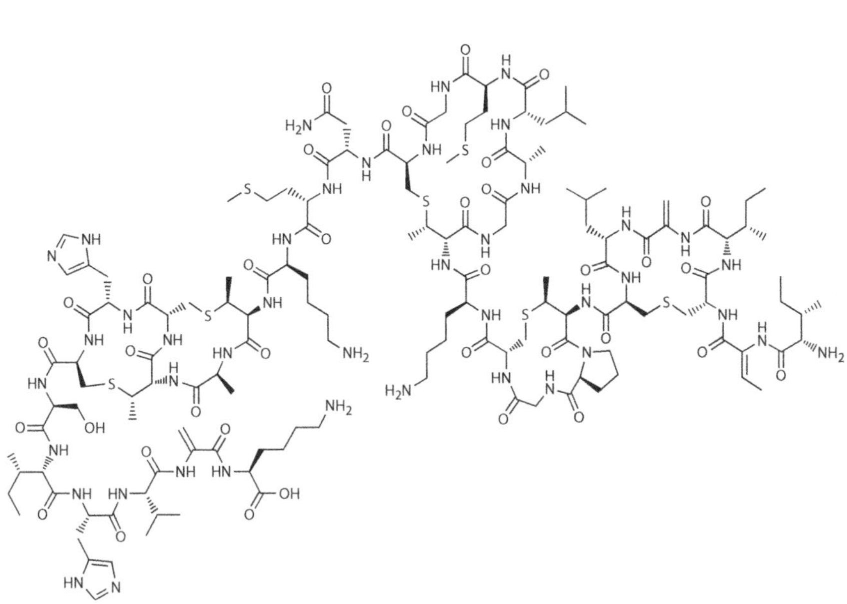

FIGURE 1.5
Structure of nisin. (From http://en.wikipedia.org/wiki/File:Nisin.png.)

enzymes, and perhaps acting as an iron scavenger (Parodi, 2007). Although milk components have been researched for the presence of antitumorigenic or antimitogenic agents, components derived from other sources such as cereals, meat, and fish that may possess anticancer properties may also possess antimitogenic action, and thus would be worth researching.

1.3 Bioavailability of Bioactive Proteins and Peptides

Bioavailability by definition is a measurement of the rate and extent of a therapeutically active drug that reaches the systemic circulation and is available at the site of action (Shargel and Yu, 1999). It has been understood that by medication a therapeutically administered drug or agent will be 100% bioavailable. It has been understood that through therapeutic administration a drug or an agent known to present 100% bioactivity could be 100% bioavailable, while drugs administered via different routes may not present as 100% bioavailable. It is thus essential to investigate the bioavailability of each drug or agent based on bioactivity, dosage, and its physicochemical properties.

Gastrointestinal resistance is a major player in dictating the extent of bioavailability to a drug or agent. For bioavailability and for human consumption, peptides or hydrolysates derived from food sources and possessing bioactivities need to be resistant to physical (pH), proteolytic (enzymes), and microbial (microbial flora present in the gut) degradation. Soluble peptides that are resistant to the gastrointestinal environment can thus render bioactivity more effectively than nonresistant peptides, and can thus extrapolate as bioavailable nutraceutical ingredients. Bioactive peptides that affect bodily function are generally inactive within their parent protein, but exert a specific function upon hydrolysis of the parent protein. Gastrointestinal (GI) enzymes help to release the peptides, and even a fraction of the compound is sufficient to exert the specific function for normal physiology (Mine and Shahidi, 2006).

There are three phases of assimilation for dietary proteins: (1) initiation of proteolysis in the stomach by enzymes such as pepsin and highly efficient endo- and carboxy-terminal cleavage in the upper small intestine cavity (duodenum) by secreted pancreatic proteases and carboxypeptidases; (2) further processing of the resulting oligopeptide fragments by exo- and endopeptidases anchored in the brush-border surface membrane of the upper small intestinal epithelium (jejunum); and (3) final facilitated transport of the resulting amino acids, di-, and tripeptides, across the epithelial cells into the lamina propria. The nutrients from lamina propria enter capillaries for distribution throughout the body (Hausch et al., 2002). Schuppan and others showed that enzymatic hydrolysis does not actively hydrolyze substituted and conformationally restricted amide bond of proline-rich peptides, and hence might lead to abundance of proline residues in gliadins and related proteins from wheat, rye, and barley (Schuppan, 2000). This, in turn, would lead to an increased concentration of relatively stable gluten-derived oligopeptides in the gut, some of which may even cross the intestinal barrier to gain access to the subepithelial lymphocytes.

There have been relatively few studies examining bioactive potentials of peptides to evaluate the gastrointestinal resistances of the peptides. The complexity of the human gut has provided a predicament to enable selection of parameters (physical), proteolytic (enzymes) and microbial (microbial flora present in the gut), for evaluation of gastrointestinal resistance. Ideally animal models closely representing the human gut are most suited for determining gastric and intestinal digestive resistances of compounds. Hausch et al. (2002) studied the ability of certain gliadin peptides to resist intestinal digestion using pancreatic protease digestion and determining the fate of the peptide. This and several other studies have used *in vitro* simulatory proteolytic digestion as primary presumptive models for conducting gastrointestinal evaluation of compounds.

Bioactive peptides may generally be liberated from food proteins via the whole intestine, and then display bioactivity in the small and large bowel. The digestion of proteins into specific peptides to impart bioactivities is rendered by the presence of proteases and microbial enzymes found in the large intestine. Trypsin has been shown to release effectively a number of bioactive peptides, including FFVAPFPEVFGK and TTMPLW from α_{s1}-casein and AVPYPQR from β-casein (Yamamoto, 1997). There is evidence that some proteins such as lactoferrin or immunoglobulins may partly escape degradation in the small intestine (Drescher et al., 1999; Roose et al., 1995). Plasmin from milk or enzymes from bacterial starter cultures may also degrade proteins. In comparison to digestive enzymes, microbial enzymes, either in the gut or in food, use different cleavage sites. Thus, peptides liberated by these enzymes may differ from those liberated by digestive enzymes. They may also present as precursors for peptides released in the intestinal tract. For example, Yamamoto (1997) found better antihypertensive activity using pancreatic digestion compared to initial tryptic digestion to generate KVLPVPQ from food protein.

Bioavailability is thus one of the essential parameters in pharmacokinetics, as bioavailability must be considered when calculating dosages for nonintravenous routes of administration. Several factors influence bioavailability; some of the important ones are physical properties of the drug or agent, route of administration, gastrointestinal resistance, rate of metabolism, interactions with other foods and drugs, and formulation. It is thus essential to investigate bioavailability of a drug or agent before clinical trials can be recommended, particularly for food-derived proteins and peptides with bioactive properties inasmuch as the route of administration and formulation may vary with each source.

1.4 Commercialization of Bioactive Proteins and Peptides

There is increasing commercial interest in the production of bioactive peptides from various sources. Industrial-scale production of such peptides is limited due to lack of scalable technologies. There is a need to develop technologies that retain or even enhance the activity of bioactive peptides in food systems. Considerable progress has been made over the last 20 years in technologies aimed at separation, fractionation, and isolation in a purified form of many interesting proteins occurring in bovine colostrum and milk (Korhonen and Pihlanto, 2007). Industrial-scale methods have been developed for native whey proteins such as immunoglobulins, lactoferrin, lactoperoxidase, α-lactalbumin, and β-lactoglobulin. Commercially viable products from bioactive peptides particularly those derived from milk proteins have interested markets. Several bioactive peptides have also been identified via fermentation of dairy products, some of which are already under commercial production. Bioactive peptides from milk, particularly antihypertensive peptides, can be produced by starter cultures and are generally used in the manufacture of fermented milks or cheeses (Korhonen and Pihlanto, 2003). A few of these have attained commercial value as fermented milk. Advances here will enable recovery of bioactive peptides with minimal destruction thus enabling utilization by returning these active peptides to functional food or specific nutraceutical applications. Table 1.3 lists some of the commercially available food products containing bioactive peptides.

Approaches toward large-scale production of proteins from cereal grains have also been undertaken. For example, rice grain can be processed to obtain oil, starch, and protein for various food and industrial applications that may require further processing or conversion to the desired product. A biocatalysis process to produce starch-derived and high-protein products from rice has been successfully developed (Casimir et al., 2008) and is described in Chapter 2 of this book.

In general, enzymatic hydrolysis has been shown to be an alternative successful approach for the production of protein-enriched flour from cereal grains with improved nutrition (Casimir et al., 2008). Use of enzymes as biocatalysts offers many advantages: (1) high efficiency and specificity, (2) natural and green (low pollution) process, and (3) economical instrumentation cost requirement. The development of immobilized enzymes would eliminate interfering complexed substances such as starch in cereals and produce to desired products with lower overall processing cost. Biotechnology via gene cloning, recombinant enzyme technology, and transgenic manipulations, are expected to further develop a simpler, more efficient, and inexpensive approach to produce value-added products with an easily accessible and important nutrition source.

1.5 Conclusion

Food proteins and peptides are being identified as bioactive compounds. In the last decade, research has intensified and new bioactive protein and peptide sequences are being deciphered. The preparation and production of these bioactive compounds and their incorporation into new products that can be labeled as functional foods pose several technological challenges as well as regulatory and marketing issues. Investigation of bioavailability, toxicity, and shelf-life becomes of paramount importance in fully classifying these compounds as "bioactive" in addition to their commercial value and preparation in new food products. Although several challenges are present, there is increasing evidence for discovery and identification of new bioactive proteins and peptides.

References

Armstrong, W.B., Kennedy, A.R., Wan, X.S., Atiba, J., McLaren, C.E., and Meyskens, F.L. 2000. *Cancer Epidemiol. Biomarkers Prev.* 9: 43–47.

Azuma, B., Machida, K., Saeki, Y., Kanamoto, R., and Iwami, K. 2000. Preventive effect of soybean resistant proteins against experimental tumorigenesis in rat colon. *J. Nutr. Sci. Vitaminol.* 46: 23–29.

Beaulieu, J., Dubuc, R., Beaudet, N., Dupont, C., and Lemieux, P. 2007. Immunomodulation by a malleable matrix composed of fermented whey proteins and lactic acid bacteria. *J. Med. Food* 1: 67–72.

Bukowski, R.M. 2003. AE-941, a multifunctional antiangiogenic compound: Trials in renal cell carcinoma. *Expert Opin. Investig. Drugs* 12(8): 1403–1411.

Bylund, A., Zhang, J.X., Bergh, A., Damber, J.E., Widmark, A., Johansson, A., Adlercreutz, H., Aman, P., Shepherd, M.J., and Hallmans, G. 2000. Rye bran and soy protein delay growth and increase apoptosis of human LNCaP prostate adenocarcinoma in nude mice. *Prostate* 42: 304–314.

Carey, J.L., McCoy, J.P., and Keren, D.F. (Eds.) 2007. *Flow Cytometry in Clinical Diagnosis*, 4, Chicago: ASCP Press.

Carlson, C.B., Mowery, P., Owen, R.M., Dykhuizen, E.C., and Kiessling, L.L. 2007. Selective tumor cell targeting using low-affinity, multivalent interactions. *ACS Chem. Biol.* 2: 119–127.

Casimir, C., Akoh, S.-W., Chang, G.-C. L., and Shaw, J.-F. 2008. Biocatalysis for the production of industrial products and functional foods from rice and other agricultural produce. Review. *J. Agric. Food Chem.* 56: 10445–10451.

Chan, J.C.K. and Li-Chan, E.C.Y. 2006. Antimicrobial peptides. In: *Nutraceutical Proteins and Peptides in Health and Disease* (Y. Mine and F. Shahidi, Eds.), Boca Raton, FL: Taylor & Francis, pp. 99–136.

Chiang, W., Tsou, M., Tsai, Z., and Tsai, T. 2006. Angiotensin I-converting enzyme inhibitor derived from soy protein hydrolysate and produced by using membrane reactor. *Food Chem.* 98: 725–732.

Costa, E.L., Almeida, A.R., Netto, F.M., and Gontijo, J.A. 2005. Effect of intraperitoneally administered hydrolyzed whey protein on blood pressure and renal sodium handling in awake spontaneously hypertensive rats. *Braz. J. Med. Biol. Res.* 38: 1817–1824.

Cudic, M. and Otvos, L.J. 2002. Intracellular targets of antibacterial peptides. *Curr. Drug. Targets* 3: 101–106.

Dalsgaard, T.K., Heegaard, C.W., and Larsen, L.B. 2008. Plasmin digestion of photo-oxidized milk proteins. *J. Dairy Sci.* 91: 2175–2183.

Dathe, M. and Wieprecht, T. 1999. Structural features of helical antimicrobial peptides: Their potential to modulate activity on model membranes and biological cells. *Biochim. Biophys. Acta* 1462: 71–87.

de Lumen, B.O. 2005. Lunasin: A cancer-preventive soy peptide. *Nutr. Rev.* 63: 16–21.

Drescher, K., Roos, N., Pfeuffer, M., Seyfert, H.M., Schrezenmeir, J., and Hagemeister, H. 1999. Recovery of 15N-lactoferrin is higher than that of 15N-casein in the small intestine of suckling, but not adult miniature pigs. *J. Nutr.* 129: 1026–1030.

Duvick, J.P., Rood, T., Rao, A.G., and Marshak, D.R. 1992. Purification and characterization of a novel antimicrobial peptide from maize (Zea mays L.) kernels. *J. Biol. Chem.* 267: 18814–18820.

Egorov, T.A., Odintsova, T.I., Pukhalsky, V.A., and Grishin, E.V. 2005. Diversity of wheat anti-microbial peptides. *Peptides* 11: 2064–2073.

Elgie, A.W., Sargent, J.M., Taylor, C.G., and Williamson, C. 1996. An *in vitro* study of blast cell metabolism in acute myeloid leukaemia using the MTT assay. *Leukemia Res.* 20: 407–413.

Engelman, H.M., Alekel, D.L., Hanson, L.N., Kanthasamy, A.G., and Reddy, M.B. 2005. Blood lipid and oxidative stress responses to soy protein with isoflavones and phytic acid in postmenopausal women. *Am. J. Clin. Nutr.* 81: 590–596.

Farzamirad, V. and Aluko, R.E. 2008. Angiotensin-converting enzyme inhibition and free-radical scavenging properties of cationic peptides derived from soybean protein hydrolysates. *Int. J. Food Sci. Nutr.* 59: 428–437.

FitzGerald, R.J., Murray, B.A., and Walsh D.J. 2004. Hypotensive peptides from milk proteins. *J. Nutr.* 134: 980S–988S.

Franken, N.A., Rodermond, H.M., Stap, J., Haveman, J., and van Bree, C. 2006. Clonogenic assay of cells *in vitro*. *Nat. Protoc.* 1(5): 2315–2319.

Galvez, A., Abriouel, H., Lucas-Lopez, R., and ben-Omar, N. 2007. Bacteriocin-based strategies for food biopreservation. *Int. J. Food Microbiol.* 120: 1–2.

Galvez, A.F., Chen, N., Macasieb, J., and de Lumen, B.O. 2001. Chemopreventive property of a soybean peptide that binds to deacetylated histones and inhibits histone acetylation. *Cancer Res.* 61: 7473–7478.

Garcia-Olmedo, F., Molina, A., Alamillo, J.M., and Rodriguez-Palenzuela, P. 1998. Plant defense peptides. *Biopolymers* 47: 479–491.

Gibbs, B.F., Zougman, A., Masse, R., and Mulligan, C. 2004. Production and characterization of bioactive peptides from soy hydrolysate and soy-fermented food. *Food Res. Int.* 37: 123–131.

Gill, H.S., Doull, F., Rutherfurd, K.J., and Cross, M.L. 2000. Immunoregulatory peptides in bovine milk. *Br. J. Nutr.* 84: S 111–S 117.

Guang, C. and Phillips, R.D. 2009. Plant food-derived Angiotensin I converting enzyme inhibitory peptides. *J. Agric. Food Chem.* 24;57(12): 5113–5120.

Ha, E. and Zemel, M.B. 2003. Functional properties of whey, whey components, and essential amino acids: Mechanisms underlying health benefits for active peptide (review). *J. Nutr. Biochem.* 14: 261–258.

Hakansson, A., Zhivotovsky, B., Orrenius, S., Sabharwal, H., and Svanborg, C. 1995. Apoptosis induced by a human milk protein. *Proc. Natl. Acad. Sci.* 92: 8064–8068.

Hamel, V., Kielwein, G., and Teschemacher, H. 1985. Casomorphin immunoreactive materials in cows' milk incubated with various bacterial species. *J. Dairy Res.* 52: 139–148.

Hartmann, R. and Meisel, H. 2007. Food-derived peptides with biological activity: From research to food applications. *Curr. Opin. Biotechnol.* 2: 163–169.

Haubner, R. and Decristoforo, C. 2009. Radiolabelled RGD peptides and peptidomimetics for tumour targeting. *Front Biosci.* 1: 872–886.

Hausch, F., Shan, L., Santiago, N.A., Gray, G.M., and Khosla, C. 2002. Intestinal digestive resistance of immunodominant gliadin peptides. *Am. J. Physiol. Gastrointest. Liver Physiol.* 283: G996–G1003.

Jauregi, P. 2008. Bioactive peptides from food proteins: New opportunities and challenges. *Food Sci. Tech. Bull.* 5: 11–25.

Kanamoto, R., Azuma, B., Miyamoto, Y., Saeki, T., Tsuchihashi, Y., and Iwami, K. 2001. Soybean resistant proteins interrupt an enterohepatic circulation of bile acids and suppress liver tumorigenesis induced by azoxymethane and dietary deoxycholate in rats. *Biosci. Biotechnol. Biochem.* 65: 999–1002.

Kannan, A., Hettiarachchy, N., and Narayan, S. 2009. Colon and breast anti-cancer effects of peptide hydrolysates derive from rice bran. *Open Bioact. Comp.* 2: 17–20.

Kannan, A., Hettiarachchy, N., Johnson, M.G., and Nannapaneni, R. 2008. Human colon and liver cancer cell proliferation inhibition by peptide hydroly-sates derived from heat-stabilized defatted rice bran. *J. Agric. Food Chem.* 56: 11643–11647.

Katayama, S., Xu, X., Fan, M.Z., and Mine, Y. 2006. Antioxidative stress activity of oligophosphopeptides derived from hen egg yolk phosvitin in Caco-2 cells. *J. Agric. Food Chem.* 2006 54: 773–778.

Kim, H.D., Lee, H.J., Shin, Z.I., Nam, H.S., and Woo, H.J. 1995. Anticancer effects of hydrophobic peptides derived from a cheese slurry. *Food Biotechnol.* 4: 268–272.

Kim, J.Y., Woo, H.J., Ahn, C.W., Nam, H.S., Shin, Z.I., and Lee, H.J. 1999. Cytotoxic effects of peptides fractionated from bromelain hydrolyzates of soybean pro-tein. *Food Sci. Biotechnol.* 8: 333–337.

Kim, S.E., Kim, H.H., Kim, J.Y., Kang, Y.I., Woo, H.J., and Lee, H.J. 2000. Anticancer activity of hydrophobic peptides from soy proteins. *BioFactors* 12: 151–155.

Kim, S.H., Kim, J.Y., Park, S.W., Lee, K.W., Kim, K.H., and Lee, H.J. 2003. Isolation and purification of anticancer peptides from Korean ginseng. *Food Sci. Biotechnol.* 12: 79–82.

Kobayashi, H., Suzuki, M., Kanayama, N., and Terao, T. 2004. A soybean Kunitz trypsin inhibitor suppresses ovarian cancer cell invasion by blocking urokinase upregulation. *Clin. Exp. Metastasis* 21: 159–166.

Korhonen, H. and Pihlanto, A. 2003. Food-derived bioactive peptides–opportunities for designing future foods. *Curr. Pharm. Des.* 9: 1297–1308.

Korhonen, H. and Pihlanto, A. 2007. Technological options for the production of health-promoting proteins and peptides derived from milk and colostrum. *Curr. Pharm. Des.* 13: 829–843.

Kuwata, H., Yip, T.T., Yip, C.L., Tomita, M., and Hutchens, T.W. 1998. Direct detec-tion and quantitative determination of bovine lactoferricin and lactoferrin frag-ments in human gastric contents by affinity mass spectrometry. *Adv. Exp. Med. Biol.* 443: 23–32.

Lam, Y., Galvez, A.F., and de Lumen, B.O. 2003. Lunasin suppresses E1A-mediated transformation of mammalian cells but does not inhibit growth of immortalized and established cancer cell lines. *Nutr. Cancer* 47: 88–94.

Langer, M. and Beck-Sickinger, A.G. 2001. Peptides as carrier for tumor diagnosis and treatment. *Curr. Med. Chem. Anticancer Agents.* 1: 71–93.

Leung, E.H.W. and Ng, T.B. 2007. A relatively stable antifungal peptide from buck-wheat seeds with antiproliferative activity toward cancer cells. *J. Pept. Sci.* 13: 762–767.

Loponen, J. 2004. Angiotensin converting enzyme inhibitory peptides in Finnish cere-als: A database survey. *Agric. Food Sci.* 13: 39–45.

Lovati, M.R., Manzoni, C., Corsini, A., Granata, A., Frattini, R., Fumagalli, R., and Sirtori, C.R. 1992. Low density lipoprotein receptor activity is modulated by soybean globulins in cell culture. *J. Nutr.* 122: 1971–1978.

Mahaffey, S.M., Copeland, E.M. 3rd, Economides, E., Talbert, J.L., Baumgartner, T.G., and Sitren, H.S. 1987. Decreased lung metastasis and tumor growth in parenter-ally fed mice. *J. Surg. Res.* 42:159–165.

Maina, T., Nock, B., and Mather, S. 2006. Targeting prostate cancer with radiolabelled bombesins. *Cancer Imaging* 6:153–157.

Manzoni, C., Duranti, M., Eberini, I., Scharnag, H., Marz, W., Castiglioni, S., and Lovati, M. 2003. Subcellular localization of soybean 7S globulin in HepG2 cells and LDL receptor up-regulation by its α' constituent subunit. *J. Nutr.* 133: 2149–2155.

Matoba, N., Usui, H., Fujita, H., and Yoshikawa, M. 1999. Novel anti-hypertensive peptide derived from ovalbumin induces nitric oxide-mediated vasorelaxation in an isolated SHR mesenteric artery. *FEBS Lett.* 1999 452: 181–184.

Matsui, T., Li, C.H., and Osajima, Y. 1999. Preparation and characterization of novel bioactive peptides responsible for angiotensin I-converting enzyme inhibition from wheat germ. *J. Pept. Sci.* 5: 289–297.

Mauriello, G., De Luca, E., La Storia, A., Villani, F., and Ercolini, D. 2005. Antimicrobial activity of a nisin-activated plastic film for food packaging. *Lett. Appl. Microbiol.* 41: 464–469.

McCarty, M.F. 1999. Vegan proteins may reduce risk of cancer, obesity, and cardiovascular disease by promoting increased glucagon activity. Review. *Med. Hypotheses* 53: 459–485.

Meisel, H. and FitzGerald, R.J. 2003. Biofunctional peptides from milk proteins: Mineral-binding and cytomodulatory effects. *Curr. Pharm. Des.* 9: 1289–1295.

Meisel, H. and Frister, H. 1989. Chemical characterization of bio-active peptides from *in vivo* digests of casein. *J. Dairy Res.* 56: 343–349.

Meisel, H., Walsh, D.J., Murray, B., and FitzGerald, R.J. 2006. ACE inhibitory peptides. In: *Nutraceutical Proteins and Peptides in Health and Disease*, Boca Raton, FL: CRC Press.

Mellander, O. 1950. The physiological importance of the casein phosphopeptide calcium salts. II. Peroral calcium dosage of infants. *Acta Soc. Med. Uppsala* 55: 247.

Migliore-Samour, D. and Jolles, P. 1988. Casein, a prohormone with an immunomodulating role for the newborn? Review. *Experientia* 15:188–193.

Mine, Y. and Shahidi F. (Eds.) 2006. Nutraceutical proteins and peptides in health and disease. Boca Raton, FL: CRC Press.

Nagao, J.I., Asaduzzaman, S.M., Aso, Y., Okuda, K.I., Nakayama, J., and Sonomoto, K. 2006. Lantibiotics: Insight and foresight for new paradigm. *J. Biosci. Bioeng.* 102: 139–149.

Nagaoka, S., Futamura, Y., Miwa, K., Awano, T., Yamauchi, K., Kanamaru, Y., Tadashi, K., and Kuwata, T. 2001. Identification of novel hypocholesterolemic peptides derived from bovine milk β-lactoglobulin. *Biochem. Biophys. Res. Commun.* 281:11–17.

Ng, T.B., Lam, Y.W., and Wang, H. 2003. Calcaelin, a new protein with translation-inhibiting, antiproliferative and antimitogenic activities from the mosaic puffball mushroom Calvatia caelata. *Planta Med.* Mar; 69(3):212–217.

Oguro, T., Ohaki, Y., Asano, G., Ebina, Y., and Watanabe, K. 2001. Ultrastructural and immunohistochemical characterization on the effect of ovomucin in tumor angiogenesis. *Jpn. J. Clin. Electron. Microsc.* 33: 89–99.

Okmato, A., Hanagata, H., Kawamura, Y., and Yanagida, F. 1995. Anti-hypertensive substances in fermented soybean, natto plant foods. *Hum. Nutr.* 47: 39–47.

Otani, H. and Hata, I. 1995. *J Dairy Res.* 62: 339–348.

Park, J.H., Jeong, H.J., and Lumen, B.D. 2007. *In vitro* digestibility of the cancer-preventive soy peptides lunasin and BBI. *J. Agric. Food Chem.* 55: 10703–10706.

Parodi, P.W. 2007. A role for milk proteins and their peptides in cancer prevention. *Curr. Pharm. Des.* 13: 813–828.

Pedroche, J., Yust, M.M., Giron-Calle, J., Alaiz, M., Millan, F., and Vioque, J. 2002. Utilization of chickpea protein isolates for production of peptides with angiotensin I-converting enzyme (ACE)-inhibitory activity *J. Sci. Food Agric.* 82: 960–965.

Picot, L., Bordenave, S., Didelot, S., Fruitier-Arnaudin, I., Sannier, F., Thorkelsson, G., Berge, J.P., Guerard, F., Chabeaud, A., and Piot, J.M. 2006. Antiproliferative activity of fish protein hydrolysates on human breast cancer cell lines. *Proc. Biochem.* 41: 1217–1222.

Pihlanto, A., Akkanen, S., and Korhonen, H.J. 2008. ACE-inhibitory and antioxidant properties of potato (*Solanum tuberosum*). *Food Chem.* 109: 104–112.

Pihlanto-Leppala, A. 2001. Bioactive peptides derived from bovine whey proteins: Opioid and ACE inhibitory peptides. *Trends Food Sci. Technol.* 11: 347–356.

Pihlanto-Leppala, A., Koskinen, P., Piilola, K., Tupasela, T., and Korhonen, H. 2000. Angiotensin I-converting enzyme inhibitory properties of whey protein digests: Concentration and characterization of active peptides. *J. Dairy Res.* Feb; 67(1):53–64.

Potter, S.M. 1995. Overview of proposed mechanisms for the hypocholesterolemic effect of soy. *J. Nutr.* 125: 606S–611S.

Quist, E.E., Phillips, R.D., and Saalia, F.K. 2009. Angiotensin converting enzyme inhibitory activity of proteolytic digests of peanut (*Arachis hypogaea* L.) flour. *Food Sci. Technol.* 42: 694–699.

Recio, I. and Visser, S. 1999. Two ion-exchange chromatographic methods for the isolation of antibacterial peptides from lactoferrin. In situ enzymatic hydrolysis on an ion-exchange membrane. *J. Chromatogr. A.* 831: 191–201.

Roberts, K., Bhatia, K., Stanton, P., and Lord, R. 2004. Proteomic analysis of selected prognostic factors of breast cancer. *Proteomics* 3:784–792.

Roos, N., Mahe, S., Benamouzig, R., Sick, H., Rautureau, J., and Tome, D. 1995. 15N-labeled immunoglobulins from bovine colostrum are partially resistant to digestion in human intestine. *J. Nutr.* 125: 1238–1244.

Rutherfurd-Markwick, K.J. and Moughan, P.J. 2005. Bioactive peptides derived from food. *J. AOAC Intl.* 88: 955–966.

Rydlo, T., Miltz, J., and Mor, A. 2006. Eukaryotic antimicrobial peptides: Promises and premises in food safety. *J. Food Sci.* 71: R125–R135.

Sagara, M., Kanda, T., NJelekera, M., Teramoto, T., Armitage, L., Birt, N., Birt, C., and Yamori, Y. 2004. Effects of dietary intake of soy protein and isoflavones on cardiovascular disease risk factors in high risk, middle-aged men in Scotland. *J. Am. Coll. Nutr.* 23: 85–91.

Saito, T. 2008. Antihypertensive peptides derived from bovine casein and whey proteins. *Adv. Exp. Med. Biol.* 606: 295–317.

Satué-Gracia, M.T., Frankel, E.N., Rangavajhyala, N., and German, J.B. 2000. Lactoferrin in infant formulas: Effect on oxidation, *J. Agric. Food Chem.* 48: 4984–4990.

Schibli, D.J., Hwang, P.M., and Vogel, H.J. 1999. Structure of the antimicrobial active center of lactoferricin B bound to sodium dodecyl sulfate micelles. *FEBS Lett.* 446: 213–217.

Schuppan, D. 2000. Current concepts of celiac disease pathogenesis. *Gastroenterology* 119: 234–242.

Shahidi, F. and Zhong, Y. 2008. Bioactive peptides. *J. AOAC Intl.* 91(4): 914–931.

Shargel, L. and Yu, A.B. 1999. *Applied Biopharmaceutics & Pharmacokinetics* (4th ed.). New York: McGraw-Hill.

Sirtori, C.R., Agradi, E., Mantero, O., Conti, F., and Gatti, E. 1977. Soybean-protein diet in the treatment of type-II hyperlipoproteinaemia. *Lancet* 1: 275–277.

Suetsuna, K., Ukeda, H., and Ochi, H. 2000. Isolation and characterization of free radical scavenging activities of peptides derived from casein. *J. Nutr. Biochem,* 11:128–131.

Sutas, Y., Hurme, M., and Isolauri, E. 1996. Down-regulation of anti-CD3 antibody-induced IL-4 production by bovine caseins hydrolysed with Lactobacillus GG-derived enzymes. *Scand. J. Immunol.* 43: 687–689.

Szeto, H.H. 2006. Cell-permeable, mitochondrial-targeted, peptide antioxidants. *AAPS J.* 8: E277–E283.

Takano, T. 1998. Milk derived peptides and hypertension reduction. *Int. Dairy J.* 8: 375–381.

Takenaka, A., Annaka, H., Kimura, Y., Aoki, H., and Igarashi, K. 2003. Reduction of paraquat-induced oxidative stress in rats by dietary soy peptide. *Biosci. Biotechnol. Biochem.* 67: 278–283.

Tsujita, T., Takaichi, H., Takaku, T., Aoyama, S., and Kiraki, J. 2006. *J. Lipid Res.* 47: 1852–1858.

U.S. Food and Drug Administration. 1999. *Fed. Regist.* 64: 57688–57733.

van der Ven, C., Gruppen, H., de Bont, D.B.A., and Voragen, A.G.J. 2002. Optimization of the angiotensin converting enzyme inhibition by whey protein hydrolysates using response surface methodology. *Int. Dairy J.* 12: 813–820.

Vermeirssen, V., Van Camp, J., and Verstraete, W. 2002. Optimisation and validation of an angiotensin-converting enzyme inhibition assay for the screening of bioactive peptides. *J. Biochem. Biophys. Meth.* 51: 75–87.

Wang, W. and de Mejia, E.G. 2005. A new frontier in soy bioactive peptides that may prevent age-related diseases. *Comp. Rev. Food Sci. Food Safety* 4: 63–78.

Wang, Y., Jones, P.J., Ausman, L.M., and Lichtenstein, A.H. 2004. Soy protein reduces triglyceride levels and triglyceride fatty acid fractional synthesis rate in hypercholesterolemic subjects. *Atherosclerosis* 173: 269–275.

Yamamoto, N. 1997. Antihypertensive peptides derived from food proteins. *Biopolymers* 43: 129–134.

Yang, W., Meng, L., Wang, H., Chen, R., Wang, R., Ma, X., Xu, G., Zhou, J., Wang, S., Lu, Y., and Ma, D. 2006. Inhibition of proliferative and invasive capacities of breast cancer cells by arginine-glycine-aspartic acid peptide *in vitro. Oncol. Rep.* 1:113–117.

Yang, Y., Tao, G., Liu, P., and Liu, J. 2007. Peptide with anigiotensin I-converting enzyme inhibitory activity from hydrolyzed corn gluten meal. *J. Agric. Food Chem.* 55: 7891–7895.

Yi, H.J., Kim, J.Y., Kim, K.H., Lee, H.J., and Lee, H.J. 2003. Anticancer activity of peptide fractions from egg white hydrolysate against mouse lymphoma cells. *Food Sci. Biotechnol.* 12: 224–227.

Yoo, Y.C., Watanabe, R., Koike, Y., Mitobe, M., Shimazaki, K., Watanabe, S., and Azuma, I. 1997. Apoptosis in human leukemic cells induced by lactoferricin, a bovine milk protein-derived peptide: Involvement of reactive oxygen species. *Biochem. Biophys. Res. Commun.* 237: 624–628.

Yoshinori, M. and Shahidi, F. 2000. In *Nutraceutical Proteins and Peptides in Health and Disease.* Boca Raton, FL: CRC Press.

Zhang, J., Tatsumi, E., Ding, C., and Li, L. 2006. Angiotensin I-converting enzyme inhibitory peptides in douchi, a Chinese traditional fermented soybean product. *Food Chem.* 98: 551–557.

2

Proteins and Peptides as Allergens

Soichi Tanabe and A. Wesley Burks

CONTENTS

2.1 Introduction

2.1.1 Food-Induced Allergies

Food-induced allergies include IgE-mediated, type I allergies and celiac disease, an autoimmune reaction. In type I hypersensitivity reactions, allergens are recognized by IgE antibodies and inflammatory mediators trigger tissue responses. IgE-mediated food allergies are triggered soon after antigen ingestion. Symptoms can include itching in the mouth, difficulty swallowing or breathing, nausea, vomiting, diarrhea, abdominal pain, hives, eczema, and asthma. Severe allergic reactions can result in anaphylaxis. Celiac disease is an immune-mediated disorder resulting from permanent gluten intolerance and is characterized by chronic inflammation of the small bowel's mucosa and submucosa. In untreated IgE celiac disease, intestinal villi become flattened and impaired in their ability to absorb nutrients. Celiac disease presentation ranges from no symptoms to severe malnourishment. The most common clinical manifestations of celiac disease include abdominal pain with moderate to severe distension, gastroesophageal reflux, recurrent episodes of altered bowel habits (diarrhea or constipation), weight loss, bone disease, anemia, and weakness.

More than 170 foods have been reported to cause IgE-mediated reactions, although 8 foods account for 90% of all food-allergic reactions: milk, eggs, peanuts, tree nuts, fish, shellfish, soy, and wheat (Food Allergy & Anaphylaxis Network, 2010). The prevalence of food allergy has risen throughout the past 10 to 20 years (Branum and Lukacs, 2009; Sicherer, Muñoz-Furlong, and Sampson, 2003). Approximately 3% of children and adults have an allergy to milk, eggs, peanuts, fish, or crustacean shellfish, as assessed by a double-blind, placebo-controlled food challenge or by self-reported symptoms plus sensitization (Rona et al., 2007). The allergenic regions of protein recognized by the binding sites of IgE molecules are called IgE-binding epitopes. These epitopes can be classified into two categories: (1) conformational epitopes where residues are distantly separated in the sequence and brought into physical proximity by protein folding; and (2) linear or sequential epitopes composed of a single continuous stretch of amino acids within a protein sequence that can react with antiprotein antibodies. Both conformational and sequential epitopes may be responsible for allergic reactions. However, in this chapter, the focus is mainly on linear epitopes.

The overall prevalence of celiac disease in the general population appears to be approximately 1/160 (6.2‰), but this figure varies on the basis of diagnostic criteria (Biagi et al., 2010). The triggers for celiac disease are specific immunogenic peptides that are present only and exclusively in dietary gluten proteins from wheat and similar structural cereals such as rye and barley. Tissue transglutaminase-2 (tTG), appears to be an important component of the pathogenesis of celiac disease (Rodrigo, 2006), and active celiac disease is characterized by intestinal or extraintestinal symptoms, villous atrophy, crypt hyperplasia, and strongly positive tTG autoantibodies. A duodenal biopsy is considered to be the gold standard for diagnosis, but the assessment is limited by variations in interpretation.

Treatment for food allergies involves strict avoidance of the trigger allergen. Similarly, treatment of celiac disease is a life-long, gluten-free diet, which leads to remission for most individuals.

2.2 Allergenicity of Wheat Gluten

Wheat is one of the world's most important grains, and a variety of wheat-derived products are consumed throughout the world. However, wheat is one of the most common food allergens. The adverse reactions to wheat flour may present clinical outcomes such as (1) atopic dermatitis, (2) food-dependent exercise-induced anaphylaxis (FDEIA), (3) celiac disease, and (4) baker's asthma (Tanabe, 2008). In these cases, certain wheat proteins induce abnormal allergic reactions in sensitized individuals.

2.2.1 Atopic Dermatitis

The wheat allergens associated with atopic dermatitis are heterogeneous. Using serum from patients allergic to wheat, Tanabe (2008) evaluated the allergenicity of salt-soluble and salt-insoluble (gluten) fractions by means of enzyme-linked immunosorbent assay (ELISA) and found most patients were sensitive to gluten.

The primary allergenic structure of gluten is a 30-mer peptide with the sequence (Ser-Gln-Gln-Gln-(Gln-)Pro-Pro-Phe)$_4$ (Table 2.1A). This allergenic peptide shows high sequence similarity (almost 90%) to precursors of low-molecular-weight glutenin, a urea-, detergent-, or KOH-soluble fraction of gluten. Similarities of about 70% also exist between the sequence of gluten's allergenic peptide and those of low-molecular-weight glutenin precursors from durum wheat.

Repeat sequences in allergenic peptides such as (Ser-Gln-Gln-Gln-(Gln-)Pro-Pro-Phe)$_4$ may be favorable for crosslinking IgE antibodies and triggering the release of chemical mediators from mast cells in the body. Indeed, there are 44 and 25 Gln-Gln-Gln-Pro-Pro sequences in high-molecular-weight glutenin subunit x and y, respectively (accession numbers, CAC40686 and CAC40687). Like Gln-Gln-Gln-Pro-Pro repeat sequences, cod allergen (*Gad c* 1, allergen M) is well known for containing three homologous IgE-binding tetrapeptides (Elsayed et al., 1982).

The peptides listed in Table 2.1 were synthesized by the solid phase method (Tanabe et al., 1996). As shown in Table 2.1A, (Ser-Gln-Gln-Gln-(Gln-)Pro-Pro-Phe)$_4$, (Ser-Gln-Gln-Gln-(Gln-)Pro-Pro-Phe)$_2$, and Ser-Gln-Gln-Gln-(Gln-)Pro-Pro-Phe bind to IgE almost equally. There is no difference between the relative ELISA values of Ser-(Gln)$_4$-Pro-Pro-Phe and Ser-(Gln)$_3$-Pro-Pro-Phe. These data suggest that the Ser-Gln-Gln-Gln-Pro-Pro-Phe motif is involved in binding to IgE antibodies.

When any of the asterisked amino acid residues in the sequence Ser-Gln*-Gln-Gln-Pro*-Pro*-Phe are replaced by Gly, the ELISA value drops below the limit of detection (Table 2.1B). These amino acid residues are therefore thought to be indispensable for binding to IgE. Within Gln-Gln-Gln-Pro-Pro, the N-terminal glutamine residue and the two proline residues are essential for binding to IgE (Table 2.1C). The IgE-binding epitope of the allergenic peptide thus comprises Gln-Xaa-Xaa-Pro-Pro, where Xaa are replaceable amino acid residues. Indeed, data from inhibition ELISA

TABLE 2.1

IgE-Binding Abilities of Wheat Low Molecular Weight Glutenin Peptides

Peptides	Relative ELISA Value
A:	
acetyl-SQQQQPPF SQQQPPF SQQQQPPF SQQQPPF	1.0
acetyl-SQQQQPPF SQQQPPF	1.1
acetyl-SQQQQPPF	1.1
acetyl-SQQQPPF	1.0
B:	
acetyl-GQQQPPF	1.1
acetyl-SGQQPPF	nd
acetyl-SQGQPPF	0.8
acetyl-SQQGPPF	1.0
acetyl-SQQQGPF	nd
acetyl-SQQQPGF	nd
acetyl-SQQQPPG	0.9
C:	
acetyl-QQQPP	0.9
acetyl-GQQPP	nd
acetyl-QGQPP	0.7
acetyl-QQGPP	1.0
acetyl-QQQGP	nd
acetyl-QQQPG	nd
QQQPP	0.6

Source: Tanabe, S., Arai, S., Yanagihara, Y. et al. (1996). *Biochem. Biophys. Res. Commun.* 219:290–293 Table 2.2. With permission from Elsevier.

assays indicate that Ac-Gln-Gln-Gln-Pro-Pro binds to wheat-specific IgE in the serum of patients.

Recombinant low-molecular-weight glutenins with many Gln-Gln-Gln-Pro-Pro motifs were expressed in *Escherichia coli* by a pET vector system and confirmed to possess IgE-binding ability (Maruyama et al., 1998). In addition, several allergens were identified from the water-soluble "albumin" and salt-soluble "globulin" fractions, such as alpha-amylase inhibitors. However, these soluble allergens are beyond the scope of this chapter.

2.2.2 Food-Dependent Exercise-Induced Anaphylaxis (FDEIA)

FDEIA is a severe, life-threatening form of allergy (IgE-mediated) in which the ingestion of a specific food before physical exercise induces symptoms of anaphylaxis. Although various foods (including shellfish, celery, hazelnuts, peanuts, soy, peas, and bananas) have been associated with FDEIA, the most frequently reported cause of these reactions seems to be wheat (Palosuo et al., 1999). Wheat-dependent, exercise-induced anaphylaxis is not as rare as previously thought, but instead is a rather poorly recognized disorder. Four IgE-binding epitopes, Gln-Gln-Ile-Pro-Gln-Gln-Gln (QQIPQQQ), QQFPQQQ, QQSPEQQ, and QQSPQQQ in omega-5 gliadin; an ethanol-soluble fraction of gluten; and three IgE-binding epitopes, QQPGQ, QQPGQGQQ, and QQSGQGQ, in high-molecular-weight glutenin have been identified (Morita, Kunie, and Matsuo, 2007).

2.2.3 Celiac Disease

Gluten-sensitive enteropathy is caused by ingestion of gliadin. The minimum epitope structures responsible for symptoms are PSQQ and QQQP (Sturgess et al., 1994). Although the disease is mediated by T lymphocyte-driven immunological activation in the gastrointestinal mucosa, the levels of total and wheat-specific IgE antibodies in celiac patients are usually not elevated, and celiac disease should not be classified as an allergic disorder (Bahna, 1996) but as an autoimmune disorder.

2.2.4 Baker's Asthma

The inhalation of wheat flour often causes asthma (Gómez et al., 1990), which is known as baker's asthma, a typical occupational allergic disease that has been known since ancient Roman times. Extensive studies have identified some proteins as allergens associated with asthma. Among them, α-amylase inhibitors from globulin fraction were identified as major allergens (Amano et al., 1998; Gómez et al., 1990). The IgE-binding epitope structures of an α-amylase inhibitor (known as the 0.28 wheat AI) have already been determined (Walsh and Howden, 1989). Acyl-CoA oxidase

(Weiss, Huber, and Engel, 1997), peroxidase (Sánchez-Monge et al., 1997), and fructose-bisphosphate aldolase (Posch et al., 1995) have been identified as other allergens.

2.3 Structure of Proteinaceous Allergens in Plants (Cereal, Beans, Vegetables, Fruits)

Allergenic plants include nuts, seeds, grains, and a variety of fresh fruits and vegetables. The allergens are designated by the Latin name of the plant, with the first three letters of the genus followed by the first letter of the species and an Arabic number. Thus, an allergen from *Arachis hypogaea* (peanut) is prefaced Ara h followed by a number, which usually represents the order in which allergens are identified. Approximately 65% of plant food allergens belong to just four protein superfamilies: the prolamin, cupin, Bet v 1, and profilin superfamilies (Chapman et al., 2007). Within three of these super-families (prolamin, cupin, and profilin) are allergens found in the legume family of plants, which are important in global nutrition but common sources of food allergens (Table 2.2).

2.3.1 Prolamins

The prolamin superfamily is characterized by a conserved pattern of cyste-ine residues found in seed storage prolamins, α-amylase/trypsin inhibitors of monocotyledonous cereal seeds, and 2S storage albumins. Soybean hydro-phobic protein, nonspecific lipid transfer proteins, and α-globulins also belong to the prolamin superfamily. The conserved pattern comprises a core of eight cysteine residues that includes a Cys-Cys and Cys-Xaa-Cys motif, (Xaa representing any other residue; Shewry et al., 2002). Two additional cysteine residues are found in the α-amylase and trypsin inhibitors. Except for seed storage prolamins, members of this superfamily share a common three-dimensional structure: a bundle of four α-helices stabilized by disul-fide bonds (Shewry et al., 2002). The bundle maintains the three-dimensional structure of many of these proteins after heating, which may maintain aller-genicity after cooking.

2.3.1.1 2S Albumins

The 2S albumins are seed storage proteins usually synthesized in the seed as single chains of M_r 10,000–15,000 (Bewley, Black, and Halmer, 2006). The sin-gle chains may be posttranslationally processed to give small and large sub-units that usually remain joined by disulphide bonds. Several 2S albumins

TABLE 2.2

Legume Allergens and Protein Classification

| Crop | Cupins | | Prolamins | | | Profilins | Allergen/Other Proteins Families |
	7S Globulins	11S Globulins	2S Albumins	ns-LTPS	PR-Proteins		
Peanuts	Ara h 1	Ara h 3 Ara h 4	Ara h 2 Ara h 6 Ara h 7	Ara h LTP	Ara h 8 (PR-10)	Ara h 5	Ara h oleosin
Soybeans	Gly m Bd 28K Gly m 5	Gly m 6	Gly m 2S Albumin	Gly m 1	Gly m 4 (PR-10)	Gly m 3	P34/Cysteine protease Gly m TI/Kunitz trypsin inhibitor Seed biotynilated protein
Lentils	Len c 1						Len c 2/Seed biotynilated protein
Mung beans					Vig r 1 (PR-10)	Vig r profilin	
Common beans					Pha v chitinase (PR-3)		
Cowpeas			Vig u				
Chickpeas			Cic a 2S albumin				
Peas	Pis s 1 Pis s 2					Pis s profilin	
White lupin	Lup a gamma_conglutin Lup a vicilin	Lup a 11S Globulin					
Locust tree					Rob p glucanase (PR-2)	Rob p profilin	Rob p 4/Calcium-binding protein

Source: Reprinted from Riascos, J.J., Weissinger, A.K., Weissinger, S.M. et al. 2010. *J. Agric. Food Chem.* 58: 20–27. With permission.

have been identified as important allergens in nuts and seeds. Examples include allergens in walnuts (Jug r 1), Brazil nuts (Ber e 1), peanuts (Ara h 2 and Ara h 6), cashews (Ana o 3), mustard (Bra j 1), and sunflower seeds (SFA-8); (Radauer and Breiteneder, 2007).

2.3.1.2 Lipid Transfer Proteins

Lipid transfer proteins are located in the outer epidermal tissues of plants, such as the peel of peaches or apples. They were named for their ability to transfer lipids *in vitro*. They have been termed pan-allergens because they are found in a variety of plant organs, including seeds, fruit, and vegetative tissues.

2.3.1.3 Seed Storage Prolamins

Prolamins are the major seed storage proteins of wheat, barley, and rye. In their native state, prolamins are insoluble in water or dilute salt solutions. They have characteristic amino acid compositions that include high contents of proline, and glutamine, as well as methionine, aromatic amino acids, and glycine (Shewry and Tatham, 1990). The prolamins in wheat form the large, disulfide-linked polymers of gluten.

2.3.1.4 Bifunctional Inhibitors

The α-amylase/trypsin inhibitor family is associated with baker's asthma or food allergies to flour of: wheat, barley, and rye which are members of the grass tribe Triticeae (Breiteneder and Ebner, 2000). Members of the Triticeae inhibitor family have 12–16 kDa polypeptides, are usually rich in glutamine plus asparagine and proline residues, and have four or five disulphide bridges essential for their inhibitory activity (Salcedo et al., 2004).

2.3.2 Bet v 1 Homologues

The major birch pollen allergen, Bet v 1, is a β-barrel protein that can bind plant steroids in a central tunnel. Bet v 1 and its homologues belong to family 10 of the pathogenesis-related proteins (Radauer and Breiteneder, 2007) and may have a role in plant protection, although this has not been confirmed. The conservation of both the amino acid sequence and molecular surfaces of Bet v 1 and its homologues explain the cross-reactivity of IgE and hence the widespread cross-reactive allergies to fresh fruits and vegetables frequently observed in individuals with birch pollen allergy.

2.3.3 Cupins

Cupins are characterized by a β-barrel structure (Chapman et al., 2007). The 11S and 7S seed storage proteins, which belong to the cupin family, are major allergens in nuts and seeds (Radauer and Breiteneder, 2007). The 11S globulins, sometimes termed legumins, are hexameric proteins (Adachi et al., 2003) of $M_r \sim$ 300,000–450,000 (Bewley et al., 2006). The 7/8S globulins comprise three subunits of $M_r \sim$ 40,000–80,000 (Bewley et al., 2006). Both 11S and 7S seed storage globulins have been reported as allergens in peanuts (Ara h 1 and Ara h 3), cashew nuts (Ana o 1 and Ana o 2), walnuts (Jug r 2 and Jug r 4), and sesame seeds (Ses i 1, Ses i 3, Ses i 6); (Radauer and Breiteneder, 2007).

2.3.4 Profilins

Profilins, cytosolic proteins of 12 to 15 kDa (Radauer and Breiteneder, 2007), are involved in the pollen-fruit allergy syndrome. Profilins are thought to regulate actin polymerization by binding to monomeric actin and a number of other proteins (Witke, 2004). Profilins are found in all eukaryotic cells (Radauer and Breiteneder, 2007), but only those found in plants have been described as allergens. Profilin-specific IgE cross-reacts with homologues from virtually every plant source, and sensitization to these allergens has been considered a risk factor for multiple pollen allergies and pollen-associated food allergy. However, the clinical relevance of cross-reactivity remains unclear.

2.4 Structure of Proteinaceous Allergens in Animal Products (Meat, Milk, Eggs)

2.4.1 Meat Allergens

The major beef allergens are bovine serum albumin (BSA, Bos d 6, 66-67 kDa) (Fiocchi et al., 2000) and gamma globulin (BGG, Bos d 7, 160 kDa) (Besler, Fiocchi, and Restani, 2001). Actin, myoglobin, and tropomyosin are minor allergens, but myosin is rarely allergenic (Besler, et al., 2001; Fiocchi, Restani, and Riva, 2000). BSA and BGG are also present in milk, and Ayuso et al. (1999) have suggested sensitization to beef may be secondary to milk allergy, because milk is introduced first in the diet.

Serum albumin is one of the most widely studied and applied proteins in biochemistry. The most abundant protein in the circulatory system, serum albumin accounts for 60% of total serum protein, with a concentration of approximately 40 mg/mL. Serum albumins comprise approximately 580 amino acid residues and are characterized by a low tryptophan content and a high cysteine content, and charged amino acids such as aspartic and glutamic acids, lysine, and arginine. The tertiary structure comprises three domains, I, II, and III (aa1–190, 191–382, 383–581), which assemble to form a heart-shaped molecule. The fact that a major component of serum acts as an allergen is very surprising. It is remarkable that albumins from animals, which are very similar in sequence, structure, and function to human serum albumin, do not induce tolerance.

The precise regions of IgE-binding epitopes of BSA have been identified (Tanabe et al., 2002; Beretta et al., 2001). It was hypothesized that BSA-specific antibodies and T cells react primarily with sequential epitopes in which the amino acid sequences differ greatly between bovine and human albumin. Sixteen peptides (P-1 through P-16) corresponding to such regions were synthesized as candidate epitopes. Among them, aa336-345 (P-7) and aa451-459 (P-11) were found to be major IgE-binding epitopes (Figure 2.1a). In inhibition ELISA (enzyme-linked immunosorbent assay), EYAV (aa338-341) bound to patient IgE antibodies and was found to be the core of the IgE-binding epitopes (Tanabe et al., 2002).

Among eight IgE-binding epitopes, three were found to contain an EXXV motif (HPEYAVSVLL [P-7], PVESKVT [P-12], and VMENFVAF [P-15]). The corresponding sequences in HSA are HPDYSVVLLL, PVSDRVT, and VMDDFAAF, respectively. Comparing two epitope sequences (P-7 and P-15) in BSA with the corresponding sequences in HSA, it appeared likely that E residues (E338 and E547) are important for recognition by IgE-antibodies, because the corresponding residues in HSA are D in both peptide P-7 and P-15. Therefore, two analogue peptides, E338D (HPDYAVSVLL) and E547D (VMDNFVAF), with amino acid substitutions from E to D, were characterized for their IgE-binding abilities (Tanabe, Shibata, and Nishimura, 2004). The substitution of glutamic acid in the EXXV sequence with aspartic acid led to a remarkable reduction in IgE-binding ability (Figure 2.1b). Thus, [338]E and [547]E in BSA were thought to be important for recognition by patient IgE antibodies. In other words, the difference between D (human type) and E (bovine type) at positions 338 and 547 seems to be a major cause of the allergenicity of BSA.

Restani et al. (1997) reported the cross-reactivity between serum albumins from different animal species. In their study, a skin prick test (SPT) and immunoblotting were performed for each of seven serum albumins (beef, sheep, pig, horse, rabbit, turkey, and chicken). As a result, a clear relationship was found between the sequence homology of different serum albumins with BSA and the percentage of positive SPT and immunoblotting in the serum of beef-allergic children. Therefore, the use of alternative meats in meat-allergic patients must be carefully evaluated on an individual basis (Fiocchi et al., 2000).

The cross-reactivity of porcine meat and cat epithelia/dander has been reported and called "pork–cat syndrome" (Hilger et al., 1997). Nearly all patients with IgE antibodies to pork also have IgE antibodies to cat epithelia/dander. Similarly, there is cross-reactivity between chicken and hen's egg yolk, which is known as "bird–egg syndrome" (Quirce et al., 2001). Egg yolk contains significant quantities of serum proteins. In bird–egg syndrome, the cross-reacting allergen has been reported to be α-livetin in egg yolk, which

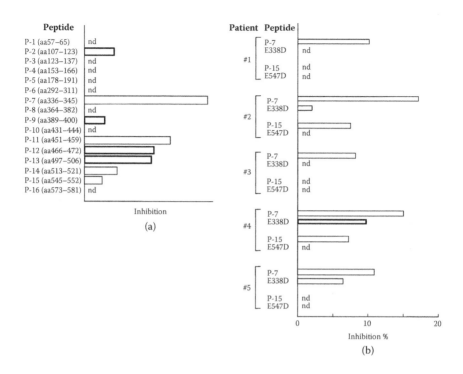

FIGURE 2.1

IgE-binding abilities of the synthetic BSA peptides. (a) Solid-phage ELISA of 16 kinds of BSA peptides. (From Tanabe, S., Kobayashi, Y., Takahata, Y. et al. 2002. *Biochem. Biophys. Res. Commun.* 293: 1348–53. Reprinted with modifications with permission from Elsevier.) (b) Inhibition ELISA of P-7, P-15, E338D, and E547D. (From Tanabe, S., Shibata, R., Nishimura, T. 2004. *Mol. Immunol.* 41: 885–890. Reprinted with modifications with permission from Elsevier.)

is identical to chicken serum albumin (Gal d 5). Chicken serum albumin can act as both an inhalant and a food allergen in patients with bird–egg syndrome (Quirce et al., 2001). In this regard, sensitization to meat may be secondary to hair and dander sensitization (especially in veterinarians) as well as secondary to milk allergy, as described earlier.

2.4.2 Milk Allergens

Cow's milk is one of the major causes of food allergies in children throughout the world. Milk proteins are classified into caseins and whey proteins. Casein and whey proteins constitute approximately 80% and 20% of the total milk protein, respectively. Casein is composed of four different proteins: α_{S1}-, α_{S2}-, β-, and κ-casein in proportions of 40:10:40:10, respectively. The major milk allergens are β-lactoglobulin (BLG, Bos d 5, 18 kDa), the main component (approximately 50%) of whey proteins, and caseins (Bos d 8, 19-25 kDa); (Cerecedo et al., 2008; Restani et al., 2009). A recent study of 113 patients with milk allergy suggested that 64.6% had detectable

IgE to α-casein, 75.2% to β-casein, 47.8% to κ-casein, and 36.2% to both α-lactalbumin and BLG (Shek et al., 2005).

Milk has numerous IgE-binding epitopes. Recently, Cerecedo et al. (2008) used a peptide microarray-based immunoassay to map IgE and IgG4 sequential epitopes in 31 children with IgE-mediated milk allergy. The 31 children included 16 with positive oral milk challenge results (reactive group). A library of peptides, consisting of 20 amino acids overlapping by 17 (3-offset), corresponding to the primary sequences of α_{S1}-, α_{S2}-, β-, and κ-caseins, and BLG, was printed on epoxy-coated slides. A region was defined as an epitope if it was statistically associated with the reactive group and recognized by at least 75% of reactive patients. A total of 10 epitopes were identified. These were aa28-50 in α_{S1}-casein; aa1-20, aa13-32, aa67-86, and aa181-207 in α_{S2}-casein; aa25-50, aa52-74, aa154-173 in β-casein; aa34-53 in κ-casein; and aa58-77 in BLG (Cerecedo et al., 2008).

Cow's milk is commonly subjected to pasteurization, usually ultra-high-temperature (UHT) processing. Sensitization to pasteurized cow's milk indicates that the allergenicity of the milk allergen is heat-resistant; linear epitopes are not affected by structural changes and maintain their IgE-binding ability after heating (Restani et al., 2009). Protease treatment has been successfully used to develop hypoallergenic formulas, which is described in Section 2.8.

2.4.3 Egg Allergens

The major egg white allergens are ovomucoid (OVM, Gal d 1, 28 kDa) and ovalbumin (OVA, Gal d 2, 45 kDa). Minor allergens include ovotransferrin (OVT, Gal d 3, 77 kDa) and lysozyme (LZ, Gal d 4, 14 kDa) (Mine and Yang, 2008). Although egg whites are the primary source of major egg allergens, egg yolk also contains allergenic proteins such as apovitellenins I and VI and phosvitin (Walsh et al., 1988).

IgE-binding epitopes of OVM have been reported in both human egg-allergic patients and murine experimental models. Probing with pooled sera from seven egg-allergic patients led to the identification of five IgE-binding epitopes and seven IgG-binding regions (Cooke and Sampson, 1997). Interestingly, the strongest IgE-binding activity was to OVM domain II, aa65-120. Another study determined instead that the strongest binding activity was to OVM domain III with immunodominant epitopes being mainly linear (Zhang and Mine, 1998). *In vitro* experiments with targeted chemical modifications indicated that hydrophilic residues were important for IgG isotype binding, whereas hydrophobic residues were important for IgE isotype binding (Zhang and Mine, 1999). Subsequently, a detailed study using pooled sera from eight egg-allergic patients and overlapping decapeptides synthesized on cellulose membranes (simple precise original test system) led to the identification of three IgE epitopes in domain I, four IgE epitopes in domain II, and two IgE epitopes in domain III (Mine and Rupa, 2003).

OVM is characterized by its high heat stability, and this feature has been credited for its dominance as a major egg allergen and its role in the prognosis of egg allergy. OVM is known to exhibit trypsin inhibitor activity (Mine and Yang, 2008). The allergenic and antigenic properties of OVM are maintained after peptic digestion (Kovacs-Nolan et al., 2000). As evaluated by ELISA with sera from human egg-allergic patients, pepsin digestion of ovomucoid does not alter its IgE-binding capacity (Kovacs-Nolan et al., 2000).

OVA is the major protein in avian egg white and composes 54% of its total protein content. OVA has been widely used for preparing allergic models in experimental animals. OVA has 386 amino acid residues with a single carbohydrate chain linked to Asn293 and a single disulfide between Cys74 and Cys121. Studies involving both egg-allergic patients and murine model sera have examined IgE-binding epitopes of OVA (Mine and Yang, 2008). Elsayed, Holen, and Haugstad (1988) first reported human IgE-binding activity in the fragment aa1-10 of the N terminus. Two subsequent papers showed that residues aa11-19 and aa34-70 (Elsayed and Stavseng, 1994) and also fragments aa41-172 and aa301-385 (Kahlert et al., 1992) had human IgE-binding activity. In *in vitro* experiments, peptide sequence aa347-385 inhibited histamine release from human basophils (Honma et al., 1996).

In comparison with OVM, OVA is relatively heat labile. Exposure of OVA to a temperature of 80°C for 3 min decreases its IgE-binding activity by 90% (Honma et al., 1994). However, the antigenicity of OVA can resist heat treatment under certain conditions (Elsayed et al., 1986). Disparities in the above results may be explained by variations inherent to patient cohort characteristics (Mine and Yang, 2008). Currently, it has been recognized that OVA is relatively stable in simulated gastric and intestinal fluid; however, preheating could lead to a decrease of its resistance to proteolysis (Takagi et al., 2003).

The epitopes of OVT and LZ relevant to egg allergy have not as yet been identified.

2.5 Structure of Proteinaceous Allergens in Seafood

The major allergens of Crustacea and Mollusca invertebrates are tropomyosins. Together with the proteins actin and myosin, tropomyosins regulate contraction in both muscle and nonmuscle cells. Tropomyosins, which are ubiquitous in animal cells, comprise a repetitive sequence of heptapeptide repeats that form two α-helixes assembled into double-stranded coils. The monomers form head-to-tail polymers along the length of an actin filament.

Data from IgE-epitope mapping indicate that sequences unique to invertebrate tropomyosins, located in the C-terminal region of the protein, play an important role in their allergenic potential. The lack of homology between vertebrate and invertebrate tropomyosins means there is no cross-reactivity between shellfish and animal muscle tropomyosins.

2.6 Structural Features of Allergenic Proteins and Peptides

Protein allergens possess a wide range of physical characteristics, but none are unique to protein allergens as a class. Allergens contain B-cell epitopes to which IgE can bind and T cell epitopes capable of inducing a type 2 T-lymphocyte response; however, the presence of appropriate epitopes alone is insufficient to impart allergenic potential. There have been efforts to use bioinformatic methods to predict what makes some proteins allergens; however, prediction of allergenic activity of proteins is not currently possible.

Several biochemical and physiochemical factors contribute to allergenicity of a food. Stability of the protein or epitope structure can be an important factor in maintaining allergenicity. Many allergens are exceptionally heat stable and retain allergenicity after heating. Others resist attack by proteases, such as those encountered in the gastrointestinal tract. Stability can be modified by ligands, such as lipids and metal ions. Intramolecular disulfide bonding may also contribute to allergenicity; induction of IgE antibody by wheat agglutinin can be lessened by reducing disulfide bonds with thioredoxin (Buchanan et al., 1997). It is not clear, however, whether disulfide bonds make a specific qualitative contribution to allergenicity of wheat agglutinin, or simply make a quantitative contribution to overall antigenicity.

2.7 Characterization of Allergenic Proteins and Peptides

The process of characterizing allergenic proteins and peptides involves analysis of reactive IgE from sera of allergic patients. Multiple methods can be used to isolate and identify proteins and peptides. Food extracts can be separated by SDS-PAGE gel and subjected to IgE immunoblotting using patient sera. Selected IgE reactive bands can be subjected to N-terminal amino acid sequencing. Proteins can also be purified by chromatography and *de novo* sequenced using matrix-associated laser desorption ionization and tandem mass spectrometry. Purification and characterization of allergens can also be done by using in-gel tryptic digestion mass spectrometry to confirm their identities, and circular dicroism and Fourier-transform infrared spectroscopies to indicate authentic folding.

2.8 Hypoallergenic Treatment

The allergenicity of food allergens can be changed by the treatment of physical (heating, high pressure, etc.) and biochemical (proteolysis, fermentation, etc.) processes. Heating (baking, boiling, cooking, drying, grilling, roasting, pasteurization, and sterilization) is the easiest way to change allergenicity. As heat generally induces protein denaturation, heat treatment reduces allergenicity of heat-liable allergens by altering protein conformation in such a way that destroys IgE-reactive conformational epitopes.

However, there are many heat-stable allergens. Examples are allergens in the prolamin superfamily, which include the 2S albumin allergens such as the Brazil nut allergen Ber e 1 and the nonspecific lipid transfer protein (ns LTP) allergens from apple Mal d 3 and grape Vit v 1 (Mills et al., 2009). In a small number of cases, heat treatment increases allergenicity. IgE immunoreactivity of peanut allergens Ara h 1 (7S globulin) and Ara h 2 (2S albumin) prepared from roasted peanuts has been reported to be higher than that of their counterparts prepared from raw and boiled peanuts (Mondoulet et al., 2005).

Biochemical food processing methods for preparing hypoallergenic foods often involve use of enzymes such as proteases, oxidases, or transglutaminase (Paschke, 2009). For example, treatment of milk (Bahna, 2008), rice (Watanabe et al., 1990), wheat flour (Tanabe, 2008), or soybeans (Yamanishi et al., 1996) with proteases decreases allergenicity of these foods. In contrast, IgE-binding epitopes of major peanut allergen Ara h 1 are apparently protected from degradation (Maleki et al., 2000).

Hypoallergenic infant formulas are categorized into at least three types: extensively hydrolyzed, partially hydrolyzed, and elemental formulas. The former two products are processed by enzymatic hydrolysis of not only bovine casein/whey but also soy. Because elemental formulas are prepared from synthesized free amino acids, they are well tolerated by practically all cow's milk allergic individuals. Extensively hydrolyzed formulas (casein, whey, and mixed) are also tolerated by most allergic individuals. Partially hydrolyzed whey formula causes allergy in one third to one half of milk allergic individuals and is not considered hypoallergenic. Unfortunately, the palatability of these formulas is usually less than satisfactory because of bitter peptides derived through hydrolysis.

A method for producing hypoallergenic wheat flour by enzymatic modification has been developed. The goal was to hydrolyze wheat's Gln-Xaa-Xaa-Pro-Pro epitope. Enzyme screening revealed actinase has a high ability to hydrolyze peptide bonds with low amylase activities. Wheat flour treated with actinase in water at 40°C for 1 hr yields a product that is hypoallergenic in most cases. By gelatinizing the starch present in the hypoallergenic batter prior to processing and increasing the viscosity of the batter by heating, the resulting product is a partially or exhaustively gelatinized batter suitable for making various wheat products such as cupcakes, pizza, cookies, wafers, pastalike noodles, and puffed items.

2.9 Production, Design, and Commercialization of Antiallergenic Proteins and Peptides

A strategy for preventing hypersensitivity reactions involves eliminating allergenic proteins from crop plants. Currently, no hypoallergenic crops are commercially available, although several attempts at reducing allergenic proteins in various crops have been made (Riascos et al., 2010). In nonleguminous crops, such as rice, tomato, apple, and ryegrass, several allergens have been genetically targeted for reduction. These include RAP (α-amylase-inhibitor) (Tada et al., 2003), Lyc e 1 (Le et al., 2006a) and Lyc e 3 (Le et al., 2006b) (profilin and ns-LTP, respectively), Mal d 1 (PR-10) (Gilissen et al., 2005), and Lol p 5 (Group 5/6 grass pollen allergens) (Bhalla, Swoboda, and Singh, 1999). In legume crops, most efforts have been focused on downregulating the cysteine protease P34 (Gly m Bd 30K) in soybeans and Ara h 2 (a 2S albumin) in peanuts.

Although cultivars of various crops with reduced allergenic proteins have been produced, there are several concerns regarding their use in foods labeled as hypoallergenic. Reduction of an allergenic protein could lead to compensatory synthesis of similar proteins that could be allergenic. Reversion of a suppressed trait could occur through the silencing of transgenes involved in its suppression. Reappearance of a significant allergen could be dangerous even if reversion involves only a portion of crop populations. In addition, individuals may be allergic to a minor allergen that was not eliminated during cultivation.

2.10 Conclusion

Common allergens among plant foods include seed storage proteins, α-amylase inhibitors, and defense proteins. In animal products, albumin, casein, and ovomucoid are common allergens. Isolation and characterization of allergens is important for understanding the mechanics of allergic reactions induced by food and for designing and evaluating potential therapeutic strategies.

References

Adachi, M., Kanamori, J., Masuda, T. et al. 2003. Crystal structure of soybean 11S globulin: Glycinin A3B4 homohexamer. *Proc. Natl. Acad. Sci. USA* 100: 7395–7400.

Amano, M., Ogawa, H., Kojima, K. et al. 1998. Identification of the major allergens in wheat flour responsible for baker's asthma. *Biochem. J.* 330: 1229–1234.

Ayuso, R., Lehrer, S.B., Tanaka, L. et al. 1999. IgE antibody response to vertebrate meat proteins including tropomyosin. *Ann. Allergy Asthma. Immunol.* 83: 399–405.

Bahna, S.L. 1996. Celiac disease: A food allergy? Contra! In *Highlights in Food Allergy* (B. Wuthrich and C. Ortolani, Eds.), Basel: Karger, pp. 211–215.

Bahna, S.L. 2008. Hypoallergenic formulas: Optimal choices for treatment versus prevention. *Ann. Allergy Asthma Immunol.* 101: 453–459.

Beretta, B., Conti, A., Fiocchi, A. et al. 2001. Antigenic determinants of bovine serum albumin. *Int. Arch. Allergy Immunol.* 126: 188–195.

Besler, M., Fiocchi, A., and Restani, P. 2001. Allergen data collection: Beef. *Internet Symp. Food Allerg.* 3: 171–184.

Bewley, D., Black, M., and Halmer, P. 2006. *The Encyclopedia of Seeds: Science, Technology and Uses.* Trowbridge, UK: Cromwell Press, p. 668.

Bhalla, P.L., Swoboda, I., and Singh, M.B. 1999. Antisense-mediated silencing of a gene encoding a major ryegrass pollen allergen. *Proc. Natl. Acad. Sci. USA* 96: 11676–11680.

Biagi, F., Klersy, C., Balduzzi, D. et al. 2010. Are we not over-estimating the prevalence of coeliac disease in the general population? *Ann. Med.* 42: 557–561.

Branum, A.M. and Lukacs, S.L. 2009. Food allergy among children in the United States. *Pediatrics* 124: 1549–1555.

Breiteneder, H. and Ebner, C. 2000. Molecular and biochemical classification of plant-derived food allergens. *J. Allergy Clin. Immunol.* 106: 27–36.

Buchanan, B.B., Adamidi, C., Lozano, R.M. et al. 1997. Thioredoxin-linked mitigation of allergic responses to wheat. *Proc. Natl. Acad. Sci. USA* 94: 5372–5377.

Cerecedo, I., Zamora, J., Shreffler, W.G. et al. 2008. Mapping of the IgE and IgG4 sequential epitopes of milk allergens with a peptide microarray-based immunoassay. *J. Allergy Clin. Immunol.* 122: 589–594.

Chapman, M.D., Pomés, A., Breiteneder, H. et al. 2007. Nomenclature and structural biology of allergens. *J. Allergy Clin. Immunol.* 119: 414–420.

Cooke, S.K. and Sampson, H.A. 1997. Allergenic properties of ovomucoid in man. *J. Immunol.* 159: 2026–2032.

Elsayed, S., Hammer, A.S., Kalvenes, M.B. et al. 1986. Antigenic and allergenic determinants of ovalbumin. I. Peptide mapping, cleavage at the methionyl peptide bonds and enzymatic hydrolysis of native and carboxymethyl OA. *Int. Arch. Allergy Appl. Immunol.* 79: 101–107.

Elsayed, S., Holen, E., and Haugstad, M.B. 1988. Antigenic and allergenic determinants of ovalbumin. II. The reactivity of the NH2 terminal decapeptide. *Scand. J. Immunol.* 27: 587–591.

Elsayed, S., Sornes, S., Apold, J. et al. 1982. The immunological reactivity of the three homologous repetitive tetrapeptides in the region 41-64 of allergen M from cod. *Scand. J. Immunol.* 16: 77–82.

Elsayed, S. and Stavseng, L. 1994. Epitope mapping of region 11-70 of ovalbumin (Gal d I) using five synthetic peptides. *Int. Arch. Allergy Immunol.* 104: 65–71.

Fiocchi, A., Restani, P., and Riva, E. 2000. Beef allergy in chidren. *Nutrition* 16: 454–457.

Food Allergy & Anaphylaxis Network. Allergens. Available at: http://www.foodallergy.org/section/allergens. Accessed December 1, 2010.

Gilissen, L.J., Bolhaar, S.T., Matos, C.I. et al. 2005. Silencing the major apple allergen Mal d 1 by using the RNA interference approach. *J. Allergy Clin. Immunol.* 115: 364–369.

Gómez, L., Martin, E., Hernandez, D. et al. 1990. Members of the alpha-amylase inhibitors family from wheat endosperm are major allergens associated with baker's asthma. *FEBS Lett.* 261: 85–88.

Hilger, C., Kohnen, M., Grigioni, F. et al. 1997. Allergic cross-reactions between cat and pig serum albumin. *Allergy* 52: 179–187.

Honma, K., Kohno, Y., Saito, K. et al. 1994. Specificities of IgE, IgG and IgA antibodies to ovalbumin. Comparison of binding activities to denatured ovalbumin or ovalbumin fragments of IgE antibodies with those of IgG or IgA antibodies. *Int. Arch. Allergy Immunol.* 103: 28–35.

Honma, K., Kohno, Y., Saito, K. et al. 1996. Allergenic epitopes of ovalbumin (OVA) in patients with hen's egg allergy: Inhibition of basophil histamine release by haptenic ovalbumin peptide. *Clin. Exp. Immunol.* 103: 446–453.

Kahlert, H., Petersen, A., Becker, W.M. et al. 1992. Epitope analysis of the allergen ovalbumin (Gal d II) with monoclonal antibodies and patients' IgE. *Mol. Immunol.* 29: 1191–1201.

Kovacs-Nolan, J., Zhang, J.W., Hayakawa, S. et al. 2000. Immunochemical and structural analysis of pepsin-digested egg white ovomucoid. *J. Agric. Food Chem.* 48: 6261–6266.

Le, L.Q., Lorenz, Y., Scheurer, S. et al. 2006b. Design of tomato fruits with reduced allergenicity by dsRNAi-mediated inhibition of ns-LTP (Lyc e 3) expression. *Plant Biotechnol. J.* 4: 231–242.

Le, L.Q., Mahler, V., Lorenz, Y. et al. 2006a. Reduced allergenicity of tomato fruits harvested from lyc e 1-silenced transgenic tomato plants. *J. Allergy Clin. Immunol.* 118: 1176–1183.

Maleki, S.J., Kopper, R.A., Shin, D.S. et al. 2000. Structure of the major peanut allergen Ara h 1 may protect IgE-binding epitopes from degradation. *J. Immunol.* 164: 5844–5849.

Maruyama, N., Ichise, K., Katsube, T. et al. 1998. Identification of major wheat allergens by means of the *Escherichia coli* expression system. *Eur. J. Biochem.* 255: 739–745.

Mills, E.N.C., Sancho, A.I., Rigby, N.M. et al. 2009. Impact of food processing on the structural and allergenic properties of food allergens. *Mol. Nutr. Food Res.* 53: 963–969.

Mine, Y. and Rupa, P. 2003. Fine mapping and structural analysis of immunodominant IgE allergenic epitopes in chicken egg ovalbumin. *Protein Eng.* 16: 747–752.

Mine, Y. and Yang, M. 2008. Recent advances in the understanding of egg allergens: Basic, industrial, and clinical perspectives. *J. Agric. Food Chem.* 56: 4874–4900.

Mondoulet, L., Paty, E., Drumare, M.F. et al. 2005. Influence of thermal processing on the allergenicity of peanut proteins. *J. Agric. Food Chem.* 53: 4547–4553.

Morita, E., Kunie, K., and Matsuo, H. 2007. Food-dependent exercise-induced anaphylaxis. *J. Dermatol. Sci.* 47: 109–117.

Palosuo, K., Alenius, H., Varjonen, E. et al. 1999. A novel wheat gliadin as a cause of exercise-induced anaphylaxis. *J. Allergy Clin. Immunol.* 103: 912–917.

Paschke, A. 2009. Aspects of food processing and its effect on allergen structure. *Mol. Nutr. Food Res.* 53: 959–962.

Posch, A., Weiss, W., Wheeler, C. et al. 1995. Sequence analysis of wheat grain allergens separated by two-dimensional electrophoresis with immobilized pH gradients. *Electrophoresis* 16: 1115–1119.

Quirce, S., Maranon, F., Umpierrez, A. et al. 2001. Chicken serum albumin (*Gal d* 5) is a partially heat-labile inhalant and food allergen implicated in the bird-egg syndrome. *Allergy* 56: 754–762.

Radauer, C. and Breiteneder, H. 2007. Evolutionary biology of plant food allergens. *J. Allergy Clin. Immunol.* 120: 518–525.

Restani, P., Ballabio, C., Di Lorenzo, C. et al. 2009. Molecular aspects of milk allergens and their role in clinical events. *Anal. Bioanal. Chem.* 395: 47–56.

Restani, P., Fiocchi, A., Beretta, B. et al. 1997. Meat allergy: III—Proteins involved and cross-reactivity between different animal species. *J. Am. Coll. Nutr.* 16: 383–389.

Riascos, J.J., Weissinger, A.K., Weissinger, S.M. et al. 2010. Hypoallergenic legume crops and food allergy: Factors affecting feasibility and risk. *J. Agric. Food Chem.* 58:20–27.

Rodrigo, L. 2006. Celiac disease. *World J. Gastroenterol.* 12: 6585–6593.

Rona, R.J., Keil, T., Summers, C. et al. 2007. The prevalence of food allergy: A meta-analysis. *J. Allergy Clin. Immunol.* 120: 638–646.

Salcedo, G., Sanchez-Monge, R., Garcia-Casado, G. et al. 2004. The cereal α-amylase/trypsin inhibitor family associated with bakers' asthma and food allergy. In: *Plant Food Allergens*, Cornwall, UK: Wiley-Blackwell, pp. 70–84.

Sánchez-Monge, R., Garcia-Casado, G., Lopez-Otin, C. et al. 1997. Wheat flour peroxidase is a prominent allergen associated with baker's asthma. *Clin. Exp. Allergy* 27: 1130–1137.

Shek, L.P., Bardina, L., Castro, R. et al. 2005. Humoral and cellular responses to cow milk proteins in patients with milk-induced IgE-mediated and non-IgE-mediated disorders. *Allergy* 60: 912–919.

Shewry, P.R., Beaudoin, F., Jenkins, J. et al. 2002. Plant protein families and their relationships to food allergy. *Biochem. Soc. Trans.* 30: 906–910.

Shewry, P.R. and Tatham, A.S. 1990. The prolamin storage proteins of cereal seeds: Structure and evolution. *Biochem. J.* 267: 1–12.

Sicherer, S.H., Muñoz-Furlong, A., and Sampson, H.A. 2003. Prevalence of peanut and tree nut allergy in the United States determined by means of a random digit dial telephone survey: A 5-year follow-up study. *J. Allergy Clin. Immunol.* 112: 1203–1207.

Sturgess, R., Day, P., Ellis, H.J. et al. 1994. Wheat peptide challenge in coeliac disease. *Lancet* 343: 758–761.

Tada, Y., Akagi, H., Fujimura, T. et al. 2003. Effect of an antisense sequence on rice allergen genes comprising a multigene family. *Breeding Sci.* 53: 61–67.

Takagi, K., Teshima, R., Okunuki, H. et al. 2003. Comparative study of *in vitro* digestibility of food proteins and effect of preheating on the digestion. *Biol. Pharm. Bull.* 26: 969–973.

Tanabe, S. 2008. Analysis of food allergen structures and development of foods for allergic patients. *Biosci. Biotechnol. Biochem.* 72: 649–659.

Tanabe, S., Arai, S., Yanagihara, Y. et al. 1996. A major wheat allergen has a Gln-Gln-Gln-Pro-Pro motif identified as an IgE-binding epitope. *Biochem. Biophys. Res. Commun.* 219: 290–293.

Tanabe, S., Kobayashi, Y., Takahata, Y. et al. 2002. Some human B and T cell epitopes of bovine serum albumin, the major beef allergen. *Biochem. Biophys. Res. Commun.* 293: 1348–1353.

Tanabe, S., Shibata, R., and Nishimura, T. 2004. Hypoallergenic and T cell reactive analogue peptides of bovine serum albumin, the major beef allergen. *Mol. Immunol.* 41: 885–890.

Walsh, B.J., Barnett, D., Burley, R.W. et al. 1988. New allergens from hen's egg white and egg yolk. *In vitro* study of ovomucin, apovitellenin I and VI, and phosvitin. *Int. Arch. Allergy Appl. Immunol.* 87: 81–86.

Walsh, B.J. and Howden, M.E. 1989. A method for the detection of IgE binding sequences of allergens based on a modification of epitope mapping. *J. Immunol. Meth.* 121: 275–280.

Watanabe, M., Miyakawa, J., Ikezawa, Z. et al. 1990. Production of hypoallergenic rice by enzymatic decomposition of constituent proteins. *J. Food Sci.* 55: 781–783.

Weiss, W., Huber, G., and Engel, K.H. 1997. Identification and characterization of wheat grain albumin/globulin allergens. *Electrophoresis* 18: 826–833.

Witke, W. 2004. The role of profilin complexes in cell motility and other cellular processes. *Trends Cell Biol.* 14: 461–469.

Yamanishi, R., Tsuji, H., Bando, N. et al. 1996. Reduction of the allergenicity of soybean by treatment with proteases. *J. Nutr. Sci. Vitaminol.* 42: 581–587.

Zhang, J.W. and Mine, Y. 1998. Characterization of IgE and IgG epitopes on ovomucoid using egg-white-allergic patients' sera. *Biochem. Biophys. Res. Commun.* 253: 124–127.

Zhang, J.W. and Mine, Y. 1999. Characterization of residues in human IgE and IgG binding site by chemical modification of ovomucoid third domain. *Biochem. Biophys. Res. Commun.* 261: 610–613.

3

Properties and Applications of Antimicrobial Proteins and Peptides from Milk and Eggs

Hannu J. Korhonen and Susanna Rokka

CONTENTS

3.1 Introduction

Bioactive compounds derived from various natural sources have attracted growing interest in the food and pharmaceutical industries in view of their potential to be exploited in many industries. They can find applications in nutraceuticals promoting health, in special dietary preparations preventing microbial infections in humans and farm animals, in biopreservation of foods and feeds, as well as in pharmaceutical preparations replacing synthetic drugs. This review focuses on antimicrobial proteins and peptides derived from animal sources, namely bovine colostrum and milk and hen eggs, as they have shown to be rich and readily available sources for bioactive compounds and scientific research over the past decades has revealed many exciting opportunities for their commercial applications.

Colostrum is not merely a source of nutrients for the newborn mammal, but it also provides effective protection against microbial and viral infections. For more than 100 years, antibodies (immunoglobulins, Igs) present in colostrum and milk have been known to protect the neonate against specific pathogens. In addition, a number of components of colostrum and milk constitute a nonspecific protective system, in which various proteins seem to play a dominant role. The best characterized among these are lactoferrin, lysozyme, and lactoperoxidase, which consists of lactoperoxidase–thiocyanate–hydrogen peroxide as a system, and are now generally recognized as nonantibody antibacterial factors and termed "innate factors." In addition to secretory IgA, they are acknowledged to be part of the mucosal associated lymphoid tissue (MALT). Table 3.1 provides a brief overview of the established or putative mechanisms of action of these bioactive components. Recent research has expanded our knowledge about bioactive properties of many other milk proteins, not only intact proteins but also peptides released by enzymatic digestion. A wide range of bioactive milk peptides has now been identified and many of them exhibit antimicrobial activity (Floris et al., 2003; Pellegrini, 2003; Pihlanto and Korhonen, 2003; Korhonen, 2009a).

Similar to mammary secretions, the avian egg is an important source of nutrients and defense factors to protect the embryo against microbial infections. The best-known antimicrobial components in hen eggs are lysozyme

TABLE 3.1

Antimicrobial Mechanisms of Milk Proteins

Protein	Mechanism(s)
Immunoglobulins (Ig) (antibodies)	Agglutination
	Bacteriolysis
	Bacteriostatis
	Opsonization
	Neutralization of viruses and toxins
Lactoferrin (LF)	Binding of iron from environment
	Destabilization of outer membranes of bacteria and molds
	Prevention of binding of viruses to host cells
Lactoperoxidase (LP) ($LP/SCN^-/H_2O_2$)	Oxidation of SH-groups of cell membranes
	Bacteriolysis of Gram-negative bacteria
	Bacteriostasis of Gram-positive bacteria
Lysozyme (LZM)	Hydrolysis of glycopeptides of bacterial cell walls (bacteriolysis)
Glycomacropeptide (GMP)	Inhibition of adhesion of bacteria and viruses to epithelial dental plaque

and immunoglobulin Y present in the albumin (egg white) and yolk, respectively. Recent research has identified many other native bioactive egg protein-based components and peptides with novel activities, including antimicrobial, antiadhesive, immunomodulatory, anticancer, and antihypertensive (Kovacs-Nolan, Mine, and Hatta, 2006). Technologies for recovery of bioactive components from milk and eggs are being developed continuously and the number of commercial applications in global markets is expected to grow quickly.

3.2 Enzyme-Based Antimicrobial Proteins

3.2.1 Lactoperoxidase

Lactoperoxidase (LP) (EC 1.11.1.7), is a peroxidase with a broad substrate specificity. Chemically, LP is a glycoprotein with a molecular weight of about 78 kD. It occurs naturally in colostrum, milk, and many other human and animal secretions. LP represents the most abundant enzyme in bovine milk (approx. 30 mg/l) and can be recovered in substantial quantities from whey using chromatographic techniques. Bovine LP is relatively heat-resistant, retaining about 50% of its original activity after high temperature short time (HTST) pasteurization (Kussendrager and Hooijdonk, 2000). The antimicrobial properties catalyzed by LP have been studied since 1924 when Hanssen (1924) traced the bactericidal properties of cow's milk to its peroxidase activity. LP catalyzes peroxidation of thiocyanate anion and some halides in the presence of a hydrogen peroxide source to generate short-lived oxidation products of SCN^-, primarily hypothiocyanate ($OSCN^-$), which kill or inhibit the growth of many species of micro-organisms. The hypothiocyanate anion causes oxidation of sulphydryl groups (SH) in microbial enzymes and other membrane proteins leading to intermediary inhibition of growth or killing of susceptible micro-organisms. The complete antimicrobial LP/SCN/H_2O_2 system was originally characterized in milk by Reiter and Oram (1967). Nowadays this system is considered to be an important part of the natural host defense system in mammals, the protective function being mediated by several mechanisms (Boots and Floris, 2006). LP is resistant to pH as low as 3, and to human gastric juice. The thiocyanate ion is widely distributed in animal secretions and tissues. The concentration of SCN^- is related to nutrition, plants of the *Cruciferae* family being a good source of cyanogenic glucosides as precursors for this ion. The third component, hydrogen peroxide, may be formed endogenously, as many lactobacilli, lactococci, and streptococci produce sufficient amounts of H_2O_2 under aerobic conditions to activate the LP system. H_2O_2 can also be produced in situ, for example, by adding glucose oxidase and glucose to the medium (Reiter and Perraudin, 1991).

In *in vitro* studies, the LP system has been shown to be active against a wide range of micro-organisms, including bacteria, viruses, fungi, molds, and protozoa (Seifu et al., 2005). The LP system is known to be bactericidal against Gram-negative pathogenic and spoilage bacteria, such as *E. coli, Salmonella* spp., *Pseudomonas* spp., and *Campylobacter* spp. This antibacterial system is particularly effective against cariogenic bacteria *Streptococcus mutans* and periodontitis-associated bacteria, such as *Actinobacillus actinomycetemcomitans, Porhyromonas gingivalis, and Fusobacterium nucleatum* (Tenovuo et al., 1991). On the other hand, the system is bacteriostatic against many Gram-positive bacteria, such as *Listeria* spp., *Staphylococcus* spp., and *Streptococcus* spp. Also, the LP system prevents the growth of *Helicobacter pylori* and generation of urease by this bacterium (Shin et al., 2002) and prevents bacterial colonization of the airway epithelium (Gerson et al., 2000). Oral administration of LP in combination with lactoferrin has been demonstrated to attenuate symptoms of pneumonia in influenza-virus-infected mice through the suppression of infiltration of inflammatory cells in the lung (Shin et al., 2005). Furthermore, this system has been shown to be inhibitory to *Candida* spp. and the protozoan *Plasmodium falciparium* and it can inactivate *in vitro* the HIV type 1 and polio virus (Seifu et al., 2005).

3.2.2 Lysozyme

Lysozyme (LZM), also known as muramidase (peptidoglycan N-acetylmuramoyl hydrolase; EC 3.2.1.17) is a potent antibacterial enzyme (molecular weight about 15 kD) that catalyzes the hydrolysis of ß-(1-4) linkage between N-acetylmuramic acid and N-acetylglucosamine of bacterial peptidoglycan. This structural component is particularly abundant in the cell wall of Gram-positive bacteria (Farkye, 2002a). Discovered by Alexander Fleming in the early 1920s (Fleming, 1922), LZM has been known for a long time as an effective antimicrobial agent. LZM is widely distributed in various biological fluids and tissues, including avian eggs, plants, bacteria, and animal secretions, but its concentration varies considerably in different sources. Hen egg white (1–3 g/l), human milk (up to 0.4 g/l), and mare's milk (0.4–1 g/l) are abundant sources of LZM. Also, tears and saliva contain significant amounts of LZM, whereas its concentration in bovine colostrum and milk is relatively low (0.0004 g/l) (Floris et al., 2003).

Lysozymes from different sources vary with regard to their antibacterial spectrum and specificity toward different types of mucopolysaccharides. LZM shows high activity against mesophilic and thermophilic spore-forming bacteria, such as *Bacillus stearothermophilus, Clostridium thermosaccharolyticum*, and *Clostridium tyrobutyricum* (Johnson, 1994). The antibacterial spectrum of LZM can be broadened to include other spoilage and pathogenic micro-organisms, for example, *C. botulinum, Listeria monozytogenes, Staphylococcus aureus*, and *Candida albicans*, as well as Gram-negative bacteria, when used in conjunction with compounds such as

EDTA, organic acids, or nisin (Losso, Nakai, and Charter, 2000), or by coupling it with a hydrophobic carrier or compounds lethal to the bacterial membrane (Ibrahim, Thomas, and Pellegrini, 2001). LZM, lactoferrin, and antibodies have a synergistic bactericidal effect on many micro-organisms (Pan et al., 2007a).

In addition to antibacterial activity, LZM has been demonstrated to have many other functions, including inactivation of certain viruses, enhancement of antibiotic effects, anti-inflammatory and antihistamine actions, activation of immune cells, and antitumor activity (Floris et al., 2003). The antiviral action of LZM has partly been explained by its role in the precipitation of viral particles and by its immune-enhancing action on the host together with its interaction with the pathogens (Sava, 1996).

An interesting characteristic of LZM is that it can be efficient after oral administration, because it is rather resistant to digestive enzymes (Sava, 1996). Oral and topical applications of LZM were found to be effective in preventing and controlling several viral skin infections, including herpes simplex and chicken pox (Sava, 1996), and Lee-Huang et al. (1999) found that egg white LZM also possessed activity against HIV type 1.

Egg white LZM has been found to affect some components of the immune system in the host which may explain part of its reported activity, for example, in reduction of thymus hyperplasia, activation of host immunity against subchronic enterocolitis, child viral hepatitis and poliomyelitis, increase in absolute granulocyte counts after antiblastic therapy, and modification of the lymphocyte responses. This last property suggests that LZM is of interest in cancer treatments for recovering from immune suppression due to anticancer therapy. LZM may also control the imbalances of immunity during auto-immune diseases. These activities have been mainly attributed to human LZM, but hen egg white LZM has been shown to be effective on human immunocompetent cells (Sava, 1996).

3.3 Nonenzymatic Antimicrobial Proteins

3.3.1 Immunoglobulins

Immunoglobulins (Igs), also called antibodies, are a homogenic group of conserved proteins that are able to bind foreign structures in a specific manner. They are present in blood, but also milk and colostrum of all lactating species. Maternal colostrum and milk offer passive protection to a newborn infant against enteric pathogens, primarily via the transfer of Igs and associated bioactive factors (Ehrlich, 1892). Ruminant neonates are born virtually without Igs, and therefore, the colostral Igs are essential for survival. Probably due to this unique function the Igs represent the major protein fraction of colostrum accounting for 70–80% of its total protein content, whereas in milk they account for only 1–2% of total protein (Butler, 1998; Marnila and Korhonen, 2002). Avian species accumulate large amounts of Ig in the yolk. An egg yolk contains from 50 to 100 mg of Igs per yolk.

Igs are produced by B-lymphocytes and plasma cells diversified from B cells. The basic structure of an immunoglobulin molecule is composed of two identical nonglycosylated light chains and two identical glycosylated heavy chains joined together with disulphide bonds. Both the light and heavy chains contain domains referred to as constant (C_L, C_H) and variable (V_L, V_H) regions. The antigen binding sites are located in the Fab region in the N-terminal "arms" of the Y-shaped flexible molecule. Generally, each antibody can bind specifically to only one antigen. The Fc portion of mammalian Igs interacts with cells of the immune system and activates the complement-mediated bacteriolytic reactions whereas avian Ig is not able to activate the complement. The molecular weight of the basic structure of an Ig molecule is about 160 kDa.

The Ig classes and subclasses are determined by genes encoding the constant regions of heavy chains (Butler, 1998). In mammals, five known classes of Igs have been characterized: IgG, IgM, IgA, IgD, and IgE. Avian IgY is homologous to mammalian Igs.

IgM is a pentamer of the basic structure. It is the first class of antibodies to appear in the serum after injection of an antigen, and is especially efficient in complement fixation, opsonization, and agglutination of bacteria. IgG is the major antibody in serum and bovine milk. It has four subclasses differing mainly in the number of disulphide bonds. IgGs are able to inactivate toxins, aggregate and opsonize parasites, and activate the complement system.

Monomeric IgA occurs in serum but in milk it is present as a dimer. This secretory IgA (SIgA) has a molecular weight of about 380 kDa and is more resistant to proteolysis, and therefore, more stable in the gastrointestinal tract than antibodies without the secretory component. The dimeric IgA is the major class of immunoglobulins in external secretions such as saliva, tears, intestinal mucus, and milk with the ruminal milk being an exception with IgG1 as the major class of immunoglobulins. It agglutinates antigens, binds viruses and toxins, and prevents the adhesion of enteropathogenic bacteria to mucosal epithelial cells. The other two classes of immunoglobulins are IgD, a B cell surface immunoglobulin known to be involved in B cell/T cell cognate activation, and IgE, involved in protection against parasites, as well as in allergic reactions. Avian IgY is secreted by B cells and transferred from blood to egg yolk where it is accumulated to protect the offspring.

Specific Igs are rarely cidal to micro-organisms but may disturb cellular metabolism by blocking receptors or enzymes, and cause structural alterations leading to immobilization, increased membrane permeability, and impaired cell growth. Igs may contribute to the killing of microbes by activating the classical complement pathway. In blood and tissues, the activation of complement-mediated bacteriolytic reactions is one of the most important functions of Igs. Bovine and human colostrum contain an active complement system participating in the immune defense of the udder (Butler, 1998).

Prevention of microbial adhesion to epithelial linings may be the most important mechanism of antibodies in protecting the host. Specific antibodies are capable of binding to adhesins, and microbes coated with Igs cannot establish a colony. The ability of Igs to form cross-links between surface antigens results in a network of cells that can be mechanically removed from the body. This agglutination of microbes reduces their capability to adhere to surfaces. Agglutinated microbes are usually not able to release toxins or to colonize the host. Normal colostrum and milk are known to contain natural antibodies that can agglutinate a large number of pathogenic and non-pathogenic micro-organisms (Bostwick, Steijns, and Braun 2000; Korhonen, Marnila, and Gill, 2000b).

Many bacterial toxins must first be actively transported via receptors inside the host cells to cause cell death. Igs can bind bacterial toxins that can then be recognized more effectively by phagocytic leukocytes. Specific Igs can inhibit the effect of toxins by blocking the toxin and preventing its internalization in host cells. Also, Igs can inhibit or reduce the production of toxins and other harmful components by inhibiting bacterial metabolism and by blocking enzymes and receptors.

Specific Igs can protect against viral infections by binding viruses and preventing virus replication, and by blocking the receptor-mediated internalization of viruses in the host cells. Specific Igs augment recognition and phagocytosis of antigens by leukocytes (opsonization). Divalent or polyvalent binding of an Ig to an antigenic structure results in conformational change of the Ig molecule which again enables the Fc portion to bind to the corresponding Fc receptors on the leukocyte surface. This receptor binding leads to various immune cell effector's functions depending on the cell Ig and leukocyte type. Leukocytes are an integral part of normal milk and colostrum and of vital importance in defending the mammary gland against pathogens (Korhonen et al., 2000b).

The milk Igs have also been found to exert a synergistic effect on activity of nonspecific antimicrobial factors in milk, such as LF and LZM as well as the lactoperoxidase–thiocyanate–hydrogen peroxide system (Loimaranta, Tenovuo, and Korhonen, 1998b; Bostwick et al., 2000). The antibody titer against certain antigenic pathogens or structures (e.g., virulence factor) can be raised up to several hundred times by immunizing the cow before parturition with vaccines containing antigens (Korhonen et al., 1995). Both milk and colostrum have been used as a source of specific bovine antibodies. Colostrum is generally preferred because of its high concentration of IgGs, but, on the other hand, colostrum is secreted only for a limited time. Avian eggs are another rich source for specific antibodies. Immunized hens produce specific IgY in the water-soluble part of egg yolks for months. The eggs are easy to collect and store until processed. Other advantages of using specific chicken antibodies are low costs, large yield, and scaleable production. IgYs can be concentrated from the yolk by simple water extraction and ammonium sulphate precipitation methods (Akita and Nakai, 1992; Kokko, Kuronen, and Kärenlampi, 1998). The material can be further purified by chromatographic methods. If specific purified IgY is needed, immune-affinity chromatography can be used.

The progress in understanding the mechanisms of Ig-mediated immune functions and the rapid development of industrial fractionation technologies have raised interest in developing formulations supplemented with bovine colostral or cheese whey-derived Igs. Most of the current commercial Ig products are prepared from colostrum of nonimmunized cows by removing the fat followed by microfiltration or pasteurization under conditions that retain biological activity of Igs. These products are usually in the form of spray-dried and freeze-dried powders whereas some are in the form of filtered colostral whey liquids or concentrates. Immunoglobulins can be isolated from colostral or milk whey on an industrial scale by a number of methods based on ultrafiltration or a combination of ultrafiltration and chromatography. The recovery rate of immunoglobulins has varied from 40 to 70% (Elfstrand et al., 2002). With microfiltration (0.1 m), recovery has been reported to be over 90% (Korhonen and Pihlanto, 2007b).

Thermal treatment during processing may influence the stability of immunoglobulins. During storage in ambient or cold temperatures, the freeze-dried immunoglobulin fractions retain their specific activity for years (Korhonen et al., 2000b; Pant et al., 2007). The shelf life of bovine milk-based products containing antibodies can be prolonged by membrane filtration (Fukumoto et al., 1994) or by fermentation processing (Wei et al., 2002).

The concept of immune milk preparations dates back to the 1950s when Petersen and Campbell first suggested that orally administered bovine colostrum from hyperimmunized cows could provide passive immune protection for humans (Campbell and Petersen, 1963). Since then, a great number of animal and human studies have been carried out to demonstrate that these preparations can be effective in the prevention or treatment of human and animal diseases caused by different pathogenic microbes (Weiner et al., 1999; Korhonen et al., 2000a; Lilius and Marnila, 2001; Hoerr and Bostwick, 2002; Korhonen and Marnila, 2006; Mehra, Marnila, and Korhonen, 2006).

Colostral and immune milk preparations against enteropathogenic *E. coli*, rotavirus, *Cryptosporidium*, *Shigella flexneri*, *Helicobacter pylori*, *Vibrio cholerae*, and caries-promoting streptococci have been studied in controlled clinical trials with variable results (see recent reviews by Hammarström and Weiner, 2008 and Korhonen and Marnila, 2009). In general, promising results have been obtained in prevention studies. For example, immune bovine colostral preparations provided good protection against infection caused by enterotoxigenic *E. coli* (ETEC; Tacket et al., 1988; Casswall et al., 2000). The therapeutic efficacy of immune products, however, seems limited, although encouraging results have been recorded in clinical studies on *Clostridium difficile* diarrhea (van Dissel et al., 2005).

Immune bovine colostrum containing specific antibodies against S. *mutans* and S. *sobrinus* was capable of inhibiting the bacterial enzymes producing sticky capsule glycoproteins (Loimaranta et al., 1997), adherence of S. *mutans* cells to saliva-coated hydroxyapatite beads (Loimaranta et al., 1998a), and augmenting the recognition, phagocytosis, and killing of S. *mutans* by human leukocytes (Loimaranta et al., 1999b). In a short-term human study, immune colostrum as a mouth rinse resulted in a higher resting pH in dental plaque and a lower proportion of streptococci caries in plaque microbial flora (Loimaranta et al., 1999a).

Oona et al. (1997) studied the effect of bovine immune whey preparation with specific H. *pylori* antibodies in treatment of H. *pylori*-positive children. The severity of symptoms and degree of gastric inflammation decreased. The degree of H. *pylori* was reduced but total eradication was not observed. These results were consistent with those of Casswall et al. (1998). A prophylactic effect of a colostral immune preparation having specific anti-H. *pylori* antibodies was demonstrated in a mouse model. Efficacy of the therapeutic treatment seemed to be related to concentration of antibodies administered (Casswall et al., 2002). Specific IgY inhibited Helicobacter infection in 70% of infected mice with a significant decrease in the degrees of gastritis (Attallah et al., 2009).

Rotavirus infections have also been successfully prevented and treated with milk preparations from hyperimmunized cows (Ebina et al., 1985; Davidson et al., 1989; Sarker et al., 1998) or egg yolk antibodies (Yolken et al., 1988; Hiraga et al., 1990; Hatta et al., 1993; Sarker et al., 2001).

Cystic fibrosis patients often suffer from chronic *Pseudomonas aeruginosa* infection. Clinical experiments show that the infection can be prevented by specific IgY (Kollberg et al., 2003; Nilsson et al., 2007, 2008).

In animal models bovine immune colostrum preparations have been effective in the treatment of experimental C. *difficile* diarrhea. Van Dissel et al. (2005) gave immune whey protein concentrate to 16 patients suffering from *Clostridium difficile* diarrhea after antibiotic treatment. In all but one case, C. *difficile* toxins disappeared from the feces upon completion of treatment. During a follow-up period none of the patients suffered another episode of C. *difficile* diarrhea. The safety of this product has been demonstrated (Young et al., 2007). Numan et al. (2007) assessed the efficacy of C. *difficile* immune milk preparation for aiding the prevention of relapses in C. *difficile* patients. Mattila et al. (2008) conducted a controlled double-blind randomized study where patients with C. *difficile*-associated diarrhea received orally colostral immune whey preparation with specific anti-C. *difficile* IgG. The authors concluded that the immune colostrum treatment is somewhat less effective than standard treatment with antibiotics, but on the other hand, it does not cause antibiotic resistance problems and does not alter the normal colonic bacterial flora as antibiotics do.

In humans, the oral administration of bovine milk immunoglobulins is generally well tolerated. Local passive immunization with bovine milk antibodies has the advantage of bypassing the host's own immune system. Orally administered immunoglobulins are degraded by intestinal proteases (Reilly, Domingo, and Sandhu, 1997) but IgG1 from bovine milk is quite resistant to pepsin cleavage and retains part of its immunological activity in the human intestine (Petschow and Talbott, 1994; Roos et al., 1995; Warny et al., 1999). Further, pepsin cleaves IgGs to F(ab') fragments, but F(ab') also is able to neutralize antigens.

Wei et al. (2002) studied the combined effect of specific colostral Ig preparation against streptococci caries with the probiotic bacterium *Lactobacillus rhamnosus* GG (LGG). LGG added in milk was earlier shown to reduce the risk of caries in daycare children (Näse et al., 2001). The LGG-bacteria and specific Igs in LGG-fermented milk synergistically inhibited adhesion of *S. mutans* to saliva-coated hydroxyapatite particles. Pant et al. (2007) combined *L. rhamnosus* GG with specific bovine colostrum-derived immunoglobulins. In an infant mouse model this combination had a significant prophylactic effect against rotavirus diarrhea.

A few colostral preparations from nonimmunized cows have boasted specific health or nutrition function claims, such as boosting immunity against microbial infections or speeding recovery from physical endurance exercises, and so on (Scammel, 2001; Tripathi and Vashishtha, 2006).

The efficacy of IgY has been demonstrated in several applications, for example, the treatment and prevention of fatal enteric colibacillosis in neonatal piglets and calves, viral diarrhea in infants, dental caries, canine parvovirus, and snake venom as well as *Salmonella, H. pylori, Pseudomonas aeruginosa, Campylobacter jejuni, E. coli,* rotavirus, corona virus, TNF-α, and various fish and other animal pathogens (Mine and Kovacs-Nolan, 2002; Schade et al., 2005; Shin et al., 2002; Kovacs-Nolan, Mine, and Hatta, 2006; Chalghoumi et al., 2009a,b). There are also numerous applications using IgY for immunodiagnostic purposes. The phylogenetic distance between birds and mammals enables antibody production in eggs against many conserved mammalian antigens.

Egg yolk Igs obtained from hens immunized with *S. mutans* glucan binding protein had a clear protective effect against *S. mutans* infection and caries development in a rat model (Smith, King, and Godiska, 2001). The immune colostrum preparation with specific antibodies against the saliva-binding region of a cell surface protein antigen and a glucan binding domain of a cell surface protein from *S. mutans* prevented dental caries development effectively in a rat model (Mitoma et al., 2002) and humans (Shimazaki et al., 2001).

Specific IgY against ETEC antigens has been reported to inhibit the binding of *E. coli* to porcine epithelial cells and mucus, and resulted in 100% survival and reduced diarrhea of neonatal ETEC-infected piglets (reviewed by Kovacs-Nolan et al., 2006). Similarly, specific IgY protected neonatal calves against fatal enteric colibacillosis (Ikemori et al., 1992).

Orally administered egg yolk and colostrum powders protected against bovine corona virus-induced diarrhea in neonatal calves. On a titer basis, the egg yolk used provided a higher degree of protection compared to colostrum powder (Ikemori et al., 1997).

Anti-*S. enteritis* IgYs, administered as whole egg powder in hen feed, reduced the rate of *Salmonella*-contaminated eggs (Gürtler et al., 2004). However in a recent study, the specific IgY in feed failed to reduce the level of cecal colonization of *Salmonella* in challenged broiler chickens (Chalghoumi et al., 2009b).

Table 3.2 presents examples of recent clinical studies performed using specific antibody preparations made from bovine colostral, milk whey, or hen egg yolk.

3.3.2 Complement System

The main functions of the complement system are to eliminate foreign pathogens by causing cytolytic membrane lesions, to opsonize microbes, and to mediate an inflammatory reaction. The complement system is present in serum, and also in bovine colostrum (Korhonen et al., 2000b). The complement system is composed of more than 20 proteins and regulatory components. Activation of the complement system can occur via three pathways in a sequential manner by proteolytic cleavages and association of precursor molecules (Figure 3.1).

The classical pathway of complement uses antibodies, mainly IgM, for the recognition of foreign cells. The binding of the C1 complex to antigen–antibody complexes initiates proteolytic cleavage of complement components C4 and C2 by C1s, leading to the formation of a C4b2a enzyme complex, the C3 convertase of the classical pathway (Markiewski and Lambris, 2007), (Figure 3.1). The lectin pathway begins with binding of the complex mannose-binding lectin (MBL) and mannose-binding lectin-associated proteases 1 and 2 to a bacterial cell wall which leads to the formation of the C3 convertase. The alternative pathway is initiated by binding to surfaces lacking complement regulatory proteins. Spontaneous hydrolysis of C3 leads to the formation of $C3(H_2O)$ which forms a complex with factor B, followed by the cleavage of factor B within this complex by factor D. The final product of these enzymatic reactions is the $C3(H_2O)Bb$ complex. It is stabilized by properdin and works as a C3 convertase.

C3 convertases generated through various pathways cleave C3 to C3a and C3b. C3b contributes to the formation of the C5 convertase, which cleaves C5 to C5a and C5b. Generated cleavage products function as opsonins augmenting phagocytosis, and recruit leucocytes to the sites of inflammation and activate them. The complement cascade leads to membrane attack complex (MAC), the pores generated leading to the death of the bacterial cell (Born and Bhakdi, 1986).

All Gram-positive and several Gram-negative organisms are resistant to complement-mediated lysis. In these cases the opsonophagocytosis is the main defense mechanism (Rautemaa and Meri, 1999). Opsonization means

TABLE 3.2

Efficacy of Orally Administered Immunoglobulin Preparations

Preparation/Product	Antigen Used in Immunization	Disease or Condition	Treatment Regimen	Treatment Effect	Reference
Bovine Colostrum, Milk, and Blood Preparations					
Bovine immune whey preparation	*Clostridium difficile* toxin and *C. difficile* whole cells	*C. difficile* diarrhea	2 weeks as supportive treatment after antibiotic treatment during 11-month follow-up period	*C. difficile* toxins eradicated from 15 of 16 human subjects and no relapses	Van Dissel et al. 2005
Bovine colostral preparation ColoPlus®	No immunization	HIV-associated diarrhea	3–4 g of Ig twice a day for four weeks	Substantial decrease in stool frequency and in fatigue and increase in body weight	Florén et al. 2006
Bovine colostral preparation IMMULACT®	No immunization	EHEC-colitis	Fed ad libitum (around 300 mg per day) for 3 weeks	Decreased rapidly EHEC colonization and decreased attachment to cecum walls and mortality in mice	Funatogawa et al. 2001
IgG supplemented baby formula	Polyvalent or monovalent *E. coli* vaccine	Diarrhea	Daily dose 0.5 g of IgG per kg of body weight	Lower incidence of diarrhea and shorter duration of diarrhea in human infants during follow-up episode period for 6 months	Tawfeek et al. 2003
Bovine immune colostral preparation	Cocktail of 17 strains of pathogenic diarrhea bacteria	*E. coli* and *Salmonella* diarrhoea	20 mg of specific Ig per day for 10 days starting 6 or 8 days before infection	Prevented enteroinvasive *E. coli* and normalized immunological parameters in mice	Xu et al. 2006
Bovine immune whey preparation	Cocktail of 17 strains of pathogenic diarrhea bacteria	*E. coli*, *Salmonella* and *Shigella* diarrhea	5 mg/ml per day for 10 days starting 6 or 8 days before infection	Decreased clinical signs of diarrhea and supported splenic NK-cell functions in mice	Huang et al. 2008

Product	Immunization	Condition	Dosage	Result	Reference
Bovine colostral product Lactobin®	No immunization	Endotoxemia due to abdominal surgery	52 g daily in 4 doses for 3 days before surgery	Reduced levels of endotoxin and endotoxin neutralizing capacity in blood suggesting reduced endotoxemia after surgery of human subjects	Bölke et al. 2002a
Bovine colostral product Lactobin®	No immunization	Endotoxemia due to coronary surgery	42 g daily doses for 2 days before surgery	Reduced levels of CRP but no effect on perioperative endotoxemia in human subjects	Bölke et al. 2002b
Bovine immune colostral preparation	*Shigella dysenteriae* 1	Shigellosis antigen 1	100 ml orally 3 times/day for 3 days in combination with antibiotics	No significant difference in any clinical parameter in infected children	Ashraf et al. 2001
Immune milk	Virulence factors of *S. mutans*	Dental caries	Mouth rinse twice per day for 14 days with 10 ml	Inhibited recolonization of *S. mutans* after antibiotic treatment in human volunteers	Shimazaki et al. 2001
Bovine colostral immune whey (7.5% IgG)	*Helicobacter felis* whole cell vaccine	Gastritis	0.2 ml before infection or 3 times daily for 4 weeks in treatment of infected mice	Prevented infection in noninfected mice and decreased gastric inflammation and colonization in readily infected mice	Marnila et al. 2003
Milk IgG fraction	No immunization	Murine rotavirus infection in mouse	Orally at time of infection and 12 hours post infection	Decreased rotavirus shedding in stools in mice	Bojsen et al. 2007
Bovine colostral preparation	No immunization	Upper respiratory tract infections	60 g of colostral protein daily for eight weeks	Reduced significantly incidence of self-estimated symptoms of respiratory infections but no difference in duration	Brinkworth and Buckley 2003
Bovine blood Ig preparation	No immunization	Mild hyper-cholesterolemia	Orally 5 g of blood derived IgG daily for 3 or 6 weeks	Both total cholesterol and LDL levels decreased from baselines in human subjects	Earnest et al. 2005

(continued)

TABLE 3.2 (continued)
Efficacy of Orally Administered Immunoglobulin Preparations

Preparation/Product	Antigen Used in Immunization	Disease or Condition	Treatment Regimen	Treatment Effect	Reference
Avian Egg Yolk Preparations					
Supracox® egg powder containing hyper-immune egg yolk IgY	Eimeria oocysts	Avian coccidiosis	Fed in standard diet supplemented with 0.01–0.05%	Significant protection against avian coccidiosis in challenged chicken	Lee et al. 2009
Yogurt with 1% avian IgY and probiotic bacteria	H. pylori urease	Gastritis	150 ml 3 times daily for 4 weeks	Decreased values in urea breath test indicating decrease in colonization in human subjects	Horie et al. 2004
Purified IgY	H. pylori whole-cell lysate	Gastritis	1 or 10 mg/ml fed to infected Mongolian gerbils	Decreased gastric mucosal injury as determined by the degree of lymphocyte and neutrophil infiltration	Shin et al. 2002
Defatted egg powder	H. pylori urease	Gastritis	25 mg/g for 10 weeks fed to infected Mongolian gerbils	No significant differences in the level of H. pylori colonization	Nomura et al. 2005
Defatted egg powder	H. pylori urease	Gastritis	25 mg/g IgY and 0.16 mg/g famotidine orally for prevention 1 week before and 8 weeks after infection of Mongolian gerbils	Inhibited H. pylori colonization and suppressed elevated gastric mucosal MPO activity	Nomura et al. 2005
Purified IgY	H. pylori whole cell lysate and Hp58 antigen	Gastritis	BALB/c mice	Inhibited infection in 70% of infected mice and a significant decrease in the degrees of gastritis	Attallah et al. 2009

Aqueous IgY antibody preparation	Pseudomonas aeruginosa		1 or 2 minutes mouth rinse	Next morning still active antibodies detected in the saliva	Carlander et al. 2002
IgY	Helicobacter pylori urease	Gastritis	Asymptomatic volunteers	UBT values were significantly decreased although no case showed H. pylori eradication	Suzuki et al. 2004
Purified IgY	Pseudomonas aeruginosa strains	Cystic fibrosis	50 mg IgY per day	Period between the first and second colonization with PA was significantly prolonged and none of the patients became chronically colonized with PA in 10-year follow-up	Kollberg et al. 2003, Nilsson et al. 2007
Purified IgY	Pseudomonas aeruginosa strains	Cystic fibrosis	50 mg IgY per day	Reduction in the number of infections; pseudomonas IgY has great potential to prevent P. aeruginosa infections	Nilsson et al. 2008
Solution of polyclonal yolk antibodies	Recombinant human tumor necrosis factor	Inflammatory bowel disease	Before and after infection 600 mg/kg/day twice per day for 5 days	Decreased significantly all inflammatory end points in rats	Worledge et al. 2000
Hyperimmune egg yolk	Human rotavirus strains	Rotavirus diarrhea	10 g daily in four doses for 4 days	Modest improvement of diarrhea associated with earlier clearance of rotavirus from stools of human subjects	Sarker et al. 2001
Egg yolk powder	Salmonella Typhimurium antigens	Salmonella Typhimurium	Treatment before and after infection of pigs	Not effective in controlling shedding of Salmonella in pigs	Mathew et al. 2009
Freeze-dried egg yolks	Salmonella Enteritidis and Salmonella Typhimurium	Salmonella infection	Fed to infected chicken 0–5% for 28 days	No effect observed in challenged chickens	Chalghoumi et al. 2009b

(continued)

TABLE 3.2 (continued)

Efficacy of Orally Administered Immunoglobulin Preparations

Preparation/Product	Antigen Used in Immunization	Disease or Condition	Treatment Regimen	Treatment Effect	Reference
IgY	Freeze-dried *Y. ruckeri* cells	Enteric redmouth disease of salmonid fish	400 mg of IgY pellet for rainbow trout once before and four times after exposure to *Y. ruckeri*	Marginal reductions in mortalities and intestine infection; prevented infection when given intraperitoneally	Lee et al. 2000
IgY solution	Porcine epidemic diarrhea virus	Neonatal diarrhea in piglets	2–4 ml twice per day	Reduced mortality in piglets	Kweon et al. 2000
Lyophilized IgY	*E. coli* O78:K80	*Escherichia coli* challenged broiler chicks	Basal diet supplemented with 0, 0.1, 0.2, or 0.4% (wt/wt) sIgY	0.2% sIgY for 3 weeks improved intestinal health indices and immunological responses	Mahdavi et al. 2010

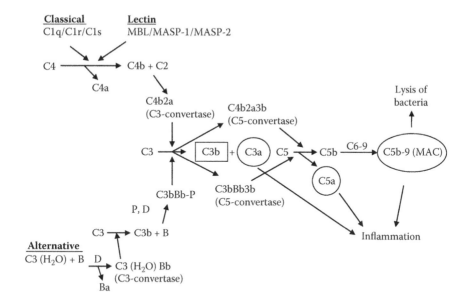

FIGURE 3.1

Schematic presentation of the complement cascade C1-C9 represent complement component molecules. MBL, mannose-binding lecting; MASP, mannose-binding lectin associated protease; P, properdin; MAC, membrane attack complex. (Modified from Thurman, J.M. and V. M. Holers. 2006. *Journal of Immunology* 176(3): 1305–10.)

"making eatable," that is, augmenting the phagocytizing leukocytes to recognize, bind, activate, ingest, and kill the pathogen. During opsonization antibodies or complement molecules bind to antigens. Phagocytic cells recognize opsonized cells and ingest or destroy them. Bovine colostral antibodies are recognized by human phagocytes and thus have opsonizing activity as well (Loimaranta et al. 1999b).

The antibody-complement system is considered as the major antimicrobial agent in bovine colostrum (Korhonen et al., 2000b). The complement system of bovine colostrum is lytic against many pathogenic bacteria *in vitro* (Reiter and Brock, 1975; Korhonen et al., 1995) but it is not known whether it is effective *in vivo* in the gut of the newborn calf (Korhonen et al., 2000b).

3.3.3 Lactoferrin

Lactoferrin (LF) is a multifunctional iron binding glycoprotein of the innate immune system. It is present in exocrine secretions such as milk, tears, nasal exudates, saliva, bronchial and cervico-vaginal mucus, seminal plasma, and gastrointestinal fluid. Bovine colostrum and milk contain LF on an average 1.5 g/l and 0.1 g/l, respectively, whereas human colostrum is a very rich source (up to 5 g/l) of this protein (Lönnerdal, 2003; Pan et al., 2007a). LF is composed of a single chain polypeptide sequence. Human LF (hLF) contains

691 and bovine LF (bLF) 689 amino acids with the degree of sequence homology of 69%. The molecular mass of hLF is between 77 and 82 kDa and that of bLF is 80–84 kDa depending on the degree of glycosylation. Fully glycosylated forms seem to be more resistant to intestinal proteases such as trypsin. The degree of glycosylation does not, however, affect the iron binding affinity.

Many features of the LF molecule define it as primarily an iron binding protein. One LF molecule can bind two ferric (Fe^{3+}) ions with the concomitant incorporation of bicarbonate ion (HCO^{3-}). The affinity of LF to ferric ions is about 300 times higher than in transferrin. In bovine milk 20–30% of bLF is in iron-saturated form (holo-LF) whereas the LF found in human secretions and mucosal surfaces is almost entirely the iron-free apo-form (apo-LF; Makino and Nishimura, 1992). A higher proportion of holo-LF gives the bovine milk LF fraction the typical reddish color (Lönnerdal and Iyer, 1995; Pan et al., 2007a; Conesa et al., 2008).

The antibacterial and antifungal activities of LF are mainly related to iron sequestration, membrane destabilization, targeting of bacterial virulence mechanisms, and host cell invasion strategies (Fernaud and Evans, 2003; Weinberg, 2007). The first recognized antibacterial mechanism of LF was its ability to prevent bacterial growth by binding iron from its medium. The majority of bacteria require iron and its availability correlates strongly with bacterial virulence. In inflammatory sites containing a high amount of activated and degranulating neutrophils, LF scavenges iron effectively. Experimental studies have shown that iron sequestering by apo-LF can strongly inhibit the growth of many bacterial species (see reviews by Weinberg, 2007; Jenssen and Hancock, 2009; Ochoa and Cleary, 2009).

Some bacterial pathogenic species encounter this iron sequestering effect. Under conditions of iron deprivation small iron-chelating molecules, siderophores are synthesized and secreted. Siderophores bind ferric ions with a high affinity and transport them into cells through specific receptors (Braun and Killmann, 1999; Andrews, Robinson, and Rodriguez-Quinones, 2003). Some pathogens like *Neisseria, Haemophilus* and *Helicobacter* species express specific outer cell membrane receptors capable of binding and removing iron directly from LF or transferrin (Andrews et al., 2003). These bacteria may benefit from the host cellular inflammatory response.

The LF molecule has large cationic patches on its surface (Baker, Baker, and Kidd, 2002) facilitating direct interaction with lipopolysaccharide (LPS) component Lipid A on the surface of Gram-negative bacteria (Appelmelk et al., 1994) leading to damage of the bacterial membrane. By damaging the bacterial membrane, LF is able to increase the antibacterial effect of many commercial drugs. The addition of apo-LF to many other antimicrobial agents enhances their antimicrobial activity, because the efficacy of the majority of antimicrobial compounds is inhibited by iron (Weinberg, 2003; Venkatesh and Rong, 2008).

LF has protease-like antimicrobial activity toward some bacterial virulence factors. For example, hLF degrades IgA1-protease and Hap-adhesin in *Haemophilus influenzae* (Plaut, Qiu, and St. Geme, 2000) and also degrades invasion plasmid antigens of *Shigella flexneri* by interaction with the type III secretion system of pathogens (Gomez et al., 2003). Furthermore, hLF has been shown to decrease invasiveness of enteropathogenic *E. coli* and *Salmonella typhimurium* by blocking the proteins that contribute to adherence and cause hemolysis (Ochoa et al., 2003; Ochoa and Cleary, 2004).

In vitro studies have shown that bLF and N-terminal peptide of LF, lactoferricin (LFcin), are able to induce apoptosis, programmed cell death, in *Listeria monocytogenes* infected THP-1 cells (Longhi et al., 2005) and in Caco-2 cells (Valenti et al., 1999). The activation of the apoptotic cell death accompanied with neutrophil chemoattractant release induced by *L. monocytogenes* infection of hepatocytes is a critical step in launching an innate immune defense action against this intracellular pathogen.

In many cases bacteria live in biofilms which are matrix-encased communities specialized for surface persistence. Biofilms notoriously resist killing by the host defense mechanisms and antibiotics. LF is able to block biofilm development of the opportunistic pathogen *Pseudomonas aeruginosa* at concentrations below those that kill or prevent growth. By chelating iron, LF stimulates surface motility, causing the bacteria to wander across the surface instead of forming cell clusters and biofilms (Singh et al., 2002). Similarly, LF effectively inhibits biofilm formation and reduces the established biofilm of periodontopathic bacteria *Porphyromonas gingivalis* and *Prevotella intermedia* (Wakabayashi et al., 2010). Another interesting finding is that hLF can prevent *in vitro* the growth of antibiotic-resistant pathogenic bacteria *S. aureus* and *Klebsiella pneumoniae* (Nibbering et al., 2001). On the other hand, bLF in combination with penicillin G has been shown to reduce the symptoms of infection caused by *S. aureus* in the mouse mastitis model (Diarra et al., 2003).

LF has a well-documented cidal activity *in vitro* against human pathogenic fungi, especially *Candida* species, including *C. albicans*. This effect is related to the adsorption of LF to the cell surface. The candidacidal effect of hLF is reported to be dependent on the extracellular cation concentration and on the metabolic state of the cell (Viejo-Diaz, Andrés, and Fierro, 2004). The occurrence of a *Candida* invasion in the mouth is associated with dry mouth symptoms. Human subjects with a dry mouth due to hyposalivation have a lowered level of salivary hLF. Some patients with an advanced stage of AIDS also have lowered salivary LF levels and have repeated periods of oral candidiosis. Oral formulations containing LF and LZM have been found effective in the treatment of oral candidiosis even in patients who fail to respond to the conventional antifungal drug therapy (Masci, 2000). Both LF and LZM are present in high concentrations in neutrophil granules and at sites of inflammation. LF can bind to the LPS layer to enable LZM to gain access to the peptidoglycan in the bacterial cell wall. The synergistic action of LF and LZM is inhibited if hLF is iron saturated (Ellison and Ghiel, 1991). Similarly, LF and LZM are synergistic against Gram-positive bacteria but the efficacy is suppressed by saturating LF with iron (Leitch and Willcox, 1998).

The oral treatment with bLF resulted in significantly decreased symptom scores in subjects with moderate vesicular or interdigital tinea pedis (Yamauchi et al., 2000a). Costantino and Guaraldi (2008) tested the clinical efficacy of a cream containing 4% of LF in the treatment of an acute vulvovaginal candidiosis. The treatment with 5 g of cream in the inflamed area twice a day for 7 days completely cured 27 of 34 patients, 5 showed good improvement, and 2 patients suffered from a vaginitis after the treatment.

bLF is more effective against viral infections than human LF. LFcin and other peptides liberated from LF are not as effective as the intact LF molecule (McCann et al., 2003). In experiments *in vitro* LF inhibited adenovirus replication in a dose-dependent manner (Arnold et al., 2002).

The antiviral activity of LF is linked to inhibition of viral host cell interaction through blocking of host cell heparin sulphate or interaction with viral surface proteins. The antiparasitic activity of LF is similar to antiviral action but is mechanistically distinct. LF inhibits the replication of a wide spectrum of RNA and DNA viruses. The list includes several enveloped viruses including herpes simplex viruses, hepatitis B and C viruses, human cytomegalovirus, hantavirus, HIV, and respiratory syncytial virus, among others, and also some naked viruses, for example, rotavirus, poliovirus, adenovirus, human papillomavirus, and enterovirus 71. The antiviral activity does not involve iron binding. LF is considered to inhibit viral replication by preventing the viral particles from infecting new host cells either by binding to host cell receptors or to the virions. Bovine LF has been shown to be more active against viruses than hLF, and apo-LF is less effective than holo-LF (Valenti and Antonini, 2005; Pan et al., 2007a).

The observations that LF is strongly antiviral when present at the time of infection but far less active if given later, suggests that antiviral therapy with LF in many cases may not be practicable (Weinberg, 2003). However, in some viral infections the immune system regulating effect may be beneficial. Shin et al. (2005) demonstrated in a mouse model that orally administered bLF did not exert any effect on influenza virus replication but the amount of inflammatory macrophages and neutrophils in the lungs was reduced significantly. Thus, orally administered bLF substantially suppressed the strong inflammatory reaction of the murine immune system against influenza virus infection, and therefore, attenuated the severity of pneumonia.

3.3.4 Other Proteins

Mammary secretions and avian eggs contain a great number of native proteins that may pose potential but less characterized antimicrobial activity. Also, many of these molecules may act as a precursor for antimicrobial peptides. Examples of such milk proteins or enzymes are xanthine oxidase (XO), alpha-lactalbumin (α-la), and beta-lactoglobulin (β-lg). In avian eggs, ovotransferrin represents similar characteristics.

Xanthine oxidase (XO), xanthine:oxygen oxidoreductase (EC 1.1.3.22), is an oxidase with a broad substrate specificity. It occurs in bovine milk at a relatively high concentration of about 35 mg/L (Farkye, 2002b) and is a dimeric metalloprotein with a molecular mass of about 300 kDa. XO is particularly abundant in the milk fat globule membrane (MFGM), accounting for about 20% of all protein present in MFGM. XO can also be considered an antimicrobial agent because it generates hydrogen peroxide as a by-product of reactions it catalyzes. H_2O_2 is generated as a result of reduction of oxygen molecule. XO has been suggested as one possible catalyst for production of H_2O_2 required for the antimicrobial LP system (Reiter and Perraudin, 1991). Further studies are warranted to explore the putative antimicrobial role of XO better.

Ovotransferrin is an iron-binding protein found in egg albumin. Similarly to LF, in addition to iron binding capacity, it has the ability to permeate bacterial outer membranes causing dissipation of electrical potential (Ibrahim, Sugimoto, and Aoki, 2000). Ovotransferrin has demonstrated antibacterial activity against a wide spectrum of bacteria, including *Pseudomonas* spp., *E.coli*, *Str. mutans*, *S. aureus*, *Bacillus cereus*, and *S. enteritidis* (reviewed by Kovacs-Nolan, Phillips, and Mine, 2005).

3.4 Antimicrobial Peptides

Peptides with antimicrobial properties continue to be the focus of growing scientific and industrial attention. The widespread distribution of broad-spectrum antimicrobial peptides in plants and animals suggests that these peptides play an important role in the host defense against micro-organisms and diseases in the natural environment (Chan and Li-Chan, 2006). In addition, their mode of action seems to elude most, if not all, known drug resistance mechanisms (Mor, 2003). Current medical practice is faced with bacterial pathogens that no longer respond to known antibiotics, for example, vancomycin-resistant enterococci (VRE) and methicillin-resistant *Staphylococcus aureus* (MRSA). This concern calls for a serious search for novel antimicrobial agents. Antimicrobial peptides share the following features (Powers and Hancock, 2003): (1) generally less than 50 amino acid residues, (2) an overall positive charge; (3) substantial (≥50%) proportion of hydrophobic residues; and (4) they may simultaneously affect multiple functions of bacterial cells. Various modes of action for antibacterial peptides have been suggested, including their effect on functions of bacterial membranes by forming pores, causing thinning of the membrane and destabilizing the lipid bilayer. As these peptides are amphipathic molecules, they act in a detergent-like manner. They also inhibit synthesis of macromolecules and interact with various components in the cell interior (van der Kraan et al., 2004). It is well documented that chemical modification of proteins often leads to changes in their antimicrobial properties. In this respect, acylation, amidation, and succinylation are the most often applied methods. In food applications, the safety aspects must, however, be taken into consideration whenever chemical modification of dietary proteins is considered.

This section focuses on the occurrence and properties of antimicrobial peptides derived from bovine milk and hen egg proteins.

3.4.1 Milk-Derived Peptides

Milk proteins have been studied for the presence of bioactive amino acid sequences since the late 1970s and are now considered the most important source of bioactive peptides. The best characterized ones include antihypertensive, antithrombotic, antimicrobial, antioxidative, immunomodulatory, and opioid peptides. More than 200 bioactive peptide sequences have been identified from caseins and whey proteins, so far. Their production, functionality, and potential applications have been reviewed recently (Gobbetti, Minervini, and Rizzello, 2007; Korhonen and Pihlanto, 2006, 2007a; Korhonen 2009a,b,c; Phelan et al., 2009). Also, antimicrobial milk peptides have been reviewed in many articles (Clare, Catignani, and Swaisgood, 2003; Floris et al., 2003; Pellegrini, 2003; Gobbetti, Minervini, and Rizzello, 2004, 2007; López-Expósito and Recio 2006; Lizzi et al., 2009). Bioactive peptides are inactive within the sequence of their parent protein molecule and can be released in the following ways: (a) hydrolysis by digestive enzymes, (b) fermentation of milk with proteolytic starter cultures, and (c) proteolysis by microbial or plant-derived enzymes (Korhonen, 2009a; Vercruysse et al., 2009). A large number of bioactive peptides with different activities have been identified in fermented dairy products, for example, sour milk, yogurt, and many cheese varieties, and the potential contribution of such peptides to the healthfulness of fermented dairy products has been speculated upon (Rizzello et al., 2005; FitzGerald and Murray, 2006; Gobbetti et al., 2007; Phelan et al., 2009). Next, examples are given of different active peptide fragments identified upon enzymatic hydrolysis of milk proteins under experimental conditions.

3.4.1.1 Peptides from Caseins

3.4.1.1.1 Isracidin (f1-23)

Isracidin (f1-23) obtained by digestion of α_{s1}-casein with chymosin, was found effective *in vivo* in a mouse model against *Staphylococcus aureus*, *Streptococcus pyogenes*, *Listeria monocytogenes*, and *C. albicans*. Isracidin had a synergistic effect *in vitro* against antibiotic-resistant *Staph. aureus*, when combined with penicillin or streptomycin (Lahov and Regelson, 1996).

McCann et al. (2006) obtained a number of peptides through peptic hydrolysis of sodium caseinate. One of these peptides, liberated from α_{s1}-casein (f99-109), had a theoretical pI of 10.46 and inhibited (MIC, as µg/mL, in brackets) pathogenic and spoilage bacteria, including: *B. subtilis* (125), *Listeria innocua* (125), *S. typhimurium* (125), *E. coli* (250), *S. enteritidis* (125), and *Citrobacter freundii* (500).

3.4.1.1.2 Casocidin-I (f150-188)

Casocidin-I (f150-188), prepared by digestion of α_{s2}-casein with trypsin, pI 8.9, inhibited the growth of *E. coli* and *S. carnosus* (Zucht et al., 1995), and fragments f164-179 and f183-207 inhibited growth of both Gram(+) and Gram(−) bacteria, with MIC values of 25–100 µM and 8–16 µM, respectively (Recio and Visser, 1999). Three cationic peptides from the C-terminal region, isolated from a chymosin digest of Na caseinate (viz. f181-207, f175-207, and f164-207), were effective in the 0.1% peptone solution against Gram(+) and Gram(−) bacteria, with MIC values (in µg/mL) of 21–168, 10.7–171, and 4.8–76.2, respectively. However, the chymosin Na caseinate digest was not effective against *S. typhimurium* in skim milk (McCann et al., 2005).

3.4.1.1.3 Kappacin

Kappacin, a nonglycosylated, phosphorylated form of caseinomacropeptide (CMP), f138-158, derived initially from κ-casein was found effective against *Str. mutans*, *E. coli*, and *Porphyromonas gingivalis* (Malkoski et al., 2001). A product of tryptic digestion, κ-caseicidin (f17-21), was found to inhibit the growth of *S. aureus*, *E. coli*, and *S. typhimurium* (Matin and Otani, 2002). Peptides liberated from κ-casein by pepsin, f18-24, f139-146, and f30-32, inhibited *Listeria innocua* and *E. coli* (López-Expósito and Recio, 2006).

3.4.1.2 Peptides from Whey Proteins

3.4.1.2.1 Lactoferricin (LFcin)

LF is nowadays commercially produced from milk or whey and has been used as an obvious source when searching for antimicrobial peptides. LF contains distinct clusters of amino acid residues with basic side chains. In bovine LF, one of these clusters occurs between residues 17 and 41, making this fragment highly cationic. This fragment, called bovine lactoferricin (bLFcin), is liberated by hydrolysis of LF with gastric pepsin (Wakabayashi, Yamauchi, and Takase, 2006). A fragment of human lactoferrin, hFcin (f1-47), also released by pepsin, contains within its sequence a region homologous with bFcin. bLFcin is a more potent antimicrobial agent than LF. LFcin binds rapidly to the surface of *E. coli* and *Bacillus subtilis* cells, upsetting their functions and killing them. The rate of killing of these bacteria corresponds to the rate of binding of LFcin to cells. The killing effect against *E. coli* is more pronounced at 20°C than at 37°C. LFcins are two- to twelvefold more active against *Escherichia coli* compared to the corresponding undigested lactoferrin molecule. Bovine LFcin is a more potent bactericidal peptide than hLFc and has antimicrobial activity against a wide range of microbes (Wakabayashi, Takase, and Tomita, 2003; Weinberg, 2007). The net positive charge of LFcin, and of other peptides derived from LF such as lactoferrampin, is essential for their interactions with the negatively charged membranes of bacteria.

LFcin binds to LPS in Gram-negative bacteria and teichoic acid in Gram-positive bacteria (Vorland et al., 1999) and translocates across the cytoplasmic membrane of both Gram-negative and Gram-positive bacteria (Haukland et al., 2001). Ulvatne et al. (2004) reported that bLFcin, on reaching the cytoplasm, inhibited protein synthesis in *E. coli* and *Bacillus subtilis*; the authors did not propose a mechanism for this inhibition. The minimum inhibitory concentrations (MIC) of an intact bovine LF and of LFcin against a strain of *E. coli* were 2 mM and 100 μM, respectively (Bellamy et al., 1992). In an animal model, bLFcin protected mice against infection by parasite *Toxoplasma gondii* (Isamida et al., 1998).

Positive effects of orally administered LF in animal models and in humans may partly be attributed to LFcin, as LF is degraded by pepsin in the stomach. Richardson et al. (2009) have recently demonstrated that the antimicrobial core (RRWQWR) of bLFcin, referred to as LfcinB6, when delivered to the cytosolic compartment by fusogenic liposomes, kills T-leukemia cells. This observation opens the way for further studies on potential applications of bLF's fragments as anticancer agents.

Lactoferrampin (LFampin), LF (f268-284), is a synthetic peptide. This particular part of the bLF sequence was selected on the basis of its putative antimicrobial properties (van der Kraan et al., 2004). Hydrolysis of bLF by pepsin, in addition to bLFcin, generates cationic peptides in the LFampin region, for example, LF (f277-288), LF (f265-285), and LF (f267-288), but LFampin cannot be obtained by proteolysis.

3.4.1.2.2 *Glycomacropeptide (GMP)*

Glycomacropeptide (GMP) is a C-terminal glycopeptide f(106-169) released from the κ-casein molecule at [105]Phe-[106]Met by the action of chymosin. GMP is hydrophilic and remains in the whey fraction in the cheese manufacturing process, whereas the remaining part of κ-casein, termed para-κ-casein, precipitates into the cheese curd. GMP has a molecular weight of about 8,000 Daltons and it contains a significant (50–60% of total GMP) carbohydrate (glycoside) fraction which is composed of galactose, *N*-acetyl-galactosamine, and *N*-neuraminic acid. The nonglycosylated form of GMP is often termed caseinomacropeptide or CMP. GMP is the most abundant protein/peptide in whey proteins produced from cheese whey amounting to 20–25% of the proteins (Abd El-Salam, El-Shibini, and Buchheim, 1996; Thomä-Worringer, Sörensen, and López-Fandino, 2006). GMP has been shown to exhibit *in vitro* a wide range of antimicrobial and other bioactivities. It inactivates microbial toxins of *E. coli* and *V. cholerae*, inhibits *in vitro* adhesion of cariogenic *Str. mutans* and *Str. sobrinus* bacteria and influenza virus, modulates immune system responses, promotes growth of bifidobacteria, suppresses gastric hormone activities, and regulates blood circulation through antihypertensive and antithrombotic activity (Brody, 2000; Manso and López-Fandino, 2004). Rhoades et al. (2005) have demonstrated that GMP effectively inhibited *in vitro* adhesion of pathogenic (VTEC and EPEC) *E. coli* strains to human HT29 colon carcinoma cells, whereas probiotic *Lactobacillus* strains were inhibited to a lesser extent. Earlier it was shown that oral GMP administration reduces *E. coli* induced diarrhea in infant rhesus monkeys (Brück et al., 2003). Conclusive evidence about the antimicrobial effects of GMP is still lacking.

3.4.1.2.3 Peptides from α-Lactalbumin

Alpha-lactalbumin (α-La) is one of the globular proteins found in bovine and human milk. Bovine α-La is quantitatively the second most important protein in whey (on an average 1, 2 g/L). It makes up approximately 20% to 25% of the whey proteins and has a MW of 14.2 kDa (Kamau et al., 2010). α-La contains a tightly bound Ca^{2+} and is not antimicrobial in its native state. However, some of the peptides (17–31) S–S (109–114) liberated by trypsin and chymotrypsin, but not those released by pepsin, have exhibited bactericidal properties against Gram-positive bacteria. Gram-negative bacteria were only weakly affected. Stabilization of the peptides' structure through disulfide bridges was essential for antibacterial effects (Pellegrini et al., 1999). Pihlanto-Leppälä et al. (1999) observed that a successive digestion of α-la and β-lg with pepsin and trypsin generated several peptides with antibacterial activity against *E. coli*. The bacteriostatic properties of the hydrolysates could be enriched by ultrafiltration through membranes having 10 kDa and 1 kDa molecular mass cut-off. If generated *in vivo*, these peptides may be of importance for the modulation of the microflora in the gut and for protection against some enteropathogenic bacteria. The combination of peptide fractions containing α-La and glycomacropeptide (GMP) with the same concentration of 0.25 to 0.05 mg/mL showed a synergistic effect in inhibiting the association of enteropathogenic *E. coli* EPEC, *S. typhimurium*, and *Shigella flexneri* with intestinal cells, and thus, may prevent infection (Brück et al., 2006).

In the study by Almaas et al. (2006), the *in vitro* digestion of goat whey proteins using human gastric and duodenal juice resulted in antibacterial peptide profiles different from those obtained using commercial enzymes; α-La was partially degraded by human enzymes, whereas treatment with commercial enzymes fully degraded the protein. α-La hydrolysates obtained with both human gastric juice and human duodenal juice strongly inhibited *L. monocytogenes* at 0.3% to 0.6% protein concentration in the bacterial culture media. Undigested goat whey and hydrolysates digested with human gastric juice demonstrated no significant effect in inhibiting *L. monocytogenes*. This suggests that antibacterial caprine whey hydrolysates are mainly obtained in the duodenum during digestion. The authors suggest that listeriosis, a bacterial infection occurring primarily in newborn infants, elderly patients, and patients who are immune-compromised, could be controlled by these peptides. Peptides with antiviral activity have been derived from α-La. Several peptide fragments with antiherpetic activity were obtained by proteolytic digestion of bovine α-La using trypsin, chymotrypsin, and pepsin (Oevermann et al., 2003).

3.4.1.2.4 Peptides from β-Lactoglobulin

β-Lactoglobulin (β-Lg, MW 18.3 kDa) is the main protein in whey, at ~3.0–3.5 g/L (~10% of milk proteins). It forms reversible dimers at the native pH of milk and binds a range of small hydrophobic and amphipathic molecules (retinol, vitamin D2, fatty acids, etc.). β-Lg is not antimicrobial in its native state, but peptides liberated by successive hydrolysis with pepsin and trypsin have been found to possess antibacterial properties (Pihlanto-Leppälä et al., 1999). Pellegrini et al. (2001) reported that digestion of β-Lg with trypsin produced bactericidal peptide fragments (f15-20), (f25-40), (f78-83), and (f92-100) which were found effective against Gram-positive bacteria tested. Apparently, more research is warranted to elucidate better the antimicrobial properties of this interesting molecule, as its biological importance is still largely unknown.

3.4.2 Egg-Derived Peptides

Enzymatic hydrolysis of hen egg white LZM produces peptides with antibacterial activity against *E. coli* and *S. aureus* (Pellegrini et al., 2000; Mine, Ma, and Lauriau, 2004). These peptides correspond to amino acid residues 98-112, 98-108, and 15-21. The action of LZM-derived peptides differs from enzymatic lysis of cell membranes and the mechanism could be explained by a helix–loop–helix structure that causes membrane permeability changes and inhibition of redox-driven bacterial respiration (Ibrahim et al., 2005). The bactericidal polypeptides derived from lysozyme also block the synthesis of DNA and RNA (Pellegrini et al., 2000). Thammasirirak et al. (2010) reported a similar helix structure in an antimicrobial peptide from goose LZM. The antimicrobial activity of LZM-derived peptides is thus not directly dependent on the amino acid sequence. Abdou et al. (2007) evaluated the antimicrobial effect of a commercial powder containing LZM peptides and found it active against *Bacillus* species suggesting that the product could be useful in controlling growth of food spoilage organisms.

Egg white ovalbumin is not known to have any antimicrobial properties. However, peptides derived from ovalbumin by enzymatic digestion with trypsin and chymotrypsin have shown strong antimicrobial activity against *Bacillus subtilis* and to a lesser extent against *E. coli*, *Bordetella bronchiseptica*, *Pseudomosas aeruginosa*, and *Serratia marcescens*. A weak fungicidal activity against *C. albicans* was also shown by some peptides (Pelligrini et al., 2004).

A 92-amino acid peptide from ovotransferrin was found to kill Gram-negative bacteria by crossing the bacterial outer membrane and causing damage to the biological function of the cytoplasmic membrane (Ibrahim, Sugimoto, and Aoki, 2000). Two fragments of ovotransferrin have shown antiviral activity towards Marek's disease virus (Giansanti et al., 2005).

Peptides derived from ovomucin by pronase treatment have demonstrated antiviral activity against Newcastle disease virus, bovine rotavirus, and human influenza virus *in vitro* (Tsuge, Shimoyamada, and Watanabe, 1996, 1997). Ovodefensins is a new group of potential antimicrobial peptides found in egg white. The biology of gallin, a member of the ovodefensins, has been examined in detail, and a clear inhibition of *E. coli* with relatively low levels of gallin was seen (Gong et al., 2010).

3.5 Commercialization of Antimicrobial Proteins and Peptides

There has been increasing industrial interest in the application of functional food proteins and peptides because of their potential as health-promoting food ingredients or as medically interesting agents of biological origin. In this respect, concrete approaches toward commercialization have been made in recent years in particular in the field of biopreservation of foods and feeds, digestive tract healthcare, and prevention of hospital infections caused by drug-resistant bacteria (Mine and Kovacs-Nolan, 2004; Korhonen, 2009b,c). Table 3.3 provides current commercial examples of applications for milk- and egg-derived antimicrobial proteins and peptides. The opportunities for commercial exploitation of bioactive components depend to a great extent on techniques available for their large-scale isolation in an active form. Industrial-scale processes are already established for the recovery of nondenatured, biologically active whey proteins such as Ig, LF, LP, and GMP, and recently also bioactive peptides (Pouliot, Gauthier, and Groleau, 2006; Vercruysse et al., 2009). These components are known to retain their activity in the regular processes applied in the dairy industry. Cheese whey, therefore, provides a natural plentiful source for their fractionation in purified or concentrated form.

Similarly, egg white is an abundant source of LZM, ovoalbumin, and ovotransferrin (Kovacs-Nolan et al., 2005; Pouliot et al., 2006). As a result of progress made in membrane separation and chromatographic techniques, it is now possible to isolate Igs also from bovine colostrum and egg yolk on a large scale (Korhonen and Pihlanto, 2007b; Chalghoumi et al., 2009a). This progress has facilitated and boosted manufacture of Ig concentrates as dietary supplements or ingredients. Some of these preparations are being marketed with health or nutrition functional claims even though the clinical evidence is limited or not reported. More scientific research is therefore needed in this

TABLE 3.3

Milk and Egg Proteins and Peptides Exploited in Commercial Applications

Protein /Peptide	Application	Comments
Immunoglobulins*	Prevention/treatment of oral or intestinal microbial infections	Commercial products against rotavirus and traveller's diarrhea, potential for *Helicobacter* and hospital infections, e.g., *C. difficile*
		Colostral Ig supplements on the market
Glycomacropeptide*	Dental caries	Efficacy not established in humans
		Cheese whey-derived supplements on the market
Lactoferrin*	Prevention of enteropathogen and viral infections	Many beneficial effects documented in animal and human studies
		Commercial products for oral hygiene and cosmetics, applied as ingredient in infant formulas and yogurt
	Prevention of pathogen contamination of raw meat	GRAS status and commercial use in USA
Lactoperoxidase system*	Preservation of raw milk	Approved by Codex Alimentarius Commission
		In use in many countries with insufficient milk cooling facilities
	Prevention of dental caries	Efficacy proven against mutans streptococci
		Commercial products for oral hygiene
Anticariogenic peptides*	Prevention of dental caries	Beneficial effect documented in human studies
		Commercial products for prevention of dental caries
Lysozyme**	Prevention of microbial defects in hard cheese	Prevention of butyric acid fermentation caused by contaminating clostridia spores

Note: * = milk derived; ** = egg-white derived.

field to substantiate the claimed health benefits. As for immune milk preparations, there are already a few commercial products on the market with quite well-established clinical evidence about their efficacy (Mehra, Marnila, and Korhonen, 2006).

The main limitation of the clinical use of bovine milk or hen egg antibodies for humans is that as proteins of another species, the Igs can be used only orally against gastrointestinal pathogens. An interesting attempt to overcome this limitation is a transchromosomic calf that has been cloned for producing humanized polyclonal Igs. As a result of five sequential genetic modifications and seven consecutive cloning events this calf was reported to have full human IgG as 10–20% of total serum. After hyperimmunization of this calf with an anthrax protective antigen, both full human IgG and chimeric IgG were found to be effective in a toxin-neutralization assay (Kuroiwa et al., 2009). The normal colostrum and preparations made from it are in most countries regarded as a food or dietary supplement, whereas the Ig containing preparations from immunized cows or hens are often regarded as pharmaceuticals, for example, in the European Union or United States, or their regulatory status is not defined (Scammel, 2001; Hoerr and Bostwick, 2002; Mehra, Marnila, and Korhonen, 2006).

A few immune milk products have been commercialized and more can be expected in the future, for example, as a supportive means in antibiotic treatments and prevention of hospital infections (see Table 3.2; Mehra et al., 2006; Korhonen and Marnila, 2009). Passive immunization with specific antibodies provides an immediate protection that is available also for persons with impaired immunity (Pant et al., 2007).

Egg antibodies (IgY from yolk) are promising for use in immunoassays (Schade et al., 2005) to quantify, for example, toxins or pathogenic viruses. The potential use of specific egg yolk-derived antibodies in clinical medicine or healthcare products is continuously being explored and applications are tested, for example, against dental caries or *Helicobacter pylori*-caused gastritis and rotavirus infections (Kovacs-Nolan, Mine, and Hatta 2006).

In view of the broad antimicrobial spectrum against a variety of spoilage and pathogenic micro-organisms, the LP system has been investigated extensively for many potential applications. Since the 1970s, various industries have investigated the possibility of utilizing the LP system as a natural antimicrobial agent in a number of diverse products (Reiter and Perraudin, 1991). Most of the applications concern the preservation of different foodstuffs, mainly dairy products, but raw fish and meat products have also been subjects of applied research. In feeding trials, a combination of the LP system and LF has proven effective in the treatment of scouring calves.

It has been suggested that the combined antimicrobial activity of these components may be more powerful than either of the components alone (van Hooijdonk, Kussendrager, and Steijns, 2000). The effectiveness of an activated LP system in extending the keeping quality of unpasteurized nonrefrigerated and refrigerated milk (bovine, camel, buffalo) has been demonstrated in many pilot studies and field trials worldwide (Anon., 2005; Seifu, Buys, and Donkin, 2005). Successful field trials conducted in the 1970s in Cuba, Kenya, Pakistan, Sri Lanka, and Mexico demonstrated that milk collected in locations remote from processing facilities could be transported at ambient temperatures (sometimes as high as 30°C) for up to eight hours, without an increase of total bacterial count. The keeping quality of refrigerated raw milk at 4°C can be prolonged up to 5–6 days, respectively. Following these trials, the LP system was recommended for use by the World Health Organisation (WHO) and the Food and Agriculture Organization of the United Nations (FAO). Since 1991, the LP system has been approved by the Codex Alimentarius Committee for preservation of raw milk under specified conditions. The method is now being utilized in practice in a number of countries where facilities for raw milk cooling are not available or are inadequate. The LP system has found other applications, for example, in dental healthcare products and animal feeds. A potential new target is gastritis caused by *H. pylori*, as this pathogen is killed *in vitro* by the LP system (Shin et al., 2002). More applications have been envisaged for preservation of different products including meat, fish, vegetables, fruits, and flowers (Boots and Floris, 2006).

Egg white LZM is largely used as a food preservative. It has found commercial application in semi-hard and hard cheeses to prevent the late blowing defect caused by Clostridia and because it does not inhibit the growth of starters and secondary cultures required for the ripening of these cheeses. LZM also prevents the growth of pathogenic bacteria on refrigerated foods: *Listeria monocytogenes*, *Clostridium botulinum*, *Clostridium jejuni*, and *Yersinia enterocolitica* (Johnson, 1994). The antibacterial properties of LZM have led to its use in oral health care products, such as toothpaste, mouthwash, and chewing gum, to protect against periodontal-causing bacteria and prevent infections in the oral mucosa (Kovacs-Nolan, Phillips, and Mine, 2005).

The industrial large-scale production of bovine LF was started in Belgium in 1985. Since then, commercial exploitation of bLF has developed relatively slowly until recently. On the other hand, Asian companies have applied bLF for more than 20 years as an ingredient of commercial foodstuffs, most of them being milk-based products (Tomita et al., 2009). In 1986, marketing of a bLF-containing infant formula, "BF-L dry milk," was started by Morinaga Milk Industry in Japan. Currently, bLF-containing infant formulas are sold in Indonesia and Korea, as well as in Japan. Over the years, scientific evidence on the benefits of supplementation of bLF in infant formulas has been accumulating. Roberts et al. (1992) reported that an infant formula containing 1 mg/ml bLF established a *Bifidobacterium*-dominant fecal flora in normal infants, whereas infant formulas containing none or 0.1 mg/ml bLF did not have a similar effect. In an intervention study with weaned Peruvian children, the test group that received bLF daily for nine months was demonstrated to have a significantly reduced rate of fecal colonization by *Giardia* species as compared to the control group (Ochoa et al., 2008). Other bLF-containing products include yogurt, skim milk, milk-type drinks, supplemental tablets, pet food, oral care products (mouth wash, gel, toothpaste, and chewing gum), and cosmetics that are mainly skin care products (Wakabayashi et al., 2006). The beneficial effects of these bLF-containing products on health have been proved in clinical and animal studies (Weinberg, 2007; Marnila and Korhonen, 2009). The therapeutic effect of LF-supplemented pet food on dermatitis in dogs and cats has also been demonstrated (Tomita et al., 2009).

The antimicrobial and antioxidative effects of bLF favor its use as a food preservative. The U.S. Food and Drug Administration (FDA) has approved the use of bLF (at not more than 2% by weight) as a spray to reduce microbial contamination on the surface of raw beef carcasses. FDA has granted to bLF a "Generally Recognized As Safe" (GRAS, GRN 67) status and this determination accounts for uses at defined levels in beef carcasses, subprimals, and finished cuts (Taylor et al., 2004). Another potential application area for bLF is the use as a component of edible coatings. Brown, Wang, and Oh (2008) found that the combination of bLF with lysozyme in the coating material in chitosan films significantly decreased the growth of *E. coli* O157:H7. In the study by Min, Harris, and Krochta (2006), edible whey protein film incorporated with the active LP system was shown to inhibit the growth of *S. enterica* and *E. coli* O157:H whereas films incorporating either LF or LZM were not effective against these pathogens.

Many microbes isolated from mastitic milk are sensitive to bLF or bLFcin whereas many pathogenic strains of *E. coli* and *S. aureus* show resistance or can develop resistance to LF. However, bLF possesses a synergistic effect with antibiotics against mastitic *S. aureus* (Diarra, Petitclerc, and Lacasse, 2002; Komine et al., 2006; Petitclerc et al., 2007; Lacasse et al., 2008). These results would suggest that bLF added to penicillin may offer an effective combination for the treatment of chronic infections caused by beta lactam antibiotic resistant *S. aureus* strains.

The well-documented antimicrobial, anti-infective, cancer preventive, and immune response regulating activities as well as effects on iron metabolism of LF have led to the development of commercial applications varying from clinical and nutraceutical uses to infant milk formulas, oral hygiene, cosmetics, and various food supplements (Tomita et al., 2002; Weinberg, 2007). Although LF is a molecule of our own immune system, and thus well tolerated, potential hazards have been documented and further research is needed to define the safe ways to exploit LF's therapeutic properties (Weinberg, 2003). Cerven, DeGeorge, and Bethell (2008) studied the safety of oral administration of purified recombinant human apo-LF and holo-LF in Wistar rats. No treatment-related changes in clinical parameters were observed in any of these preparations. Toxicological studies in rats for 13 weeks with a daily oral administration of bLF (Yamauchi et al., 2000b) and of recombinant hLF produced in transgenic cows (Appel et al., 2006) revealed that the levels of no observed adverse effect were at least 2 g/kg body weight/day.

The results of these studies can be regarded as supportive of the safety of clinical studies with bLF for use in food products and supplements (Tamano et al., 2008). Many bacteria, including many *E. coli* and *S. aureus* strains, possess some degree of resistance against bLF or bLFcin. This resistance can be due to proteases on cell outer membranes or it may be due to positively charged groups on the cell surface preventing the incorporation of cationic sequences of LF or LFcin (Ulvatne et al., 2001, 2002). Thus, a widespread use of LF in the food industry may lead to development of resistant pathogens. The risks due to possibly emerging LF-resistant microbes may be even higher than those of antibiotics because in the case of LF potential pathogens are, in practice, educated to resist the molecules of our own innate immunity. Another type of risk may emerge owing to LF's ability to stimulate the replication of human T cell leukemia virus type 1. Also, orally ingested LF may be harmful to humans with infections by gastrointestinal bacteria that can use iron from LF (Weinberg, 2003). Immunological effects of orally administered LF are diverse and, thus far, only partially known. Therefore, further experimental research concomitant with relevant safety evaluation appears necessary in those particular application areas.

GMP has been claimed to deliver many health effects, such as weight reduction and prevention of various gastrointestinal infections, including dental caries. Although conclusive clinical evidence about such benefits remains still to be substantiated, already on the global market there are a great number of healthcare products and supplements containing GMP as a claimed effective ingredient.

3.6 Conclusion

Safety of foodstuffs has arisen as a global concern for all stakeholders of the food chain from farm to consumers' table. Consumers demand higher-quality, preservative-free, safe yet mildly processed foods with an extended shelf-life. The public health authorities are forced to tighten quality and safety regulations on foods in view of past and emerging food safety crises related to microbial contamination. This trend is invigorated by the fast-growing global marketing of food raw materials and processed foods and expanding tourism as well as a result of potential climate change. These issues pose challenges also to the food industries which have to look for novel means to ensure the safety of their products. Again, humans and domestic animals are exposed to the alarming emergence of resistance among bacteria to antimicrobial agents causing difficulties in the chemotherapy of microbial diseases.

In view of the above considerations, natural antimicrobials derived from either plants or animals are likely to offer innovative solutions. The increased scientific knowledge about the mechanisms of action of such antimicrobial compounds as well as rapid progress made in novel mild or green technologies (e.g., membrane separation and chromatographic techniques) has enabled the industrial manufacture of concentrated or pure antimicrobial components from bovine colostrum, milk, and avian eggs. New emerging technologies, such as micro- or nanoencapsulation and high hydrostatic pressure may offer feasible solutions for improving stability and bioavailability of such components in various food products and during digestion (Korhonen, 2002; Kulozik, 2009). In recent years, this potential has been rapidly expanded due to the fast-growing findings about antimicrobial peptides derived from plant and animal sources (Rydlo, Miltz, and Mor, 2006).

There seems to be a wide range of potential areas for applications of these antimicrobials in various industrial sectors, for example, plant and livestock production, food processing industries, pharmaceutical, and well-being

TABLE 3.4

Areas of Application for Milk and Egg-Derived Antimicrobial Proteins and Peptides

- Contribution to natural defense mechanisms
  cow (prevention/therapy of mastitis)
  calf, pigs (modulation of gut microbiota)
  humans (modulation of gut microbiota, stimulation of mucosal immune system)
- Prevention and treatment of microbial infections of the gastrointestinal tract
  farm animals, fish
  humans
- Extension of shelf-life of foods and feeds
  milk, meat, fish
  vegetables, fruits
- Control of microbiological quality of foods and feeds
  pathogenic micro-organisms
  toxins
  microbiological defects
- Use as active ingredients for healthcare products
  oral hygiene
  ophthalmics
  cosmetics

industries as well as chemical and cosmetics industries. Table 3.4 provides a brief list of potential future applications of bovine colostrums, milk, and hen egg derived antimicrobial proteins, and their enzymatic derivatives. At present, peptides seem to present the most promising group among these compounds due to their chemical and biological diversity, the most promising applications being as natural preservatives of various foodstuffs and as replacers of or supplements to antimicrobial drugs. However, safety of novel peptides generated by new technologies has to be assessed before entering into proper product development.

References

Abd El-Salam, M.H., El-Shibini, S., and Buchheim, W. 1996. Characteristics and potential uses of the casein macropeptide. *Int. Dairy J.* 6: 327–341.

Abdou, A.M., Higashiguchi, S., Aboueleinin, A.M., Kim, M., and Ibrahim, H.R. 2007. Antimicrobial peptides derived from hen egg lysozyme with inhibitory effect against bacillus species. *Food Control* 18(2): 173–178.

Akita, E.M. and Nakai, S. 1992. Immunoglobulins from egg yolk: Isolation and purification. *J. Food Sci.* 57: 629–634.

Almaas, H., Holm, H., Langsrud, T., Flengsrud, R., and Vegarud, G.E. 2006. *In vitro* studies of the digestion of caprine whey proteins by human gastric and duodenal juice and the effects on selected microorganisms. *Brit. J. Nutr.* 96(3): 562–569.

Andrews, S.C., Robinson, A.K., and Rodriguez-Quinones, F. 2003. Bacterial iron homeostasis. *FEMS Microbiology Rev.* 27(2–3): 215–237.

Anon. 2005. Benefits and potential risks of the lactoperoxidase system of raw milk preservation. *Report of an FAO/WHO technical meeting.* FAO Headquarters, Rome, 28 November–2 December.

Appel, M.J., van Veen, H.A., Vietsch, H., Salaheddine, M., Nuijens, J.H., Ziere, B., and de Loos, F. 2006. Sub-chronic (13-week) oral toxicity study in rats with recombinant human lactoferrin produced in the milk of transgenic cows. *Food Chem. Toxicol.* 44(7): 964–973.

Appelmelk, B.J., An, Y.Q., Geerts, M., Thijs, B.G., de Boer, H.A., MacLaren, D.M., de Graaff, J., and Nuijens. J.H. 1994. Lactoferrin is a lipid A-binding protein. *Infect, Immunity* 62(6): 2628–2632.

Arnold, D., Di Biase, A.M., Marchetti, M., Pietrantoni, A., Valenti, P., Seganti, L., and Superti, F. 2002. Antiadenovirus activity of milk proteins: Lactoferrin prevents viral infection. *Antiviral Res.* 53(2): 153–158.

Ashraf, H., Mahalanabis, D., Mitra, A.K., Tzipori, S., and Fuchs, G.J. 2001. Hyperimmune bovine colostrum in the treatment of shigellosis in children: A double-blind, randomized, controlled trial. *Acta Paediatrica* 90(12): 1373–1378.

Attallah, A.M., Abbas, A.T., Ismail, H. Abdel-Raouf, M., and El-Dosoky, I. 2009. Efficacy of passive immunization with IgY antibodies to a 58-kDa H. pylori antigen on severe gastritis in BALB/c mouse model. *J. Immunoassay Immunochem.* 30(4): 359–377.

Baker, E.N., Baker, H.M., and Kidd, R.D. 2002. Lactoferrin and transferrin: Functional variations on a common structural framework. *Biochem. Cell Biol.* 80(1): 27–34.

Bellamy, W., Takase, M., Wakabayashi, H., Kawase, K., and Tomita, M. 1992. Antibacterial spectrum of lactoferricinB, a potent bacterial peptide from the N-terminal region of bovine lactoferrin. *J. Appl. Bacteriol.* 73: 472–479.

Bojsen, A., Buesa, J., Montava, R., Kvistgaard, A.S., Kongsbak, M.B., Petersen, T.E., Heegaard, C.W., and Rasmussen, J.T. 2007. Inhibitory activities of bovine macromolecular whey proteins on rotavirus infections *in vitro* and *in vivo*. *J. Dairy Sci.* 90(1): 66–74.

Bölke, E., Jehle, P.M., Hausmann, F., Daubler, A., Wiedeck, H., Steinbach, G., Storck, M., and Orth, K. 2002. Preoperative oral application of immunoglobulin-enriched colostrum milk and mediator response during abdominal surgery. *Shock* 17(1): 9–12.

Bölke, E., Orth, K., Jehle, P.M., Schwarz, A., Steinbach, G., Schleich, S., Ulmer, C., Storck, M., and Hannekum, A. 2002. Enteral application of an immunoglobulin-enriched colostrum milk preparation for reducing endotoxin translocation and acute phase response in patients undergoing coronary bypass surgery–A randomized placebo-controlled pilot trial. *Wiener Klinische Wochenschrift* 114(21–22): 923–928.

Boots, J.-W. and Floris, R. 2006. Lactoperoxidase: From catalytic mechanism to practical applications. *Int. Dairy J.* 16: 1272–1276.

Born, J. and Bhakdi, S. 1986. Does complement kill *E. coli* by producing transmural pores? *Immunology* 59(1): 139–145.

Bostwick, E.F., Steijns, J., and Braun, S. 2000. Lactoglobulins. In: *Natural Food Antimicrobial Systems* (A.S. Naidu, Ed.), Boca Raton, FL: CRC Press, 133–158.

Braun,V. and Killmann, H. 1999. Bacterial solutions to the iron-supply problem. *Trends in Biochem. Sci.* 24(3): 104–109.

Brinkworth, G.D. and Buckley, J.D. 2003. Concentrated bovine colostrum protein supplementation reduces the incidence of self-reported symptoms of upper respiratory tract infection in adult males. *Euro. J. Nutr.* 42(4): 228–232.

Brody, E.P. 2000. Biological activities of bovine glycomacropeptide. *Brit. J. Nutr.* 84: S39–S46.

Brown, C.A., Wang, B., and Oh, J.H. 2008. Antimicrobial activity of lactoferrin against foodborne pathogenic bacteria incorporated into edible chitosan film. *J. Food Protect.* 71(2): 319–324.

Brück, W.M., Kelleher, S.L., Gibson, G.R., Nielsen, K.E., Chatterton, D.E., and Lönnerdal, B. 2003. rRNA probes used to quantify the effects of glycomacropeptide and α-lactalbumin supplementation on the predominant groups of intestinal bacteria of infant rhesus monkeys challenged with enteropathogenic *Escherichia coli. J. Pediatr. Gastroent. Nutr.* 37: 273–280.

Brück, W.M., Redgrave, M., Tuohy, K.M., Lönnerdal, B., Graverholt, G., Hernell, O., and Gibson, G.R. 2006. Effects of bovine alpha-lactalbumin and casein glyco-macropeptide-enriched infant formulae on faecal microbiota in healthy term infants. *J. Pediatr. Gastroent. Nutr.* 43(5): 673–679.

Butler, J.E. 1998. Immunoglobulin diversity, B-cell and antibody repertoire development in large farm animals. *Revue Scientifique et Technique,* 17(1): 43–70.

Campbell, B. and Petersen, W.E. 1963. Immune milk—a historical survey. *Dairy Sc. Abstr.* 25: 345–358.

Carlander, D., Kollberg, H., and Larsson, A. 2002. Retention of specific yolk IgY in the human oral cavity. *BioDrugs: Clin. Immunotherapeut. Biopharmaceut. Gene Therapy* 16(6): 433–437.

Casswall, T.H., Nilsson, H.O., Björck, L., Sjöstedt, S., Xu, L., Nord, C.K., Boren, T., Wadström, T., and Hammarström, L. 2002. Bovine anti-helicobacter pylori antibodies for oral immunotherapy. *Scand. J. Gastroenter.* 37(12): 1380–1385.

Casswall, T.H., Sarker, S.A., Albert, M.J., Fuchs, G.J., Bergstrom, M., Björck, L., and Hammarström, L. 1998. Treatment of Helicobacter pylori infection in infants in rural Bangladesh with oral immunoglobulins from hyperimmune bovine colostrum. *Aliment. Pharmacol. Therapeut.* 12(6): 563–568.

Casswall, T.H., Sarker, S.A., Faruque, S.M., Weintraub, A , Albert, M.J., Fuchs, G.J., Alam, N.H. et al. 2000. Treatment of enterotoxigenic and enteropathogenic Escherichia coli-induced diarrhoea in children with bovine immunoglobulin milk concentrate from hyperimmunized cows: A double-blind, placebo-controlled, clinical trial. *Scand. J. Gastroent.* 35(7): 711–718.

Cerven, D., DeGeorge, G., and Bethell, D. 2008. 28-Day repeated dose oral toxicity of recombinant human apo-lactoferrin or recombinant human lysozyme in rats. *Reg. Toxicol. Pharmacol.* 51(2): 162–167.

Chalghoumi, R., Beckers, Y., Portetelle, D., and Thêwis, A. 2009a. Hen egg yolk antibodies (IgY), production and use for passive immunization against bacterial enteric infections in chicken: A review. *Biotechnol. Agron. Soc. Environ.* 13(2): 295–308.

Chalghoumi, R., Marcq, C., Thewis, A., Portetelle, D., and Beckers, Y. 2009b. Effects of feed supplementation with specific hen egg yolk antibody (immunoglobin Y) on Salmonella species cecal colonization and growth performances of challenged broiler chickens. *Poult. Sci.* 88(10): 2081–2092.

Chan, J.C.K. and Li-Chan, E.C.Y. 2006. Antimicrobial peptides. In *Nutraceutical Proteins and Peptides in Health and Disease* (Y. Mine and F. Shahidi, Eds.), Boca Raton, FL: Taylor & Francis, pp. 99–136.

Clare, D.A., Catignani, G.L., and Swaisgood, H.E. 2003. Biodefense properties of milk: The role of antimicrobial proteins and peptides. *Curr. Pharmaceut. Des.* 9: 1239–1255.

Conesa, C., Sánchez, L., Rota, C., Pérez, M.-D., Calvo, M., Farnaud, S., and Evans, R.W. 2008. Isolation of lactoferrin from milk of different species: Calorimetric and antimicrobial studies. *Compar. Biochem. Physiol. B*, 150: 131–139.

Costantino, D. and Guaraldi, C. 2008. Preliminary evaluation of a vaginal cream containing lactoferrin in the treatment of vulvovaginal candidosis. *Minerva Ginecologica* 60(2): 121–125.

Davidson, G.P., Whyte, P.B., Daniels, E., Franklin, K., Nunan, H., McCloud, P.I., Moore, A.G., and Moore, D.J. 1989. Passive immunisation of children with bovine colostrum containing antibodies to human rotavirus. *Lancet* 2(8665): 709–712.

Diarra, M.S., Petitclerc, D., Deschênes, É., Lessard, N., Grondin, G., Talbot, B.G., and Lacasse, P. 2003. Lactoferrin against staphylococcus aureus mastitis: Lactoferrin alone or in combination with penicillin G on bovine polymorphonuclear function and mammary epithelial cells colonisation by staphylococcus aureus. *Veterinary Immunol. Immunopathol.* 95(1–2): 33–42.

Diarra, M.S., Petitclerc, D., and Lacasse, P. 2002. Effect of lactoferrin in combination with penicillin on the morphology and the physiology of Staphylococcus aureus isolated from bovine mastitis. *J. Dairy Sci.* 85(5): 1141–1149.

Earnest, C.P., Jordan, A.N., Safir, M., Weaver, E., and Church, T.S. 2005. Cholesterol-lowering effects of bovine serum immunoglobulin in participants with mild hypercholesterolemia. *Amer. J. Clin. Nutr.* 81(4): 792–798.

Ebina, T., Sato, A., Umezu, K., Ishida, N., Ohyama, S., Oizumi, A., Aikawa, K., Katagiri, S., Katsushima, N., and Imai, A. 1985. Prevention of rotavirus infection by oral administration of cow colostrum containing antihumanrotavirus antibody. *Med. Microbiol. Immunol.* 174(4): 177–185.

Ehrlich, P. 1892. Über immunität durch verebung und zeugung. *Zeitschrift für Hygiene und Infektionskrankheiten* 12: 183–203.

Elfstrand, L., Lindmark-Månsson, H., Paulsson, M., Nyberg, L., and Åkesson, B. 2002. Immunoglobulins, growth factors and growth hormone in bovine colostrum and the effects of processing. *Int. Dairy J.* 12(11): 879–887.

Ellison, R.T. and Giehl T.J. 1991. Killing of gram-negative bacteria by lactoferrin and lysozyme. *J. Clin. Invest.* 88(4): 1080–1091.

Farkye, N.Y. 2002a. Enzymes indigenous to milk—Other enzymes. In *Encyclopedia of Dairy Sciences* (H. Roginski, J.F. Fuquay, and P.F. Fox, Eds.), London: Academic Press/Elsevier Science, pp. 943–948.

Farkye, N.Y. 2002b. Enzymes indigenous to milk—Xanthine oxidase. In *Encyclopedia of Dairy Sciences* (H. Roginski, J.F. Fuquay, and P.F. Fox, Eds.), London: Academic Press/Elsevier Science, pp. 941–942.

Fernaud, S. and Evans, R.W. 2003. Lactoferrin-amultifunctional protein with antimicrobial properties. *Molec. Immunol.* 40: 395–405.

FitzGerald, R.J. and Murray, B.A. 2006. Bioactive peptides and lactic fermentations. *Int. J. Dairy Tech.* 59: 118–125.

Fleming, A. 1922. On a remarkable bacteriolytic element found in tissues and secretions. *Proc. Roy. Soc.* B93:306.

Floren, C.H., Chinenye, S., Elfstrand, L., Hagman, C., and Ihse, I. 2006. ColoPlus, a new product based on bovine colostrum, alleviates HIV-associated diarrhoea. *Scand. J. Gastroent.* 41(6): 682–686.

Floris, R., Recio, I., Berkhout, B., and Visser, S. 2003. Antibacterial and antiviral effects of milk proteins and derivatives thereof. *Curr. Pharmaceut. Des.* 9: 1257–1275.

Fukumoto, L.R., Li-Chan, E., Kwan, L., and Nakai, S. 1994. Isolation of immunoglobulins from cheese whey using ultrafiltration and immobilized metal affinity chromatography. *Food Res. Int.* 27(4): 335–348.

Funatogawa, K., Ide, T., Kirikae, F., Saruta, K., Nakano, M., and Kirikae, T. 2002. Use of immunoglobulin enriched bovine colostrum against oral challenge with enterohaemorrhagic Escherichia coli O157:H7 in mice. *Microbiol. Immunol.* 46(11): 761–766.

Gerson, C., Sabater, J., Scuri, M., Torbati, A., Coffey, R., and Abraham, J. et al. 2000. The lactoperoxidase system functions in bacterial clearance of airways. *Amer. J. Respir. Cell Molec. Biol.* 22: 665–671.

Giansanti, F., Massucci, M.T., Giardi, M.F., Nozza, F., Pulsinelli, E., Nicolini, C. et al. 2005. Antiviral activity of ovotransferrin derived peptides. *Biochem. Biophys. Res. Commun.* 331(1): 69–73.

Gobbetti, M., Minervini, F., and Rizzello, C.G. 2004. Angiotensin I-converting-enzyme-inhibitory and antimicrobial bioactive peptides. *Int. J. Dairy Technol.* 57: 172–188.

Gobbetti, M., Minervini, F., and Rizzello, C.G. 2007. Bioactive peptides in dairy products. In *Handbook of Food Products Manufacturing* (Y.H. Hui, Ed.), Hoboken, NJ: John Wiley & Sons, pp. 489–517.

Gomez, H.F., Ochoa, T.J., Carlin, L.G., and Cleary, T.G. 2003. Human lactoferrin impairs virulence of Shigella flexneri. *J. Infect. Diseases* 187(1): 87–95.

Gong, D., Wilson, P., Bain, M., McDade, K., Kalina, J., Herve-Grepinet, V. et al. 2010. Gallin; An antimicrobial peptide member of a new avian defensin family, the ovodefensins, has been subject to recent gene duplication. *BMC Immunol.* 11(1): 12.

Gurtler, M., Methner, U., Kobilke, H., and Fehlhaber. K. 2004. Effect of orally administered egg yolk antibodies on salmonella enteritidis contamination of hen's eggs. *J. Veterinary Med. B, Infect. Diseases Veterinary Public Health* 51(3): 129–134.

Hammarström, L. and Weiner, C.K. 2008. Targeted antibodies in dairy-based products. *Adv. Exper. Med. Biol.* 606: 321–343.

Hanssen, F.W. 1924. The bactericidal property of milk. *Brit. J. Exper. Pathol.* 5:271.

Hatta, H., Tsuda, K., Akachi, S., Kim, M., Yamamoto, T., and Ebina, T. 1993. Oral passive immunization effect of anti-human rotavirus IgY and its behavior against proteolytic enzymes. *Biosci. Biotechnol. Biochem.* 57(7): 1077–1081.

Haukland, H.H., Ulvatne, H., Sandvik, K., and Vorland, L.H. 2001. The antimicrobial peptides lactoferricin B and magainin 2 cross over the bacterial cytoplasmic membrane and reside in the cytoplasm. *FEBS Lett.* 508: 389–93.

Hiraga, C., Kodama, Y., Sugiyama, T., and Ichikawa, Y. 1990. Prevention of human rotavirus infection with chicken egg yolk immunoglobulins containing rotavirus antibody in cat. *J. Jpn. Assoc. Infect. Diseases* 64(1): 118–123.

Hoerr, R.A. and.Bostwick, E.F 2002. Commercializing colostrum based products: A case study of Galagen Inc. *Int. Dairy Fed. Bull.* 375: 33–46.

Horie, K., Horie, N., Abdou, A.M., Yang, J.O., Yun, S.S., Chun, H.N., Park, C.K., Kim, M., and Hatta, H. 2004. Suppressive effect of functional drinking yogurt containing specific egg yolk immunoglobulin on helicobacter pylori in humans. *J. Dairy Sci.* 87(12): 4073–4079.

Huang, X.H., Chen, L., Gao, W., Zhang, W., Chen, S.J., Xu, L.B., and Zhang, S.Q. 2008. Specific IgG activity of bovine immune milk against diarrhea bacteria and its protective effects on pathogen-infected intestinal damages. *Vaccine* 26(47): 5973–5980.

Ibrahim, H.R., Inazaki, D., Abdou, A., Aoki, T., and Kim, M. 2005. Processing of lysozyme at distinct loops by pepsin: A novel action for generating multiple antimicrobial peptide motifs in the newborn stomach. *Biochimica et Biophysica Acta* 1726(1): 102–114.

Ibrahim, H.R., Sugimoto, Y., and Aoki, T. 2000. Ovotransferrin antimicrobial peptide (OTAP-92) kills bacteria through a membrane damage mechanism. *Biochimica et Biophysica Acta* 1523(2–3): 196–205.

Ibrahim, H.R., Thomas, U., and Pellegrini, A. 2001. A helix-loop peptide at the upper lip of the active site cleft of lysozyme confers potent antimicrobial activity with membrane permeabilization action. *J. Biol. Chem.* 276: 43767–43774.

Ikemori, Y., Kuroki, M., Peralta, R.C., Yokoyama, H., and Kodama, Y. 1992. Protection of neonatal calves against fatal enteric colibacillosis by administration of egg yolk powder from hens immunized with K99-piliated enterotoxigenic escherichia coli. *Amer. J. Veterinary Res.* 53(11): 2005–2008.

Ikemori, Y., Ohta, M., Umeda, K., Icatlo, F.C., Kuroki, M., Yokoyama, H., and Kodama, Y. 1997. Passive protection of neonatal calves against bovine coronavirus-induced diarrhea by administration of egg yolk or colostrum antibody powder. *Veterinary Microbiol.* 58(2–4): 105–111.

Isamida, T., Tanaka, T., Omata, Y., Yamauchi, K., Shimazaki, K., and Saito, A. 1998. Protective effect of lactoferricin against *Toxoplasma gondii* infection in mice. *J. Veterinary Med. Sci.* 60:241–244.

Jenssen, H. and R.E.W. Hancock. 2009. Antimicrobial properties of lactoferrin. *Biochimie* 91:19–29.

Johnson, E.A. 1994. Egg white lysozyme as a preservative for use in foods. In *Egg Uses and Processing Technologies: New Developments* (J.S. Sim and S. Nakai, Eds.), Oxon: CABI, pp. 177–191,

Kamau, S.M., Cheison, S.C., Chen, W., Liu, X.-M., and Lu, R.-R. 2010. Alpha-Lactalbumin: Its production technologies and bioactive peptides. *Comp. Rev. Food Sci. Food Safety* 9:197–212.

Kokko, H., Kuronen, I., and Kärenlampi, S. 1998. Rapid production of antibodies in chicken and isolation from eggs. In *Cell Biology: A Laboratory Handbook*, 2nd edition, Volume 2 (J. Celis, Ed.), San Diego: Academic Press, 410–417.

Kollberg, H., Carlander, D., Olesen, H., Wejaker, P.E., Johannesson, M., and Larsson A. 2003. Oral administration of specific yolk antibodies (IgY) may prevent pseudomonas aeruginosa infections in patients with cystic fibrosis: A phase I feasibility study. *Pediatr. Pulmonol.* 35(6): 433–440.

Komine, Y., Kuroishi, T., Kobayashi, J., Aso, H., Obara, Y., Kumagai, K. et al. 2006. Inflammatory effect of cleaved bovine lactoferrin by elastase on staphylococcal mastitis. *J. Veterinary Med. Sci.* 68(7): 715–723.

Korhonen, H. 2002. Technology options for new nutritional concepts. *Int. J. Dairy Technol.* 55: 79–88.

Korhonen, H. 2009a. Bioactive milk peptides: from science to applications. *J. Funct. Foods* 1(2): 177–187.

Korhonen, H. 2009b. Bioactive components in bovine milk. In *Bioactive Components in Milk and Dairy Products* (Y. Park, Ed.), Ames, IA: Wiley-Blackwell, pp. 15–42.

Korhonen, H. 2009c. Bioactive whey proteins and peptides: Functional solutions for health-promotion. *Nutrafoods* 8(4): 9–22.

Korhonen, H. and P. Marnila. 2006. Bovine milk antibodies for protection against microbial human diseases. In *Nutraceutical Proteins and Peptides in Health and Disease* (Y. Mine, and F. Shahidi, Eds.), Boca Raton, FL: Taylor & Francis, pp. 137–159.

Korhonen, H. and Marnila, P. 2009. Bovine milk immunoglobulins against microbial human diseases. In *Dairy-Derived Ingredients: Food and Nutraceutical Uses* (M. Corredig, Ed.), Oxford, UK: Woodhead, pp. 269–289.

Korhonen, H., Marnila, P., and Gill, H.S. 2000a. Bovine milk antibodies for health. *British J. Nutr.* 84 Suppl 1(Nov): S135–S146.

Korhonen, H., Marnila, P., and Gill, H.S. 2000b. Milk immunoglobulins and complement factors. *Brit. J. Nutr.* 84 Suppl 1(Nov): S75–S80.

Korhonen, H. and Pihlanto A. 2006. Bioactive peptides: Production and functionality. *Int. Dairy J.* 16(9): 945–960.

Korhonen, H. and Pihlanto, A. 2007a. Bioactive peptides from food proteins. In *Handbook of Food Products Manufacturing: Health, Meat, Milk, Poultry, Seafood, and Vegetables* (Y.H. Hui, Ed.), Hoboken, NJ: John Wiley & Sons, pp. 5–38.

Korhonen, H. and Pihlanto, A. 2007b. Technological options for the production of health-promoting proteins and peptides derived from milk and colostrum. *Curr. Pharmaceuti. Des.* 13(8): 829–843.

Korhonen, H., Syväoja, E.L., Ahola-Luttila, H., Sivelä, S., Kopola, S., Husu, J., and Kosunen, T.U. 1995. Bactericidal effect of bovine normal and immune serum, colostrum and milk against Helicobacter pylori. *J. Appl. Bacteriol.* 78(6): 655–662.

Kovacs-Nolan, J., Mine, Y., and Hatta, H. 2006. Avian immunoglobulin Y and its application in human health and disease. In *NutraceuticalProteins and Peptides in Health and Disease* (Y. Mine and F. Shahidi, Eds.), Boca Raton, FL: Taylor & Francis, pp. 161–189.

Kovacs-Nolan, J., Phillips, M., and Mine, Y. 2005. Advances in the value of eggs and egg components for human health. *J. Agricul. Food Chem.* 53(22): 8421–8431.

Kulozik, U. 2009. Novel approaches for the separation of dairy components and manufacture of dairy ingredients. In *Dairy-Derived Ingredients: Food and Nutraceutical Uses* (M. Corredig, Ed.), Oxford, UK: Woodhead, pp. 3–23.

Kuroiwa,Y., Kasinathan, P., Sathiyaseelan, T., Jiao, J.A., Matsushita, H., Sathiyaseelan, J. et al. 2009. Antigen-specific human polyclonal antibodies from hyperimmunized cattle. *Nature Biotechnol.* 27, 173–181.

Kussendrager, K.D. and van Hooijdonk, A.C.M. 2000. Lactoperoxidase: Physicochemical properties, occurrence, mechanism of action and applications. *Brit. J. Nutr.* 84(Suppl. 1): 19–25.

Kweon, C.H., Kwon, B.J., Woo, S.R., Kim, J.M., Woo, G.H., Son, D.H., Hur, W., and Lee. Y.S. 2000. Immunoprophylactic effect of chicken egg yolk immunoglobulin (Ig Y) against porcine epidemic diarrhea virus (PEDV) in piglets. *J. Veterinary Med. Sci.* 62(9): 961–964.

Lacasse, P., Lauzon, K., Diarra, M.S., and Petitclerc, D. 2008. Utilization of lactoferrin to fight antibiotic-resistant mammary gland pathogens. *J. Anim. Sci.* 86(13 Suppl.): 66–71.

Lahov, E. and Regelson, W. 1996. Antibacterial and immunostimulating casein-derived substances from milk: Casecidin, isracidin peptides. *Food Chem. Toxicol.* 34:131–145.

Lee, S. Mine, B.Y., and Stevenson, R.M.W. 2000. Effects of hen egg yolk immunoglobulin in passive protection of rainbow trout against Yersinia ruckeri. *J. Agric. Food Chem.* 48(1): 110–115.

Lee, S.H., Lillehoj, H.S., Park, D.W., Jang, S.I., Morales, A., Garcia, D., Lucio, E. et al. 2009. Induction of passive immunity in broiler chickens against Eimeria acervulina by hyperimmune egg yolk immunoglobulin Y. *Poultry Sci.* 88(3): 562–566.

Lee-Huang, S., Huang, P.L., Sun, Y., Huang, P.L., Kung, H.F., Blithe, D.L., and Chen, H.C. 1999. Lysozyme and RNases as anti-HIV components in beta-core preparations of human chorionic gonadotropin. *Proc. Nat.Acad. Sci. U.S.A.* 96:2678–2681.

Leitch, E.C. and Willcox, M.D. 1998. Synergic antistaphylococcal properties of lactoferrin and lysozyme. *J. Med. Microbiol.* 47(9): 837–842.

Lilius, E.-M. and Marnila, P. 2001. The role of colostral antibodies in prevention of microbial infections. *Curr. Opin. Infect. Diseases* 14: 295–300.

Lizzi, A.R., Carnicelli, V., Clarkson, M.M., Di Giulio, A., and Oratore. A. 2009. Lactoferrin derived peptides: Mechanisms of action and their perspectives as antimicrobial and antitumoral agents. *Mini Rev. Med. Chem.* 9: 687–695.

Loimaranta, V., Carlen, A., Olsson, J., Tenovuo, J., Syväoja, E.L., and Korhonen, H. 1998a. Concentrated bovine colostral whey proteins from Streptococcus mutans/Strep. sobrinus immunized cows inhibit the adherence of Strep. mutans and promote the aggregation of mutans streptococci. *J. Dairy Res.* 65(4): 599–607.

Loimaranta, V., Laine, M., Soderling, E., Vasara, E., Rokka, S., Marnila, P., Korhonen, H. et al. 1999a. Effects of bovine immune and non-immune whey preparations on the composition and pH response of human dental plaque. *Euro. J. Oral Sci.* 107(4): 244–250.

Loimaranta, V., Nuutila, J., Marnila, P., Tenovuo, J., Korhonen, H., and Lilius, E.M. 1999b. Colostral proteins from cows immunised with Streptococcus mutans/S. sobrinus support the phagocytosis and killing of mutans streptococci by human leucocytes. *J. Med. Microbiol.* 48(10): 917–926.

Loimaranta, V., Tenovuo, J., and Korhonen, H. (1998b). Combined inhibitory effect of bovine immune whey and peroxidase-generated hypothiocyanite against glucose uptake by Streptococcus mutans. *Oral Microbiol. Immunol.* 13(6), 378–381.

Loimaranta, V., Tenovuo, J., Virtanen, S., Marnila, P., Syväoja, E.-L., Tupasela, T., and Korhonen, H. 1997. Generation of bovine immune colostrum against Streptococcus mutans and Streptococcus sobrinus and its effect on glucose uptake and extracellular polysaccharide formation by mutans streptococci. *Vaccine* 15(11): 1261–1268.

Longhi, C., Conte, M.P., Ranaldi, S., Penta, M., Valenti, P. Tinari, A. et al. 2005. Apoptotic death of Listeria monocytogenes-infected human macrophages induced by lactoferricin B, a bovine lactoferrin-derived peptide. *Int. J. Immunopath. Pharmacol.* 18(2): 317–325.

Lönnerdal, B. 2003. Lactoferrin. In *Advanced Dairy Chemistry*, Vol. 1, *Proteins*, 3rd edition (P.F. Fox and P.L.H. McSweeney, Eds.), New York: Kluwer Academic/Plenum, pp. 449–466.

Lönnerdal, B. and Bryant, A. 2006. Absorption of iron from recombinant human lacto-ferrin in young US women. *Amer. J. Clin. Nutr.* 83(2): 305–309.

Lönnerdal, B. and Iyer, S. 1995. Lactoferrin: A general review. *Haematologica* 80: 252–267.

López-Expósito, I., Quiros, A., Amigo, L., and Recio, I. 2007. Casein hydrolysates as a source of antimicrobial, antioxydant and antihypertensive peptides. *Lait* 87: 241–249.

López-Expósito. I. and Recio, I. 2006. Antibacterial activity of peptides and folding variants from milk proteins. *Int. Dairy J.*16: 1294–1305.

López-Soto, F., León-Sicairos, N., Nazmi, K., Bolscher, J.G., and de la Garza, M. 2010. Microbicidal effect of the lactoferrin peptides Lactoferricin 17-30, Lactoferrampin 265-284, and Lactoferrin chimera on the parasite Entamoeba histolytica. *Biometals.* 23(3): 563–568.

Losso, J.N., Nakai, S., and Charter, E.A. 2000. Lysozyme. In *Natural Food Antimicrobial Systems* (A.S. Naidu, Ed.), New York: CRC Press, pp. 185–210.

Mahdavi, A.H., Rahmani, H.R., Nili, N., Samie, A.H., Soleimanian-Zad, S., and Jahanian, R. 2010. Effects of dietary egg yolk antibody powder on growth per-formance, intestinal Escherichia coli colonization, and immunocompetence of challenged broiler chicks. *Poultry Sci.* 89(3): 484–494.

Makino, Y. and Nishimura, S. 1992. High-performance liquid chromatographic sep-aration of human apolactoferrin and monoferric and diferric lactoferrins. *J. Chrom.* 579(2): 346–349.

Malkoski, M., Dashper, S.G., O'Brien-Simpson, N.M., Talbo, G.H., Macris, M., Cross, K.J., and Reynolds, E.C. 2001. Kappacin, a novel antibacterial peptide from bovine milk. *Antimicrob. Agents Chemother.* 45(8): 2309–2315.

Manso, M.A. and López-Fandino, R. 2004. K-Casein macropeptides from cheese whey: Physicochemical, biological, nutritional, and technological features for possible uses. *Food Rev. Int.* 20: 329–355.

Markiewski, M.M. and Lambris, J.D. 2007. The role of complement in inflammatory dis-eases from behind the scenes into the spotlight. *Amer. J. Pathol.* 171(3): 715–727.

Marnila, P. and Korhonen, H. 2002. Immunoglobulins. In *Encyclopedia of Dairy Sciences* (H. Roginski, J.W. Fuquay, and P.F. Fox, Eds.), London: Academic Press, pp. 1950–1956.

Marnila, P. and Korhonen, H.J. 2009. Lactoferrin for human health. In *Dairy-Derived Ingredients: Food and Nutraceutical Uses* (M. Corredig, Ed.), Oxford, UK: Woodhead, pp. 290–307.

Marnila, P., Rokka, S., Rehnberg-Laiho, L., Kärkkäinen, P., Kosunen, T.U., Rautelin, H. et al. 2003. Prevention and suppression of helicobacter felis infection in mice using colostral preparation with specific antibodies. *Helicobacter* 8(3): 192–201.

Masci, J.R. 2000. Complete response of severe, refractory oral candidiasis to mouth-wash containing lactoferrin and lysozyme. *AIDS* 14(15): 2403–2404.

Mathew, A.G., Rattanatabtimtong, S., Nyachoti, C.M., and Fang, L. 2009. Effects of in-feed egg yolk antibodies on Salmonella shedding, bacterial antibiotic resistance, and health of pigs. *J. Food Protect.* 72(2): 267–273.

Matin, M.A. and Otani, H. 2002. Cytotoxic and antibacterial activities of chemi-cally synthesized κ-casecidin and its partial peptide fragments. *J. Dairy Res.* 69: 329–334.

Mattila, E., Anttila, V.J., Broas, M., Marttila, H., Poukka, P., Kuusisto, K. et al. 2008. A randomized, double-blind study comparing Clostridium difficile immune whey and metronidazole for recurrent Clostridium difficile-associated diarrhoea: Efficacy and safety data of a prematurely interrupted trial' *Scand. J. Infect. Diseases* 40(9): 702–708.

McCann, K.B., Lee, A., Wan, J., Roginski, H., and Coventry, M.J. 2003. The effect of bovine lactoferrin and lactoferricin B on the ability of feline calicivirus (a norovirus surrogate) and poliovirus to infect cell cultures. *J. Appl. Microbiol.* 95(5): 1026–1033.

McCann, K.B., Shiell, B.J., Michalski, W.P., Lee, A., Wan, J., Roginski, H. et al. 2005. Isolation and characterisation of antibacterial peptides derived from the f(164–207) region of bovine α_{s2}-casein. *Int. Dairy J.* 15: 133–143.

McCann, K.B., Shiell, B.J., Michalski, W.P., Lee, A., Wan, J., Roginski, H. et al. 2006. Isolation and characterisation of a novel antibacterial peptide from bovine α_{s1}-casein. *Int. Dairy J.* 16: 316–323.

Mehra, R., Marnila, P., and Korhonen, H. 2006. Milk immunoglobulins for health promotion. *Int. Dairy J.* 16(11): 1262–1271.

Min, S., Harris, L.J., and Krochta, J.M. 2006. Inhibition of Salmonella enterica and Escherichia coli O157:H7 on roasted turkey by edible whey protein coatings incorporating the lactoperoxidase system. *J. Food Protect.* 69: 784–793.

Mine, Y. and Kovacs-Nolan, J. 2002. Chicken egg yolk antibodies as therapeutics in enteric infectious disease: A review. *J. Med. Food* 5(3): 159–169.

Mine, Y. and Kovacs-Nolan, J. 2004. Biologically active hen egg components in human health and disease. *J. Poult. Sci.* 41: 1–29.

Mine, Y., Ma, F., and Lauriau, S. 2004. Antimicrobial peptides released by enzymatic hydrolysis of hen egg white lysozyme. *J. Agric. Food Chem.* 52(5): 1088–1094.

Mitoma, M., Oho, T., Michibata, N., Okano, K., Nakano, Y., Fukuyama, M., and Koga, T. 2002. Passive immunization with bovine milk containing antibodies to a cell surface protein antigen-glucosyltransferase fusion protein protects rats against dental caries. *Infect. Immun.* 70(5): 2721–2724.

Mor, A. 2003. Introduction. *Peptides* 24: 1645.

Näse, L., Hatakka, K., Savilahti, E., Saxelin, M., Ponka, A., Poussa, T., Korpela, R., and Meurman, J.H. 2001. Effect of long-term consumption of a probiotic bacterium, lactobacillus rhamnosus GG, in milk on dental caries and caries risk in children. *Caries Res.* 35(6): 412–420.

Nibbering, P.H., Ravensbergen, E., Welling, M.M., van Berkel, L.A., van Berkel, P.H., Pauwels, E.K., and Nuijens, J.H. 2001. Human lactoferrin and peptides derived from its N terminus are highly effective against infections with antibiotic-resistant bacteria. *Infecti. Immun.* 69(3): 1469–1476.

Nilsson, E., Kollberg, H., Johannesson, M., Wejaker, P.E., Carlander, D., and Larsson, A. 2007. More than 10 years' continuous oral treatment with specific immunoglobulin Y for the prevention of Pseudomonas aeruginosa infections: A case report. *J. Med. Food* 10(2): 375–378.

Nilsson, E., Larsson, A., Olesen, H.V., Wejaker, P.E., and Kollberg, H. 2008. Good effect of IgY against Pseudomonas aeruginosa infections in cystic fibrosis patients. *Pediatr. Pulmonol.* 43(9): 892–899.

Nomura, S., Suzuki, H., Masaoka, T., Kurabayashi, K., Ishii, H., Kitajima, M., Nomoto, K., and Hibi, T. 2005. Effect of dietary anti-urease immunoglobulin Y on Helicobacter pylori infection in mongolian gerbils. *Helicobacter* 10(1): 43–52.

Numan, S.C., Veldkamp, P., Kuijper, E.J., van den Berg, R.J., and van Dissel, J.T., 2007. Clostridium difficile-associated diarrhoea: Bovine anti-Clostridium difficile whey protein to help aid the prevention of relapses. *Gut* 56(6): 888–889.

Ochoa, T.J., Chea-Woo, E., Campos, M., Pecho, I., Prada, A., McMahon, R.J. et al. 2008. Impact of lactoferrin supplementation on growth and prevalence of Giardia colonization in children. *Clini. Infect. Diseases* 46(12): 1881–1883.

Ochoa, T.J. and Cleary, T.G. 2004. Lactoferrin disruption of bacterial type III secretion systems. *Biometals* 17(3): 257–260.

Ochoa, T.J. and Cleary T.G. 2009. Effect of lactoferrin on enteric pathogens. *Biochimie* 91: 30–34.

Ochoa, T.J., Noguera-Obenza, M., Ebel, F., Guzman, C.A., Gomez, H.F., and Cleary, T.G. 2003. Lactoferrin impairs type III secretory system function in enteropathogenic Escherichia coli. *Infect. Immun.* 71(9): 5149–5155.

Oevermann, A., Engels, M., Thomas, U., and Pellegrini, A. 2003. The antiviral activity of naturally occurring proteins and their peptide fragments after chemical modification. *Antiviral Res.* 59: 23–33.

Oona, M., Rägo, T., Maaroos, H.I., Mikelsaar, M., Lõivukene, K., Salminen, S., and Korhonen, H. 1997. Helicobacter pylori in children with abdominal complaints: Has immune bovine colostrum some influence on gastritis? *Alpe Adria Microbiol. J.* 6: 49–57.

Pacyna, J., Siwek, K., Terry, S.J., Roberton, E.S., Johnson, R.B., and Davidson, G.P. 2001. Survival of rotavirus antibody activity derived from bovine colostrum after passage through the human gastrointestinal tract. *J. Pediatr. Gastroenter. Nutr.* 32: 162–167.

Pan, Y., Rowney, M., Guo, P., and Hobman, P. 2007a. Biological properties of lactoferrin: An overview. *Austral. J. Dairy Technol.* 62: 31–42.

Pan, Y., Wan, J., Roginski, H., Lee, A., Shiell, B., Michalski, W.P. et al. 2007b. Comparison of the effects of acylation and amidation on the antimicrobial and antiviral properties of lactoferrin. *Lett. Applied Microbiol.* 44(3): 229–234.

Pant, N., Marcotte, H., Brussow, H., Svensson, L., and Hammarström, L. 2007. Effective prophylaxis against rotavirus diarrhea using a combination of Lactobacillus rhamnosus GG and antibodies. *BMC Microbiol.* 7(1): 86.

Pellegrini, A. 2003. Antimicrobial peptides from food proteins. *Curr. Pharmaceut. Des.* 9: 1225–1238.

Pellegrini, A., Dettling, C., Thomas, U., and Hunziker, P. 2001. Isolation and characterization of four bactericidal domains in the bovine β-lactoglobulin. *Biochimica et Biophysica Acta* 1526: 131–140.

Pellegrini, A., Hülsmeier, A.J., Hunziker, P., and Thomas, U. 2004. Proteolytic fragments of ovalbumin display antimicrobial activity. *Biochimica et Biophysica Acta* 1672(2): 76–85.

Pellegrini, A., Thomas, U. Bramaz, N. Hunziker, P., and von Fellenberg, R. 1999. Isolation and identification of three bactericidal domains in the bovine alpha-lactalbumin molecule. *Biochimica et Biophysica Acta* 1426: 439–448.

Pellegrini, A., Thomas, U., Wild, P., Schraner, E., and von Fellenberg, R. 2000. Effect of lysozyme or modified lysozyme fragments on DNA and RNA synthesis and membrane permeability of Escherichia coli. *Microbiol. Res.* 155(2): 69–77.

Petitclerc, D., Lauzon, K., Cochu, A., Ster, C., Diarra, M.S., and Lacasse, P. 2007. Efficacy of a lactoferrin-penicillin combination to treat {beta}-lactam-resistant Staphylococcus aureus mastitis. *J. Dairy Sci.* 90(6): 2778–2787.

Petschow, B.W. and Talbott, R.D. 1994. Reduction in virus-neutralizing activity of a bovine colostrum immunoglobulin concentrate by gastric acid and digestive enzymes. *J. Pediatr. Gastroent. Nutr.* 19(2): 228–235.

Phelan, M., Aherne, A., Fitzgerald, R.J., and O'Brien, N.M. 2009. Casein-derived bioactive peptides: Biological effects, industrial uses, safety aspects and regulatory status. *Int. Dairy J.* 19(11): 643–654.

Pihlanto, A. and Korhonen, H. 2003. Bioactive peptides and proteins. In *Advances in Food and Nutrition Research* (S.L. Taylor, Ed.), San Diego: Elsevier, pp. 175–276.

Pihlanto-Leppälä, A., Marnila, P., Hubert, L., Rokka, T., Korhonen, H.J.T., and Karp, M. 1999. The effect of α-lactalbumin and β-lactoglobulin hydrolysates on the metabolic activity of *Escherichia coli* JM103. *J. Appl. Microbiol.* 87: 540–545.

Plaut, A.G., Qiu, J., and St. Geme, J.W. 2000. Human lactoferrin proteolytic activity: Analysis of the cleaved region in the IgA protease of Haemophilus influenzae. *Vaccine* 19(Suppl. 1): S148–S152.

Pouliot, Y., Gauthier, S.F., and Groleau, P.E. 2006. Membrane-based fractionation and purification strategies for bioactive peptides. In *Nutraceutical Proteins and Peptides in Health and Disease* (Y. Mine and F. Shahidi, Eds.), Boca Raton, FL: Taylor & Francis, pp. 639–658.

Powers, J.-P.S. and Hancock, R.E.W. 2003. The relationship between peptide structure and antibacterial activity. *Peptides* 24: 1681–1691.

Rautemaa, R. and Meri, S. 1999. Complement-resistance mechanisms of bacteria. *Microbes Infect* 1(10): 785–794.

Recio, I. and Visser, S. 1999. Identification of two distinct antibacterial domains within the sequence of bovine α_{s2}-casein. *Biochimica et Biophysica Acta* 1428:314–326.

Reilly, R.M., Domingo, R., and Sandhu, J. 1997. Oral delivery of antibodies. Future pharmacokinetic trends. *Clin. Pharmacokinet.* 32(4): 313–323.

Reiter, B. and Brock, J.H. 1975. Inhibition of Escherichia coli by bovine colostrum and post-colostral milk. I. complement-mediated bactericidal activity of antibodies to a serum susceptible strain of E. coli of the serotype O 111. *Immunology* 28(1): 71–82.

Reiter, B. and Oram J.D. 1967. Bacterial inhibitors in milk and other biological fluids. *Nature* 216: 28–30.

Reiter, B. and Perraudin, J-P. 1991. Lactoperoxidase: Biological functions. In *Peroxidases in Chemistry and Biology* (J. Everse, K.E. Everse, and M.B. Grisham, Eds.), Boca Raton, FL: CRC Press, pp. 143–180.

Rhoades, J.R., Gibson, G.R., Formentin, K., Beer, M., Greenberg, N., and Rastall, R.A. 2005. Caseinoglycomacropeptide inhibits adhesion of pathogenic Escherichia coli strains to human cells in culture. *J. Dairy Sc.* 88(10): 3455–3459.

Richardson, A., de Antueno, R., Duncan, R., and Hoskin, D.W. 2009. Intracellular delivery of bovine lactoferricin's antimicrobial core (RRWQWR) kills T-leukemia cells. *Biochem. Biophys. Res. Commun.* 388: 736–741.

Rizzello, C.G., Losito, I., Gobbetti, M., Carbonara, T., De Bari, M.D., and Zambonin, P.G. 2005. Antibacterial activities of peptides from the water-soluble extracts of Italian cheese varieties. *J. Dairy Sci.* 88: 2348–2360.

Roberts, A.K., Chierici, R., Sawatzki, G., Hill, M.J., Volpato, S., and Vigi, V. 1992. Supplementation of an adapted formula with bovine lactoferrin: 1. Effect on the infant faecal flora. *Acta Paediatrica* 81(2): 119–124.

Roos, N., Mahe, S., Benamouzig, R., Sick, H., Rautureau, J., and Tome, D. 1995. 15N-labeled immunoglobulins from bovine colostrum are partially resistant to digestion in human intestine. *J. Nutr.* 125(5): 1238–1244.

Rydlo, T., Miltz, J., and Mor, A. 2006. Eukaryotic antimicrobial peptides: Promises and premises in food safety. *J. Food Sci.* 71: R125–R135.

Samuelsen, Ø., Haukland, H.H., Jenssen, H., Krämer, M., Sandvik, K., Ulvatne, H. et al. 2005. Induced resistance to the antimicrobial peptide lactoferricin B in Staphylococcus aureus. *FEBS Lett.* 579(16): 3421–3426.

Sarker, S.A., Casswall, T.H., Juneja, L.R., Hoq, E., Hossain, I., Fuchs, G.J., and Hammarström, L. 2001. Randomized, placebo-controlled, clinical trial of hyper-immunized chicken egg yolk immunoglobulin in children with rotavirus diar-rhea. *J. Pediatr. Gastroent. Nutr.* 32(1): 19–25.

Sarker, S.A., Casswall, T.H., Mahalanabis, D., Alam, N.H., Albert, M.J., Brüssow, H., Fuchs, G.J., and Hammerström, L. 1998. Successful treatment of rotavirus diar-rhea in children with immunoglobulin from immunized bovine colostrum. *Pediatr. Infect. Disease J.* 17(12): 1149–1154.

Sava, G. 1996. Pharmacological aspects and therapeutic applications of lysozymes. *EXS* 75: 433–449.

Sava, G., Pacor, S., Dasig, G., and Bergamo, A. 1995. Lysozyme stimulates lymphocyte response to ConA and IL-2 and potentiates 5-fluorouracil action on advanced carcinomas. *Anticancer Res.* 15: 1883–1888.

Scammell, A.W. 2001. Production and uses of colostrum. *Austral. J. Dairy Technol.* 56(2): 74–82.

Schade, R., Calzado, E.G., Sarmiento, R., Chacana, P.A., Porankiewicz-Asplund, J., and Terzolo, H.R. 2005. Chicken egg yolk antibodies (IgY-technology): A review of progress in production and use in research and human and veterinary medi-cine. *Alternatives to Laboratory Animals: ATLA* 33(2): 129–154.

Seifu, E., Buys, E.M., and Donkin, E.F. 2005. Significance of the lactoperoxidase sys-tem in the dairy industry and its potential applications: A review. *Trends Food Sci. Technol.* 16: 137–154.

Shimazaki, Y., Mitoma, M., Oho, T., Nakano, Y., Yamashita, Y., Okano, K., Nakano, Y. et al. 2001. Passive immunization with milk produced from an immunized cow prevents oral recolonization by streptococcus mutans. *Clin. Vaccine Immunol.* 8(6): 1136–1139.

Shin, J.H., Yang, M., Nam, S.W., Kim, J.T., Myung, N.H., Bang, W.G., and Roe, I.H. 2002. Use of egg yolk-derived immunoglobulin as an alternative to antibiotic treatment for control of Helicobacter pylori infection. *Clin. Diagnostic Lab. Immunol.* 9(5; Sep): 1061–1066.

Shin, K., Wakabayashi, H., Yamauchi, K., Teraguchi, S., Tamura, Y., Kurokawa, M., and Shiraki, K. 2005. Effects of orally administered bovine lactoferrin and lac-toperoxidase on influenza virus infection in mice. *J. Med. Microbiol.* 54: 717–723.

Singh, K., Parsek, M.R., Greenberg, E.P., and Welsh M.J. 2002. A component of innate immunity prevents bacterial biofilm development. *Nature* 417(6888): 552–555.

Smith, D.J., King, W.F., and Godiska, R. 2001. Passive transfer of immunoglobulin Y antibody to Streptococcus mutans glucan binding protein B can confer protec-tion against experimental dental caries. *Infect. Immun.* 69(5): 3135–3142.

Suzuki, H., Nomura, S., Masaoka, T., Goshima, H., Kamata, N., Kodama, Y., Ishii, H., Kitajima, M., Nomoto, K., and Hibi, T. 2004. Effect of dietary anti-helicobacter pylori-urease immunoglobulin Y on Helicobacter pylori infection. *Aliment. Pharmacol. Therapeut.* 20 (Suppl. 1): 185–192.

Tacket, C.O., Losonsky, G., Link, H., Hoang, Y., Guesry, P., Hilpert, H., and Levine, M.M. 1988. Protection by milk immunoglobulin concentrate against oral challenge with enterotoxigenic escherichia coli. *New Engl. J. Med.* 318(19): 1240–1243.

Tamano, S., Sekine, K., Takase, M., Yamauchi, K., Iigo, M., and Tsuda, H. 2008. Lack of chronic oral toxicity of chemopreventive bovine lactoferrin in F344/DuCrj rats. *Asian Pacific J.Cancer Prevent.* 9(2): 313–316.

Tawfeek, H.I., Najim, N.H., and Al-Mashikhi, S. 2003. Efficacy of an infant formula containing anti-escherichia coli colostral antibodies from hyperimmunized cows in preventing diarrhea in infants and children: A field trial. *Int. J. Infect. Diseases* 7(2): 120–128.

Taylor, S., Brock, J., Kruger, C., Berner, T., and Murphy, M. 2004. Safety determination for the use of bovine milk-derived lactoferrin as a component of an antimicrobial beef carcass spray. *Reg. Toxicol. Pharmacol.* 39(1): 12–24.

Tenovuo, J. 2002. Clinical applications of antimicrobial host proteins lactoperoxidase, lysozyme and lactoferrin in xerostomia: Efficacy and safety. *Oral Diseases* 8: 23–29.

Tenovuo J., Lumikari, M., and Soukka, T. 1991. Salivary lysozyme, lactoferrin and peroxidases: antibacterial effects on cariogenic bacteria and clinical applications in preventive denstistry. *Proc. Finnish Dentist. Soc.* 87: 197–208.

Thammasirirak, S., Pukcothanung, Y., Preecharram, S., Daduang, S., Patramanon, R., Fukamizo, T. et al. 2010. Antimicrobial peptides derived from goose egg white lysozyme. *Compar. Biochem. Physiol. Part C: Toxicol. Pharmacol.* 151(1): 84–91.

Thomä-Worringer, C., Sörensen, J., and López-Fandino, R. 2006. Health effects and technological features of caseinomacropeptide. *Int. Dairy J.* 16: 1324–1333.

Thurman, J.M. and Holers, V.M. 2006. The central role of the alternative complement pathway in human disease. *J. Immunol.* 176(3): 1305–1310.

Tomita, M., Wakabayashi, H., Shin, K., Yamauchi, K., Yaeshima, T., and Iwatsuki, K. 2009. Twenty-five years of research on bovine lactoferrin appications. *Biochimie* 91: 52–57.

Tomita, M., Wakabayashi, H., Yamauchi, K., Teraguchi, S., and Hayasawa, H. 2002. Bovine lactoferrin and lactoferricin derived from milk: Production and applications. *Biochem. Cell Biol.* 80(1): 109–112.

Tripathi, V. and Vashishtha, B. 2006. Bioactive compounds of colostrum and its application. *Food Rev. Int.* 22: 225–244.

Tsuge, Y., Shimoyamada, M., and Watanabe, K. 1996. Differences in hemagglutination inhibition activity against bovine rotavirus and hen newcastle disease virus based on the subunits in hen egg white ovomucin. *Biosci. Biotechnol. Biochem.* 60(9): 1505–1506.

Tsuge, Y., Shimoyamada, M., and Watanabe, K. 1997. Structural features of Newcastle Disease virus- and anti-ovomucin antibody-binding glycopeptides from pronase-treated ovomucin. *J. Agric. Food Chem.* 45(7): 2393–2398.

Tsuji, S., Hirata, Y., and Matsuoka, K. 1989. Two apparent molecular forms of bovine lactoferrin. *J. Dairy Sci.* 72(5): 1130–1136.

Ueta, E., Tanida, T., and Osaki, T.2001. A novel bovine lactoferrin peptide, FKCRRWQWRM, suppresses Candida cell growth and activates neutrophils. *J. Peptide Res.* 57(3): 240–249.

Ulvatne, H., Haukland, H.H., Olsvik, O., and Vorland, L.H. 2001. Lactoferricin B causes depolarization of the cytoplasmic membrane of Escherichia coli ATCC 25922 and fusion of negatively charged liposomes. *FEBS Lett.* 492(1–2): 62–65.

Ulvatne, H., Haukland, H.H., Samuelsen, Ø., Krämer, M., and Vorland, L.H. 2002. Proteases in Escherichia coli and Staphylococcus aureus confer reduced susceptibility to lactoferricin B. *J. Antimicrob. Chemother.* 50(4): 461–467.

Ulvatne, H., Samuelsen, Ø., Haukland, H.H., Krämer, M., and Vorland, L.H. 2004. Lactoferricin B inhibits bacterial macromolecular synthesis in Escherichia coli and Bacillus subtilis. *FEMS Microbiol. Lett.* 237(2): 377–384.

Valenti, P. and Antonini, G. 2005. Lactoferrin: An important host defence against microbial and viral attack. *Cell. Molec. Life Sci.* 62(22): 2576–2587.

Valenti, P., Greco, R., Pitari, G., Rossi, P., Ajello, M., Melino G. et al. 1999. Apoptosis of Caco-2 intestinal cells invaded by Listeria monocytogenes: protective effect of lactoferrin. *Exper. Cell Res.* 250(1): 197–202.

van der Kraan, M.I.A., Groenink, J., Nazmi, K., Veerman, E.C.I., Bolscher, J.G.M., and Nieuw Amerongen, A.V. 2004. Lactoferrampin: A novel antimicrobial peptide in the N1-domain of bovine lactoferrin. *Peptides* 25: 177–183.

van Dissel, J.T., de Groot, N., Hensgens, C.M.H., Numan, S., Kuijper, E.J., Veldkamp, P. et al. 2005. Bovine antibody-enriched whey to aid in the prevention of a relapse of Clostridium difficile-associated diarrhoea: Preclinical and preliminary clinical data. *J. Med. Microbiol.* 54(2): 197–205.

van Hooijdonk, A.C.M., Kussendrager, K.D., and Steijns, J.M. 2000. *In vivo* antimicrobial and antiviral activity of components in bovine milk and colostrum involved in non-specific defence. *Brit. J. Nutr.* 84 (Suppl. 1): S127–S134.

Venkatesh, M.P. and Rong, L. 2008. Human recombinant lactoferrin acts synergistically with antimicrobials commonly used in neonatal practice against coagulase-negative staphylococci and Candida albicans causing neonatal sepsis. *J. Med. Microbiol.* 57: 1113–1121.

Vercruysse, L., Van Camp, J., Dewettinck, K., and Smagghe, G. 2009. Production and enrichment of bioactive peptides derived from milk proteins. In *Dairy-Derived Ingredients: Food and Nutraceutical Uses* (M. Corredig, Ed.), Oxford, UK: Woodhead, pp. 51–67.

Viejo-Díaz, M., Andrés, M.T., and Fierro, J.F. 2004. Modulation of *in vitro* fungicidal activity of human lactoferrin against Candida albicans by extracellular cation concentration and target cell metabolic activity. *Antimicrob. Agents Chemother.* 48(4): 1242–1248.

Vorland, L.H. 1999. Lactoferrin: A multifunctional glycoprotein. Review article. *APMIS* 107: 971–981.

Wakabayashi, H., Kondo, I., Kobayashi, T., Yamauchi, K., Toida, T., Iwatsuki, K., and Yoshie, H. 2010. Periodontitis, periodontopathic bacteria and lactoferrin. *Biometals* 23: 419–424.

Wakabayashi, H., Takase, M., and Tomita. M. 2003. Lactoferricin derived from milk protein lactoferrin. *Curr. Pharmaceut. Des.* 9(16): 1277–1287.

Wakabayashi, H., Yamauchi, K., and Takase, M. 2006. Lactoferrin research, technology and applications. *Int. Dairy J.* 16: 1241–1251.

Warny, M., Fatimi, A., Bostwick, E.F., Laine, D.C., Lebel, F., LaMont, J.T., Pothoulakis, C., and Kelly C.P. 1999. Bovine immunoglobulin concentrate-clostridium difficile retains C. difficile toxin neutralising activity after passage through the human stomach and small intestine. *Gut* 44(2): 212–217.

Wei, H., Loimaranta, V., Tenovuo, J., Rokka, S., Syväoja, E.L., Korhonen, H. et al. 2002. Stability and activity of specific antibodies against Streptococcus mutans and Streptococcus sobrinus in bovine milk fermented with Lactobacillus rhamnosus strain GG or treated at ultra-high temperature. *Oral Microbiol. Immunol.* 17(1): 9–15.

Weinberg, E.D. 2003. The therapeutic potential of lactoferrin. *Expert Opin. Investig. Drugs* 12(5): 841–851.

Weinberg, E.D. 2007. Antibiotic properties and applications of lactoferrin. *Curr. Pharmaceut. Des.* 13(8): 801–811.

Weiner, C., Pan, Q., Hurtig, M., Boren, T., Bostwick, E., and Hammarström, L. 1999. Passive immunity against human pathogens using bovine antibodies. *Clin. Exper. Immunol.* 116: 193–205.

Worledge, K.L., Godiska, R., Barrett, T.A., and Kink, J.A., 2000. Oral administration of avian tumor necrosis factor antibodies effectively treats experimental colitis in rats. *Digest. Diseases Sci.* 45(12): 2298–2305.

Xu, L.B., Chen, L., Gao, W., and Du, K.H. 2006. Bovine immune colostrum against 17 strains of diarrhea bacteria and *in vitro* and *in vivo* effects of its specific IgG. *Vaccine* 24(12): 2131–2140.

Yamauchi, K., Hiruma, M., Yamazaki, N., Wakabayashi, H., Kuwata, H., Teraguchi, S. et al. 2000a. Oral administration of bovine lactoferrin for treatment of tinea pedis. A placebo-controlled, double-blind study. *Mycoses* 43(5): 197–202.

Yamauchi, K., Toida, T., Nishimura, S., Nagano, E., Kusuoka, O., Teraguchi S. et al. 2000b. 13-Week oral repeated administration toxicity study of bovine lactoferrin in rats. *Food Chem. Toxicol.* 38(6): 503–512.

Yolken, R.H., Leister, F., Wee, S.B., Miskuff, R., and Vonderfecht, S. 1988. Antibodies to rotaviruses in chickens' eggs: A potential source of antiviral immunoglobulins suitable for human consumption. *Pediatrics* 81(2): 291–295.

Young, K.W.H., Munro, I.C., Taylor, S.L., Veldkamp, P., and van Dissel, J.T. 2007. The safety of whey protein concentrate derived from the milk of cows immunized against Clostridium difficile. *Regulatory Toxicol. Pharmacol.* 47(3): 317–326.

Zucht, H.-D., Raida, M., Adermann, K., Magert, H.J., and Forssmann, W.G. 1995. Casocidin-I: A casein-α_{s2} derived peptide exhibits antibacterial activity. *FEBS Lett.* 372: 185–188.

4

Proteins and Peptides as Antioxidants

Se-Kwon Kim, Isuru Wijesekara, Eun Young Park,
Yasuki Matsumura, Yasushi Nakamura, and Kenji Sato

CONTENTS

4.1 Introduction

In recent years, the role of protein in the diet has been acknowledged worldwide. Dietary proteins become a source of physiologically active components, which have a positive impact on the body's function after gastrointestinal digestion. Bioactive peptides are inactive in the sequences of their parent protein, but can be produced by one of three methods such as chemical hydrolysis, enzymatic hydrolysis, and microbial fermentation of food proteins (Lahl and Braun, 1994). However, especially in the food and pharmaceutical industries the enzymatic hydrolysis method is preferred in order to avoid the residual toxic chemicals in the products. The physicochemical conditions of the reaction media such as temperature and pH of the protein solution must then be adjusted in order to optimize the activity of the enzyme used. Proteolytic enzymes from microbes, plants, and animals can be used for the hydrolysis process of proteins to develop bioactive peptides (Simpson, 1998). Bioactive peptides usually contain 3–20 amino acid residues, and their activities are based on their amino acid composition and sequence

(Pihlanto-Leppala, 2001). Furthermore, some of these bioactive peptides may have greater potential for human health promotion and disease risk reduction (Shahidi and Zhong, 2008). Thus, the possible role of food-derived bioactive peptides in reducing the risk of cardiovascular disease has been well demonstrated (Erdmann, Cheung, and Schroder, 2008).

Lipid oxidation is a very important subject in food storage and in human health. Lipid oxidation is the main factor causing the deterioration of food during storage and processing, as it can induce undesirable changes in color, flavor, texture, and nutritional profile as well as produce potentially toxic reaction products. Some proteins and peptides, such as casein (Kanazawa, Ashida, and Natake, 1987), maize (Wang et al., 1991), enzymatic hydrolysates of soy (Chen, Muramoto, and Yamauchi, 1995; Park et al., 2005), gelatin (Kim et al., 2001), elastin (Hattori et al., 1998), and egg white proteins (Tsuge et al., 1991), have been shown to have potential antioxidant activities.

Reactive oxygen species are also generated in living cells during respiration, such as superoxide anion, hydroxyl radical, and hydrogen peroxide, and are highly reactive and damaging chemical species. Uncontrolled production of free radicals that attack macromolecules such as membrane lipids, proteins, and DNA may lead to many health disorders such as cancer, diabetes mellitus, and neurodegenerative and inflammatory diseases with severe tissue injuries (Butterfield et al., 2002; Frlich and Riederer, 1995; Yang et al., 2001).

To retard the peroxidation process in food systems, many synthetic antioxidants such as butylated hydroxytoluene (BHT), butylated hydroxyanisole (BHA), tert-butylhydroquinone (TBHQ), and propyl gallate (PG) have been used. However, the use of these synthetic antioxidants must be strictly controlled due to potential health issues (Hettiarachchy et al., 1996). Hence, the search for natural antioxidants as safe alternatives to synthetic products is important in the food industry (Pena-Ramos and Xiong, 2001). Recently, the use of natural antioxidants available in food and other biological substances has attracted significant interest due to their presumed safety, nutritional, and therapeutic values (Ajila et al., 2007). Therefore, bioactive protein and peptides with antioxidative properties can be introduced for the food and pharmaceutical industries as potential antioxidants with health benefits.

This chapter focuses on antioxidant proteins and peptides, and presents an overview of their value as future antioxidants in food and body systems.

4.2 Antioxidant Food Proteins and Peptides

According to recent knowledge, dairy products and marine-derived foods seem to be by far the greatest sources of food-derived bioactive proteins and peptides. However, bioactive proteins and peptides can be gained from other animal and plant sources such as eggs, gelatin, and meat products as well as wheat, rice, maize, and soy products. Bioactive protein and peptides with antioxidative properties have been detected directly or after release by hydrolyzation or fermentation, which may have other biological activities such as antihypertensive (Jung et al., 2006), anticoagulant (Jung and Kim, 2009), opioid (Haque, Chand, and Kapila, 2009) and antimicrobial (Liu et al., 2008) activities.

4.2.1 Animal-Derived Antioxidative Proteins and Peptides

Caseins and whey proteins are two major types of proteins found in milk with different biological activities such as antiviral, antimicrobial, antioxidant, and anticarcinogenic activities. Milk proteins are the most important source of bioactive peptides (Korhonen, 2009). Many milk proteins possess specific biological properties that make these components potential ingredients of health-promoting foods. Numerous products are already on the market or currently under development by food companies, exploiting the potential of milk-derived bioactive peptides. At present, casein-derived peptides have interesting applications as supplements in food and pharmaceutical preparations (Meisel and FitzGerald, 2003; Phelan et al., 2009). In addition, milk protein-derived bioactive peptides have been shown *in vivo* to exert various activities affecting body systems and are naturally found in fermented dairy products, such as yogurt, sour milk, and cheese products (Korhonen and Pihlanto, 2006).

Caseins and casein-derived peptides have been demonstrated to inhibit lipid peroxidation induced by lipoxygenase, AAPH (2,2'-azo-bis(2-amidino-propane) dihydrochloride), hemoglobin, and so on (Rival et al., 2000; Rival, Boeriu, and Wichers, 2001). Chelation of iron (lactoferrin) and copper (BSA) and inactivation of free radicals (BSA and beta-lactoglobulin) through interactions with tyrosine, cysteine, hemoglobin, and so on (Rival et al., 2000; Rival, Boeriu, and Wichers, 2001) have been suggested as underlying mechanisms for the antioxidative activity (Taylor and Richardson, 1980; Wayner et al., 1987; Meucci Mordente, and Martorana, 1991; Ostdal, Daneshvar, and Skibsted, 1996). In a Tween 20-Menhaden oil emulsion system, whey proteins isolate are thought to inhibit lipid oxidation by chelating pro-oxidant metals, inactivating free radicals or by forming a physical barrier at the lipid interface (Donnelly, Decker, and McClements, 1998).

Egg yolk hydrolysates have been shown to inhibit lipid oxidation in various muscle foods containing dispersed lipids, such as beef and tuna (Sakanaka et al., 2005; Penta-Ramos and Xiong, 2003). Sakanaka et al. demonstrated in a variety of *in vitro* assays that egg yolk protein hydrolysates are versatile antioxidants capable of scavenging superoxide, hydroxyl, and peroxyl radicals, and inhibiting lipid oxidation-induced beta-carotene bleaching.

The same hydrolysates have been demonstrated to inhibit the formation of thiobarbituric acid reactive substances (TBARS) in model food systems.

4.2.2 Plant-Derived Antioxidative Proteins and Peptides

Pulse crops and soy beans are excellent sources of bioactive proteins and peptides (Gibbs et al., 2004; Roy, Boye, and Simpson, 2010). The antioxidative property of these plant-derived proteins and peptides has been associated with many beneficial health-promoting properties such as reducing cholesterol levels, managing type-2 diabetes, and preventing various forms of cancer. However, pulse crops, like other leguminous crops, also contain many antinutritional proteins that cause vomiting, bloating, and hemaglutination after ingestion in raw form.

The antioxidative properties of soy protein hydrolysates have been ascribed the co-operative effect of a number of properties, including their ability to scavenge free radicals, to act as a metal-ion chelater, oxygen quencher, or hydrogen donor, and possibly preventing lipid oxidation initiators by forming a membrane around oil droplets (Moure, Dominguez, and Parajo, 2006; Pena-Ramos and Xiong, 2002). Soy protein isolate inhibits oxidation of ethyl esters of eicosapentaenoic acid in a maltodextrin-stabilized emulsion powder system (Park et al., 2005). Soy protein hydrolysates and fermented soy products are rich in biologically active peptides with antioxidative properties (Gibbs et al., 2004). Chen et al. (1996) reported that soy peptides have antioxidant activity against the peroxidation of linoleic acid in aqueous systems. They found that peptides with strong radical-scavenging activity have Leu-Leu-Pro-His-His motifs in sequence. They also revealed that His and Pro residues are essential for the antioxidant effect by using synthetic peptides with modified sequences. Such results suggested that some specific amino acid residues in peptides play a significant role in antioxidant activity (Chen et al., 1996).

The main component of potato protein (i.e., patatin) and a potato protein hydrolysate prepared by Amano P and pancreatic digestion were reported to exhibit antioxidant activity in an *in vitro* system (Kudoh, Mausrmoto, and Onodera, 2003; Liu et al., 2003) and in cooked beef patties (Wang and Xiong, 2005).

Sunflower protein hydrolysates have copper-chelating capacity (Megias et al., 2007), in addition to angiotensin-converting enzyme (ACE) inhibitory activity (Megias et al., 2004).

4.2.3 Marine-Derived Antioxidative Peptides

Bioactive peptides derived from marine organisms as well as marine fish processing by-products have greater potential for the development of functional foods (Ariyoshi, 1993; Yamamoto, 1997; Kim et al., 2001; Ravallec et al., 2001; Jun et al., 2004; Jung et al., 2006), which can act as potential physiological modulators of metabolism after absorption. Marine bioactive peptides have

TABLE 4.1

Antioxidant Peptides Derived from Marine Organisms[a]

Source	Amino Acid Sequence	Ref.
Jumbo squid	Phe-Asp-Ser-Gly-Pro-Ala-Gly-Val-Leu	Mendis et al. 2005
	Asn-Gly-Pro-Leu-Gln-Ala-Gly-Gln-Pro-Gly-Glu-Arg	"
	Asn-Ala-Asp-Phe-Gly-Leu-Asn-Gly-Leu-Glu-Gly-Leu-Ala	Rajapakse et al. 2005a
	Asn-Gly-Leu-Glu-Gly-Leu-Lys	"
Oyster	Leu-Lys-Glu-Glu-Leu-Glu-asp-Leu-Leu-Glu-Lys-Glu-Glu	Qian et al. 2008
Blue mussel	Phe-Gly-His-Pro-Tyr	Jung et al. 2005
	His-Phe-Gly-Asp-Pro-Phe-His	Rajpakse et al. 2005b
Hoki	Glu-Ser-Thr-Val-Pro-Glu-Arg-Thr-His-Pro-Ala-Cys-Pro-Asp-Phe-Asn	Kim et al. 2007
Tuna	Val-Lys-Ala-Gly-Phe-Ala-Trp-Thr-Ala-Asn-Glu-Glu-Leu-Ser	Je et al. 2007
Rotifer	Leu-Leu-Gly-Pro-Gly-Leu-Thr-Asn-His-Ala	Byun et al. 2009
	Asp-Leu-Gly-Leu-Gly-Leu-Pro-Gly-Ala-His	"
Microalga	Val-Glu-Cys-Tyr-Gly-Pro-Asn-Arg-Pro-Glu-Phe	Sheih et al. 2009

[a] Source and amino acid sequence.

been isolated widely by enzymatic hydrolysis of marine organisms (Je et al. 2008; Sheih, Wu, and Fang, 2009; Slizyte et al., 2009; Suetsuna and Nakano, 2000; Zhao et al., 2007). However, in fermented marine food sauces, enzymatic hydrolysis has been accomplished by micro-organisms and bioactive peptides can be induced without further hydrolysis (Je et al., 2005).

The antioxidant activity of marine-derived bioactive peptides has been reviewed previously (Kim and Wijesekara, 2010). Table 4.1 presents a summary of marine-derived antioxidant peptides. The beneficial effects of antioxidant marine bioactive peptides are well known in scavenging free radicals and reactive oxygen species or in preventing oxidative damage by interrupting the radical chain reaction of lipid peroxidation (Rajapakse et al., 2005a; Mendis et al., 2005; Qian et al., 2008; Ranathunga, Rajapakse, and Kim, 2006). The inhibition of lipid peroxidation by marine bioactive peptide, isolated from jumbo squid has been determined by a linoleic acid model system and their activity was much higher than natural antioxidant, α-tocopherol, and was close to highly active synthetic antioxidant, BHT (Mendis et al., 2005). It has been shown that this antioxidant is potent mostly due to the presence of hydrophobic amino acids in the peptide. In addition, the bioactive antioxidant peptide isolated from oyster (*Crassostrea gigas*) has exhibited a higher activity against polyunsaturated fatty acid peroxidation than antioxidant vitamin α-tocopherol (Qian et al., 2008). The antioxidant activity can be the result of specific scavenging of radicals formed during peroxidation, scavenging of oxygen-containing compounds, or metal-chelating ability. Furthermore, several studies have indicated that peptides derived from marine fish proteins have greater antioxidant properties in different oxidative systems (Jun et al., 2004). However, the exact mechanism of peptides acting as antioxidants is not clearly known but some aromatic amino acids and histidine (His) are reported to play a vital role in the activity. Gelatin peptides contain mainly nonpolar amino acids and an abundance of these amino acids favor a higher affinity to oil and better emulsifying ability. Hence, marine gelatin-derived peptides are expected to exert higher antioxidant effects among other antioxidant peptide sequences (Mendis et al., 2005).

Recently, increasing consumer knowledge of the link between diet and health has raised awareness of and demand for functional food ingredients and nutraceuticals. This has led to the avoidance of undesirable side effects associated with organically synthesized chemical food additives and drugs. Therefore, the above-stated animal-, plant-, and marine-derived bioactive peptides with antioxidative properties have great potential for use as functional ingredients in nutraceuticals and pharmaceuticals instead of synthetic antioxidants.

4.3 Methods for Evaluation of Antioxidant Activity

4.3.1 Reaction Mechanisms of Antioxidants

Various antioxidant activity methods have been used to monitor and compare the antioxidant activity of proteins and peptides. On the basis of the inactivation mechanism involved, major antioxidant activity assays can be roughly divided into two categories: hydrogen atom transfer (HAT) and single electron transfer (SET) based assays.

The end result is the same, regardless of the mechanism, but kinetics and the potential for side reactions differ. Proton-coupled electron transfer and HAT reactions may occur in parallel, and the mechanism dominating in a given system will be determined by antioxidant structure and properties, solubility and partition coefficient, and system solvent. Bond dissociation energy (BDE) and ionization potential (IP) are two major factors that determine the mechanism and efficacy of antioxidants (Prior, Wu, and Karen, 2005; Wright, Johnson, and Dilabio, 2001). HAT-based methods, such as the ORAC and TRAP assays, apply a competitive reaction scheme, in which antioxidant and substrate compete for thermally generated peroxyl radicals. SET-based methods measure an antioxidant activity to reduce an oxidant, which changes color when reduced. The degree of color change is correlated with the sample's antioxidant concentration. The SET-based method includes the total phenols assay by Folin–Cioalteu reagent, DPPH and ABTS radical scavenging activity assays, the SOD assay, and the FRAP assay (Dudonne et al., 2009; Huang, Ou, and Prior, 2005). SET and HAT mechanisms almost always occur together in all samples, with the balance determined by antioxidant structure and pH, and their difference can be made difficult.

4.3.2 Characteristics of Candidate Antioxidant Activity Methods

This chapter summarizes the principles of the most commonly used antioxidant activity assays and evaluates their advantages.

ORAC measures antioxidant inhibition of peroxyl radical-induced oxidations and thus reflects classical radical chain-breaking antioxidant activity by H atom transfer (Ou, Hampsch-Woodill, and Prior, 2001). The ORAC assay provides a controllable source of peroxyl radicals that models reactions of antioxidants with lipids in both food and physiological systems, and it can be adapted to detect both hydrophilic and hydrophobic antioxidants by altering the radical source and solvent (Prior et al., 2005). Using this technique, beta-lactoglobulin (Elias et al., 2006) and casein (Diaz and Decker, 2004) were shown to scavenge peroxyl radicals.

The TRAP assay monitors the activity of antioxidant compounds to interfere with the reaction between peroxyl radicals generated by AAPH or ABAP and a target probe (Prior et al., 2005). The basic reactions of the assay are similar to those of ORAC. The TRAP assay was designed and is most often used for measurements of *in vivo* antioxidant activity in serum or plasma because it measures nonenzymatic antioxidants, such as glutathione, ascorbic acid, alpha-tocopherol, and beta-carotene. The TRAP assay involves the initiation of lipid peroxidation by generating water-soluble peroxyl radicals and is sensitive to all known chain-breaking antioxidants (Frankel and Meyer, 2000).

Beta-carotene or crocin-bleaching assay measures the decrease in the rate of beta-carotene or crocin decay provided by antoxidants. Carotenoids bleach via autoxidation induced by light or heat or peroxyl radicals (e.g., AAPH or oxidizing lipids; Kampa et al., 2002; Ursini et al., 1998), and this decolorization can be diminished or prevented by classical antioxidants that donate hydrogen atoms to quench radicals (Prior et al., 2005). The addition of an antioxidant-containing sample, individual antioxidants, or plant extracts causes the inhibition of beta-carotene bleaching (Laguerre, Lecomte, and Villeneuve, 2007).

The total oxidant scavenging capacity (TOSC) assay has the advantage that it permits the quantification of the antioxidant activity toward three oxidants, that is, hydroxyl and peroxyl radicals and peroxynitrite. Also, other advantages of the TOSC assay are its effectiveness in detecting the scavenging activity of both hydrophilic and lipophilic antioxidants and its activity in distinguishing between fast- and slow-acting antioxidants (Regoli and Winston, 1999).

The TEAC assay has been improved and widely used in testing antioxidant activity in food samples (Cao et al., 1998; Pellegrini et al., 2003; Pietta et al., 2000). The TEAC method, which is based on the oxidation of ABTS in the presence of H_2O_2 and metomyoglobin, only measures hydrophilic antioxidants. This assay is operationally simple and can be used to study effects of pH on antioxidant mechanisms (Lemanska et al., 2001). Also, ABTS is soluble in both aqueous and organic solvents and is not affected by ionic strength, so it can be used in multiple media to determine both hydrophilic and lipophilic antioxidant activities of extracts and body fluids (Awika et al., 2003).

The DPPH radical is one of the few stable organic nitrogen radicals, and bears a deep purple color. It is commercially available and does not have to be generated before assay. This assay is based on the measurement of the reducing activity of antioxidants toward the DPPH radical. The assay is technically simple and rapid and needs only a UV-Vis spectrophotometer, which might explain its widespread use in antioxidant screening (Prior et al., 2005). However, interpretation is complicated when the test compounds have spectra that overlap DPPH at 515 nm. Carotenoids, in particular, interfere. Use of DPPH to measure antioxidant activity is plagued by many drawbacks (Noruma, Kikuchi, and Kawakami, 1997).

Folin–Cioalteu reagent (FCR) assay has for many years been used as a measure of total phenolics in natural products, but the basic mechanism is an oxidation/reduction reaction. The total phenol assay by Folin–Cioalteu reagent is convenient, simple, and reproducible (Huang et al., 2005). Several studies applied the total phenol assay by FCR and antioxidant activity assay (e.g., DPPH, FRAP TEAC, ORAC, etc.) and often found excellent linear correlations between the total phenolic profiles and the antioxidant activity (Gheldof and Engeseth, 2002; Madhujith, Izydorczyk, and Shahidi, 2006; Shahidi, Liyana-Pathirana, and Wall, 2006).

Among lipid peroxidation products used for antioxidant assays, malonaldehyde (MA) has been most widely used to evaluate the antioxidant activity of chemicals in lipid peroxidation systems (Neff and Frankel, 1984; Pryor, Stranley, and Blair, 1976). In particular, MA is a useful biomarker to investigate the final stage of lipid peroxidation. Studies of plants and their components using the thiobarbituric acid (TBA) assay that have appeared recently are studies of extracts of rosemary, green tea, grape seed, and tomato phenolic compounds (Kosar, Goeger, and Can Baser, 2008; Smet et al., 2008). The lipid peroxidation inhibition activity of the antioxidant peptide is determined by the TBA assay (Osawa and Namiki, 1985). Here, linoleic acid is subjected to oxidization in an ethanol/water emulsified model system, where transition metal ion, Fe^{2+}, accelerated lipid peroxidation. Afterward, the degree of oxidation can be evaluated by the ferric thiocyanate (FTC) method described by Mitsuda, Yasumoto, and Iwami (1996) and the capacity of anioxidative peptide to inhibit lipid peroxidation determined. The chemistry of this process involves generation of peroxyl and alkoxyl radicals from pre-existing lipid peroxide to initiate lipid peroxidation.

In this review, numerous antioxidant activity methods are reported. They differ from each other in terms of reaction mechanisms, oxidant and target/

probe species, reaction conditions, and expression of results. In addition to selecting these methods, analysis conditions, such as concentration of substrate and antioxidants should be optimized to simulate real food or biological systems as much as possible.

Generally, it is recommended to use at least two different types of assays: one to monitor the early stage of lipid peroxidation, such as beta-carotene bleaching, conjugated diene, or FTC, and the other to monitor the final stage of lipid peroxidation, such as TBA. It is also recommended that a combination of assays for scavenging electrons or radicals, such as DPPH, ABTS, or FRAP, and for the assays associated with lipids, peroxidations should be used, as these chemically distinct methods are based on different reaction mechanisms (Dudonne et al., 2009; Huang et al., 2005; Moon and Shinamoto, 2009; Prior and Cao, 1999). It is necessary to obtain comparable antioxidant values for the same sample by different methods.

4.4 Peptide Fractions with Stronger Antioxidants by Autofocusing

Bioactive peptides are usually purified by using a combination of chromatographic techniques. These may include enrichment in bioactive peptides by purification using gel filtration or ion exchange chromatographies. Most often, reversed phase high-performance liquid chromatograophy (RP-HPLC) has been used extensively for laboratory-scale isolation of the peptides in protein hydrolysates and identification of antioxidant peptides. However, the high cost of the preparative HPLC apparatus and chemicals has been a stumbling block in the isolation or preparation of antioxidant peptides in sufficient quantities to examine their value in food and *in vivo* systems. Moreover, acetonitrile, methanol, and trifluoroacetic acid, which have been extensively used for peptide preparation by the HPLC system, are not suitable solvents for food processing because of their potential toxicity (Megias et al., 2007; Park et al., 2008). Thus, it is difficult to isolate or synthesize adequate amounts of antioxidant peptides for food ingredients. Therefore, crude exzymatic hydrolysates of food proteins have been used to examine their antioxidant activities in food systems and the underlying mechanism for the antioxidant activity of peptides in foods is still unclear.

To alleviate this situation, preparative ampholyte-free isoelectro focusing, referred to as autofocusing has been developed. A large-scale apparatus that can fractionate more than 50–500 g of peptide sample has been introduced (Hashimoto et al., 2005). By using such apparatus, a sufficient amount of peptide fraction with different p*I* can be prepared by using water as the solvent.

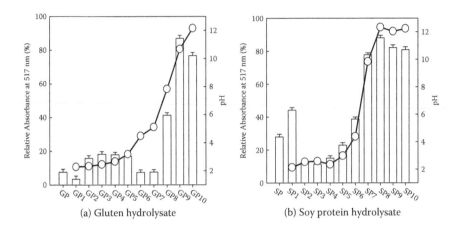

FIGURE 4.1
DPPH radical scavenging activity of gluten (a) and soy protein (b) hydrolysates with different isoelectric points (p*I*). The concentration of each sample is 1%. Each value represents the mean of four replicates and standard deviations in error bars.

Park et al. (2008) examined antioxidant activities of autofocusing fractions of SP (soy protein) and GP (gluten) hydrolysates. Peptide fractions with higher antioxidant activities than crude enzymatic hydrolysates of GP and SP were prepared without toxic solvent and reagents. By using autofocusing, adequate amounts of peptide fractions with high DPPH radical scavenging activity or high antioxidant activity against AAPH-induced linoleic acid oxidation in the emulsion system can be prepared, respectively (Figures 4.1 and 4.2).

These acidic and basic peptide fractions would be useful for examining the mechanism underlying the antioxidant activities of peptides in a food system. Also, by combining some autofocusing fractions, a peptide fraction with higher antioxidant activity against the AAPH-induced linoleic acid oxidation in the emulsion system compared to the crude hydrolysate can be prepared (Figure 4.3). If the antioxidant activity of peptides in food can be increased by autofocusing, this technique could be used for the preparation of peptide-based food additives to suppress oxidation as this technique does not require expensive and harmful solvents and has an inherent potential for further scale-up.

4.5 Conclusion

Bioactive peptides with antioxidative properties may have great potential for use as nutraceuticals and pharmaceuticals, and as substitutes for synthetic antioxidants. For example, Shahidi, Han, and Synowiecki (1995) clearly demonstrated that capelin protein hydrolysate when added to minced pork

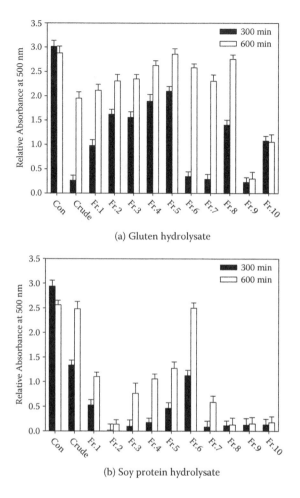

(a) Gluten hydrolysate

(b) Soy protein hydrolysate

FIGURE 4.2
Inhibitory effect of autofocusing fractions of gluten (a) and soy protein (b) hydrolysates on linoleic acid oxidation in emulsion system. The concentration of each sample is 0.75%. Each value represents the mean of four replicates and standard deviations in error bars.

muscle at a level of 0.5–3.0% reduced the formation of secondary oxidation products including thiobarbituric acid reactive substances (TBARS) in the product by 17.7–60.4%. Thus proteins and peptides have excellent antioxidant activity and great potential as food additives although they may not be suitable in all food applications. Antioxidants for food additives should satisfy the following requirements: safety, effectiveness at low concentration, no color, no odor, or no strange taste, effectiveness for processed food as well as for the raw material, ease of analysis in food, and low cost (Hattori et al., 1998).

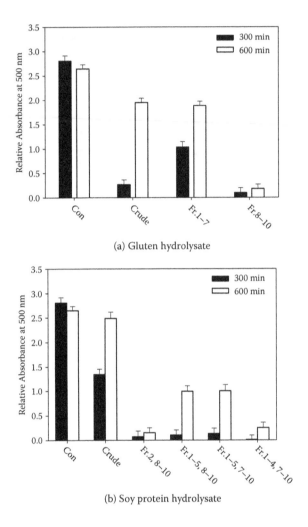

(a) Gluten hydrolysate

(b) Soy protein hydrolysate

FIGURE 4.3
Inhibitory effect of combined fractions of gluten (a) and soy protein (b) hydrolysates on linoleic acid oxidation in emulsion system. The concentration of each sample is 0.75%. Each value represents the mean of four replicates and standard deviations in error bars.

A major concern associated with the use of protein and peptides as antioxidants is the potential issue of allergenicity, especially with proteins derived from dairy, soy, nuts, and eggs. Also, proteins and peptides could cause problems by altering taste, flavor, texture, color, and so on of the food product (Ellas et al., 2008). The possible bitter taste of protein hydrolysates may prevent their use in many products as food additives (Kristinsson and Rasco, 2000) and their bioactivity may be reduced through molecular action during food processing or interaction with other food ingredients (Moller et al., 2008). To decrease bitterness, Shahidi et al. (1995) treated fish protein hydrolysate with activated carbon, which removed the bitterness. An increase of antioxidant activity utilizing food processing procedures, such as autofocusing, can reduce the peptide dose as antioxidant, consequently reducing undesirable taste and flavor.

By using *in vitro* assays, some antioxidant peptides have been isolated and identified from protein hydrolysates. However, lack of isolation methods for the preparation of enough peptide for addition to a food system, and the active peptide in a food system and its underlying mechanism have remained unclear. To break this situation, large-scale autofocusing has been developed. Autofocusing can fractionate peptides in sufficient amounts for addition to a food system. By comparing antioxidant activities of the autofocusing fractions in food and also by *in vitro* systems, active peptides in food systems and their underlying mechanism for antioxidant activity would be elucidated.

It has been speculated that oral ingestion of the antioxidant peptide might suppress oxidaton in the human body and consequently moderate oxidation-related diseases. However, *in vitro* antioxidant activity cannot be directly linked to health-promoting activity based on antioxidant activity, as the peptide can be degraded into amino acids during digestion and absorption processes. There are few data on direct antioxidant activity of food-derived peptides in the human body. Further studies are necessary to obtain direct evidence for health-promoting activities of protein and peptides through antioxidant activity.

References

Ajila, C.M., Naidu, K.A., Bhat, U.J.S., and Rao, P. (2007). Bioactive compounds and antioxidant potential of mango peel extract. *Food Chem.* 105: 982–988.

Allen, J.C. and Wrieden, W.L. (1982). Influence of milk proteins on lipid oxidation in aqueous emulsions. I. Casein, whey protein and α-lactoalbumin. *J. Dairy Res.* 49: 239–248.

Ariyoshi, Y. (1993). Angiotensin-converting enzyme inhibitors derived from food proteins. *Trends Food Sci. Technol.* 4: 139–144.

Awika, J.M., Rooney, L.W., Wu, X., Prior, R.L., and Cisneros-Zevallos, L. (2003). Screening methods to measure antioxidant activity of sorghum (*Sorghum bicolor*) and sorghum products. *J. Agric. Food Chem.* 51: 6657–6662.

Butterfield, D.A., Castenga, A., Pocernich, C.B., Drake, J., Scapagnini, G., and Calabrese, V. (2002). Nutritional approaches to combat oxidative stress in Alzheimer's disease. *J. Nutr. Biochem.* 13: 444–461.

Byun, H.G., Lee, J.W., Parl, H.G., Jeon, J.K., and Kim, S.K. (2009). Antioxidant peptides isolated from marine rotifer, *Brachionus rotundiformis*. *Process Biochem.* 44: 842–846.

Cao, G., Russell, R.M., Lischner, N., and Prior, R.L. (1998). Serum antioxidant capacity is increased by consumption of strawberries, spinach, red wine or vitamin C in elderly women. *J. Nutr.* 128: 2383–2390.

Chen, H.M., Muramoto, K., and Yamauchi, F. (1995). Structural analysis of antioxidative peptides from soybean β–conglycium. *J. Agric. Food Chem.* 43: 574–578.

Chen, H.M., Muramoto, K., Yamauchi, F., and Norihara, K. (1996). Antioxidant activity of designed peptides based on the antioxidative peptide isolated from digests of a soybean protein. *J. Agric. Food Chem.* 44, 2619–2623.

Clare, D.A. and Swaisgood, H.E. (2000). Bioactive milk peptides: A prospectus. *J. Dairy Sci.* 83: 1187–1195.

Diaz, M. and Decker, E. A. (2004). Antioxidant mechanisms of caseinophosphopeptides and casein hydrolysates and their application in ground beef. *J. Agric. Food Chem.* 52: 8208–8213.

Donnelly, J.L., Decker, E.A., and McClements, D.J. (1998). Iron-catalyzed oxidation of menhaden oil as affected by emulsifiers. *J. Food Sci.* 63: 997–1000.

Dudonne, S., Vitrac, X., Coutiere, P., Woillez, M., and Merillon, J.-M. (2009). Comparative study of antioxidant properties and total phenolic content of 30 plant extracts of industrial interest using DPPH, ABTS, FRAP, SOD and ORAC assays. *J. Agric. Food Chem.* 57: 1768–1774.

Elias, R.J., Bridgewater, J.D., Vachet, R.E., Waraho, T., McClements, D.J., and Decker, E. A. (2006). Antioxidant mechanisms of enzymatic hydrolysates of beta-lactoglobulin in food lipid dispersions. *J. Agric. Food Chem.* 54: 9565–9572.

Ellas, R.J., Kellerby, S.S., and Decker, E.A. (2008). Antioxidant activity of protein and peptides. *Crit. Rev. Food Sci. Nutr.* 48: 430–431.

Erdmann, K., Cheung, B.W.Y., and Schroder, H. (2008). The possible roles of food-derived bioactive peptides in reducing the risk of cardiovascular disease. *J. Nutr. Biochem.* 19: 643–654.

Frankel, E.N. and Meyer, A.S. (2000). The problems of using one-dimensional methods to evaluate multifunctional food and biological antioxidants. *J. Sci. Food Agric.* 80: 1925–1941.

Frlich, I. and Riederer, P. (1995). Free radical mechanisms in dementia of Alzheimer type and the potential for antioxidative treatment. *Drug Res.* 45: 443–449.

Gheldof, N. and Engeseth, N.J. (2002). Antioxidant capacity of honeys from carious floral sources based on the determination of oxygen radical absorbance capacity and inhibition of *in vitro* lipoprotein oxidation in human serum samples. *J. Agric. Food Chem.* 50: 3050–3055.

Gibbs, B.F., Zougman, A., Masse, R., and Mulligan, C. (2004). Production and characterization of bioactive peptides from soy hydrolysate and soy-fermented food. *Food Res. Int.* 37: 123–131.

Guo, Q., Zhao, B., Shen, S., Hou, J., Hut, J., and Xin, W. (1999). ESR study on the structure-antioxidant activity relationship of tea catechins and their epimers. *Biochim. Biophys. Acta* 1427: 13–23.

Haque, E., Chand, R., and Kapila, S. (2009). Biofunctional properties of bioactive peptides of milk origin. *Food Rev. Int.* 25: 28–43.

Hashimoto, K., Sato, K., Nakamura, Y., and Ohtsuki, K. (2005). Development of a large scale (50 L) apparatus for ampholyte free isoelectric focusing(Autofocusing) of peptides in enzymatic hydrolysates of food proteins. *J. Agric. Food Chem.* 53: 3801–3806.

Hattori, M., Yamaji-Tsukamoto, K.A., Kumagai, H., Feng, Y., and Takahashi, K. (1998). Antioxidative activity of soluble elastin peptides. *J. Agric. Food Chem.* 46: 2167–2170.

Hettiarachchy, N.S., Glenn, K.C., Gnanasambandan, R., and Johnson, M.G. (1996). Natural antioxidant extract from fenugreek (*Trigonella foenumgraecum*) for ground beef patties. *J. Food Sci.* 61: 516–519.

Huang, D., Ou, B., and Prior, R.L. (2005). The chemistry behind antioxidant capacity assays. *J. Agric. Food Chem.* 53: 1841–1856.

Je, J.Y., Park, J.Y., Jung, W.K., Park, P.J., & Kim, S.K. (2005). Isolation of angiotensin I converting enzyme (ACE) inhibitor from fermented oyster sauce, *Crassostrea gigas. Food Chem.* 90: 809–814.

Je, J.Y., Qian, Z.J., Byun, H.G., and Kim, S.K. (2007). Purification and characterization of an antioxidant peptide obtained from tuna backbone protein by enzymatic hydrolysis. *Process Biochem.* 42: 840–846.

Je, J.Y., Qian, Z.J., Lee, S.H., Byun, H.G., and Kim, S.K. (2008). Purification and antioxidant properties of big eye tuna (*Thunnus obesus*) dark muscle peptide on free radical-mediated oxidation systems. *J. Med. Food* 11(4): 629–637.

Jun, S.Y., Park, P.J., Jung, W.K., and Kim, S.K. (2004). Purification and characterization of an antioxidative peptide from enzymatic hydrolysates of yellowfin sole (*Limanda aspera*) frame protein. *Euro. Food Res. Technol.* 219: 20–26.

Jung, W.K. and Kim, S.K. (2009). Isolation and characterization of an anticoagulant oligopeptide from blue mussel, *Mytilus edulis. Food Chem.* 117: 687–692.

Jung, W.K., Mendis, E., Je, J.Y., Park, P.J., Son, B.W., Kim, H.C., Choi, Y.K., and Kim, S.K. (2006). Angiotensin I-converting enzyme inhibitory peptide from yellowfin sole (*Limanda aspera*) frame protein and its antihypertensive effect in spontaneously hypertensive rats. *Food Chem.* 94: 26–32.

Jung, W.K., Rajapakse, N., and Kim, S.K. (2005). Antioxidative activity of a low molecular weight peptide derived from the sauce of fermented blue mussel, *Mytilus edulis. Euro. Food Res. Technol.* 220: 535–539.

Kampa, M., Nistikaki, A., Tsaousis, V., Maliataki, N., Notas, G., and Castanas, E. (2002). A new automated method for the determination of total antioxidant capacity (TAC) of human plasma, based on the crocin bleaching assa. *BMC Clin, Patho.* 2, http://www.Biomedecentral.Com/1472–6890/1472/1473.

Kanazawa, K., Ashida, H., and Natake, M. (1987). Autoxidizing process interaction of linoleic acid with casein. *J. Food Sci.* 52: 475–478.

Kim, S.K., Kim, Y.Y., Byun, K.S., Nam, D.S.J., and Shahidi, F. (2001). Isolation and characterization of antioxidative peptides from gelatin hydrolysate of Alaska pollock skin. *J. Agric. Food Chem.* 49: 1984–1989.

Kim, S.K. and Wijesekara, I. (2010). Development and biological activities of marine-derived bioactive peptides: A review. *J. Funct. Foods* 2: 1–9.

Korhonen, H. (2009). Milk-derived bioactive peptides: From science to applications. *J. Funct. Foods* 1: 177–187.

Korhonen, H. and Pihlanto, A. (2006). Bioactive peptides: Production and functionality. *Int. Dairy J.* 16: 945–960.

Kosar, M., Goeger, F., and Can Baser, K.H. (2008). *In vitro* antioxidant properties and phenolic composition of *Salvia virgata* Jacq. from Turkey. *J. Agric. Food Chem.* 56: 2369–2374.

Kristinsson, H.G. and Rasco, B.A. (2000). Hydrolysis of salmon muscle protein by an enzyme mixture extracted from Atlantic salmon (salmo salar) pyloric caeca. *J. Food Biochem.* 24: 177–184.

Kudoh, K., Mausrmoto, M., and Onodera, S. (2003). Antioxidant activity and protective effect against ethanol-induced gastric mucosal damage of a potato protein hydrolysate. *J. Nutr. Sci. Vitaminol.* 49(6): 451–455.

Laguerre, M., Lecomte, J., and Villeneuve, P. (2007). Evaluation of the ability of antioxidants to counteract lipid oxidation: existing methods, new trends and challenges. *Progress Lipid Res.* 46: 244–282.

Lahl, W.J. and Braun, S.D. (1994). Enzymatic production of protein by hydrolysates for food use. *Food Technol.* 48: 68–71.

Lemanska, K., Szymusiak, H., Tysakowska, B., Zielinski, R., Soffer, A.E.M.F., and Rietjens, I.M.C. (2001). The influence of pH on the antioxidant properties and the mechanisms of antioxidant action of hydroxyflavones. *Free Radical Biol. Med.* 31: 869–881.

Liu, Y.W., Han, C.H., Lee, M.H., Hsu, F.L., and Hou, W.C. (2003). Patatin, the tuber storage protein of potato (*Salanum tuberosum L.*), exhibits antioxidant activity *in vitro. J. Agric. Food Chem.* 51: 4389–4393.

Madhujith, T., Izydorczyk, M., and Shahidi, F. (2006). Antioxidant properties of pearled barley fractions. *J. Agric. Food Chem.* 54: 3283–3289.

Megias, C., Pedroche, J., Yust, M.M., Giron-Calle, J., Alaiz, M., Millan, F., and Vioque, J. (2007). Affinity purification of copper-chelating peptides from sunflower protein hydrolysates. *J. Agric. Food Chem.* 55: 6509–6514.

Megias, C., Yust, M.M., Pedroche, J., Lquari, H., Giron-Calle, J., Alaiz, M., Millan, F., and Vioque, J. (2004). Purification of an ACE inhibitory peptide after hydrolysis of sunflower (*Helianthus annuus L.*) protein isolate. *J. Agric. Food Chem.* 52: 1928–1932.

Meisel, H. and FitzGerald, R.J. (2003). Biofunctional peptides from milk proteins: mineral-binding and cytomodulatory effects. *Curr. Pharmaceut. Des.* 9: 1289–1295.

Mendis, E., Rajapakse, N., Byun, H.G., and Kim, S.K. (2005). Investigation of jumbo squid (*Dosidicus gigas*) skin gelatin peptides for their *in vitro* antioxidant effects. *Life Sci.* 77: 2166–2178.

Meucci, E., Mordente, A., and Martorana, G.E. (1991). Metal-catalyzed oxidation of human serum albumin: Conformational and functional changes. *J. Biol. Chem.* 266: 4692–4699.

Mitsuda, H., Yasumoto, K., and Iwami, K. (1996). Antioxidant action of indole compounds during the autoxidation of linoleic acid. *Eiyo to Shokuryo* 29: 238–244.

Moller, N.P., Scholz-Ahrens, K.E., Roos, N., and Schrezenmeir, J. (2008). Bioactive peptides and proteins from foods: indication for health effects. *Euro. J. Nutr.* 47: 171–182.

Moon, J.K. and Shinamoto, T. (2009). Antioxidant assays for plant and food components. *J. Agric. Food Chem.* 57: 1655–1666.

Moure, A., Dominguez, H., and Parajo, J. C. (2006). Antioxidant properties of ultrafiltration-recovered soy protein fractions from industrial effluents and their hydrolysates. *Process Biochem.* 41: 447–456.

Nanjo, F., Goto, K., Seto, R., Suzuki, M., Sakai, M., and Hara, Y. (1996). Scavenging effects of tea catechins and their derivatives on 1,1,-diphenyl picrylhydrazyl radical. *Free Radical Biol. Med.* 21: 895–202.

Neff, W.E. and Frankel, E.N. (1984). Photosensitized oxidation of methyl linolenate monohydroperoxides: hydroperoxy cyclic peroxides, dihydroperoxides and hydroperoxy bis (cyclic peroxides). *Lipids* 19: 952–957.

Noruma, T., Kikuchi, M., and Kawakami, Y. (1997). Proton-donative antioxidant activity of fucoxanthin with 1,1-diphenyl-2-picrylhydrazyl (DPPH). *Biochem. Mol. Biol. Int.* 42: 361–370.

Osawa, T. and Namiki, M. (1985). Natural antioxidants isolated from eucalyptus waxes. *J. Agric. Food Chem.* 33: 770–780.

Ostdal, H., Daneshvar, B., and Skibsted, L.H. (1996). Reduction of ferrylmyglobin by beta-lactoglobulin. *Free Radical Res.* 24: 429–438.

Ou, B., Hampsch-Woodill, M., and Prior, R.L. (2001). Development and validation of an improved oxygen radical absorbance capacity assay using fluorescein as the fluorescent probe. *J. Agric. Food Chem.* 49: 4619–4926.

Park, E.Y., Morimae, M., Matsumura, Y., Nakamura, Y., and Sato, K. (2008). Antioxidant activity of some protein hydrolysates and their fractions with different isoelectric points. *J. Agric. Food Chem.* 56(19): 9246–9251.

Park, E.Y., Murakami, H., Mori, T., and Matsumura, Y. (2005). Effects of protein and peptides addition on lipid oxidation in power model system. *J. Agric. Food Chem.* 53: 137–144.

Pellegrini, P., Serafini, M., Colombi, B., Del Rio, D., Salvatore, S., Bianchi, M., and Brighenti, F. (2003). Total antioxidant capacity of plant foods, beverages and oils consumed in Italy assessed by three different *in vitro* assays. *J. Nutr.* 133: 2812–2819.

Pena-Ramos, E.A. and Xiong, Y.L. (2001). Antioxidative activity of whey protein hydrolysates in a liposomal system. *J. Dairy Sci.* 84: 2577–2583.

Pena-Ramos, E.A. and Xiong, Y.L. (2002). Antioxidant activity of soy protein hydrolyzates in a liposomial system. *J. Food Sci.* 67(8): 2952–2956.

Pena-Ramos, E.A. and Xiong, Y.L. (2003). Whey and soy protein hydrolysates inhibit lipid oxidation in cooked pork patties. *Meat Sci.* 64: 259–263.

Phelan, M., Aherne, A., FitzGerald, R.J., and O'Brien, N.M. (2009). Casein-derived bioactive peptides: Biological effects, industrial uses, safety aspects and regulatory status. *Int. Dairy J.* 19: 643–654.

Pietta, P., Simonetti, P., Gardana, C., and Mauri, P. (2000). Trolox equivalent antioxidant capacity (TEAC) of Ginkgo biloba flavonol and *Camellia sinensis* catechin metabolites. *J. Pharmaceut. Biomed. Anal.* 23: 223–226.

Pihlanto-Leppala, A. (2001). Bioactive peptides derived from bovine proteins: Opioid and ACE-inhibitory peptides. *Trends Food Sci. Technol.* 11: 347–356.

Prior, R.L. and Cao, G. (1999). *In vivo* total antioxidant capacity: Comparison of different analytical methods. *Free Radical Biol. Med.* 27: 1173–1181.

Prior, R.L., Wu, X., and Karen, S. (2005). Standardized methods for the determination of antioxidant capacity and phenolics in foods and dietary supplements. *J. Agric. Food Chem.* 53: 4290–4302.

Pryor, W.A., Stranley, J.P., and Blair, E. (1976). Autoxidation of polyunsaturated fatty acids: II. A suggested mechanism for the formation of TBA-reactive materials from prostaglandin-like endoperoxides. *Lipids* 11: 370–379.

Qian, Z.J., Jung, W.K., Byun, H.G., and Kim, S.K. (2008). Protective effect of an antioxidative peptide purified from gastrointestinal digests of oyster, *Crassostrea gigas* against free radical induced DNA damage. *Biores. Technol.* 99: 3365–3371.

Rajapakse, N., Mendis, E., Byun, H.G., and Kim, S.K. (2005a). Purification and *in vitro* antioxidative effects of giant squid muscle peptides on free radical-mediated oxidative systems. *J. Nutr. Biochem.* 16: 562–569.

Rajapakse, N., Mendis, E., Jung, W.K., Je, J.Y., and Kim, S.K. (2005b). Purification of a radical scavenging peptide from fermented mussel sauce and its antioxidant properties. *Food Res. Int.* 38: 175–182.

Ranathunga, S., Rajapakse, N., and Kim, S.K. (2006). Purification and characterization of antioxidative peptide derived from muscle of conger eel (*Conger myriaster*). *Euro. Food Res. Technol.* 222: 310–315.

Ravallec, P.R., Charlot, C., Pires, C., Braga, V., Batista, I., Wormhoudt, A.V., Gal, Y.L., and Fouchereau-Peron, M. (2001). The presence of bioactive peptides in hydrolysates prepared from processing waste of sardine (*Sardina pilchardus*). *J. Sci. Food Agric.* 81(11): 1120–1125.

Regoli, F. and Winston, G.W. (1999). Quantification of total oxidant scavenging capacity of antioxidants for peroxynitrite, peroxyl radicals, and hydroxyl radicals. *Toxicol. Appl. Pharmacol.* 156: 96–105.

Rival, S.G., Boeriu, C.G., and Wichers, H.J. (2001). Caseins and casein hydrolysates. 2. Antioxidative properties and relevance to Lipoxygenase inhibition. *J. Agric. Food Chem.* 49: 295–302.

Rival, S.G., Fornaroli, S., Boeriu, C.G., and Wichers, H.J. (2000) Caseins and casein hydrolysates. 1. Lipoxygenase inhibitory properties. *J. Agric. Food Chem.* 48: 287–294.

Rosen, G.M. and Rauckman, E.J. (1984). Spin trapping of superoxide and hydroxyl radicals. *Meth. Enzymol.* 105: 198–209.

Roy, F., Boye, J.I., and Simpson, B.K. (2010). Bioactive proteins and peptides in pulse crops: Pea, chickpea and lentil. *Food Res. Int.* 43: 432–442.

Sakanaka, S., Tachibana, Y., Ishihara, N., and Juneja, L.R. (2005). Antioxidant properties of casein calcium peptides and their effects on lipid oxidation in beef homogenates. *J. Agric. Food Chem.* 53: 464–468.

Shahidi, F. and Zhong, Y. (2008). Bioactive peptides. *J. AOAC Int.* 91: 914–931.

Shahidi, F., Han, X.-Q., and Synowiecki, J. (1995). Production and characteristics of protein hydrolysates from capelin (*Mallotus villosus*). *Food Chem.* 53: 285–293.

Shahidi, F., Liyana-Pathirana, C.M., and Wall, D.S. (2006). Antioxidant activity of white and black sesame seeds and their hull fractions. *Food Chem.* 99: 478–483.

Sheih, I.-C., Wu, T.K., and Fang, T.J. (2009). Antioxidant properties of a new antioxidative peptide from algae protein waste hydrolysate in different oxidation systems. *Biores. Technol.* 100: 3419–3425.

Simpson, B.K., Nayeri, G., Yaylayan, V., and Ashie, I.N.A. (1998). Enzymatic hydrolysis of shrimp meat. *Food Chem.* 61: 131–138.

Slizyte, R., Mozuraityte, R., Martinez-Alvarez, O., Falch, E., Fouchereau-Peron, M., and Rustad, T. (2009). Functional, bioactive and antioxidative properties of hydrolysates obtained from cod (*Gadus morhua*) backbones. *Process Biochem.* 44: 668–677.

Smet, K., Raes, K., Hughebaert, G., Haak, L., Arnouts, S., and DeSmet, S. (2008). Lipid and protein oxidation of broiler meat as influenced by dietary natural antioxidant supplementation. *Poultry Sci.* 87: 1682–1688.

Suetsuna, K. and Nakano, T. (2000). Identification of an antihypertensive peptide from peptic digest of wakame (*Undaria pinnatifida*). *J. Nutr. Biochem.* 11: 450–454.

Taylor, M.J. and Richardson, T. (1980). Antioxidant activity of cysteine and protein sulfhydryls in linoleate emulsion oxidized by hemoglobin. *J. Food Sci.* 45: 1223–1227.

Tsuge, N., Eiwaka, Y., Nomura, Y., Yamamoto, M., and Sugisawa, K. (1991). Antioxidative activity of peptides prepared by enzymatic hydrolysis of egg-white albumin. *Nippon Nogei Kagaku Kaishi* 65: 1635–1641.

Ursini, F., Zamburlini, A., Cazzolato, G., Maiorino, M., Bon, G.B., and Sevanian, A. (1998). Postprandial plasma lipid hydroperoxides: A possible link between diet and atherosclerosis. *Free Radical Biol. Med.* 25: 250–252.

Wang, J.Y., Fujimoto, K., Miyazawa, T., and Endo, Y. (1991). Antioxidative mechanism of maize zein in powder model systems against methyl linoleate: Effect of water activity and coexistence of antioxidants. *J. Agric. Food Chem.* 39: 351–355.

Wang, L.L. and Xiong, Y.L. (2005). Inhibition of lipid oxidation in cooked beef patties by hydrolyzed potato protein is related to its reducing and radical scavenging ability. *J. Agric. Food Chem.* 53: 9186–9192.

Wayner, D.D.M., Burton, G.W., Ingold, K.U., Barclay, L.R.C., and Locke, S.J. (1987). The relative contributions of vitamin E, urate, ascorbate and proteins to the total peroxyl radical-trapping antioxidant activity of human blood plasma. *Biochim. Biophys. Acta* 924: 408–419.

Wright, J.S., Johnson, E.R., and Dilabio, G.A. (2001). Predicting the activity of phenolic antioxidants: Theoretical method, analysis of substituent effects, and application to major families of antioxidants. *J. Amer. Chem. Soc.* 123: 1173–1183.

Yamamoto, N. (1997). Antihypertensive peptides derived from food proteins. *Biopolymers*, 43, 129–134.

Yang, C.S., Landau, J.M., Huang, M.T., and Newmark, H.L. (2001). Inhibition of carcinogenesis by dietary polyphenolic compounds. *Ann. Rev. Nutr.* 21: 381–406.

Zhao, Y., Li, B., Liu, Z., Dong, S., Zhao, X., and Zeng, M. (2007). Antihypertensive effect and purification of an ACE inhibitory peptide from sea cucumber gelatin hydrolysate. *Process Biochem.* 42: 1586–1591.

5

Mineral-Binding Peptides from Food

Saïd Bouhallab and Dominique Bouglé

CONTENTS

5.1 Introduction

The primary role of diet is to provide enough nutrients to meet metabolic requirements besides giving the consumer satisfaction and well-being. Recent knowledge supports the hypothesis that beyond meeting nutrition needs, diet may modulate various physiological functions in the body and may play beneficial roles in some diseases leading to the development of the "functional food" concept (Roberfroid, 2000). Food components are a potential source of bioactive substances. It was suggested that food can be designed as functional if it satisfactorily affects one or more target functions in the body, beyond the adequate nutritional status in a way that is relevant to either the state of well-being and health or the reduction in risk of a disease (Diplock et al., 1999). All food-derived components (i.e., proteins, lipids, carbohydrates, minerals, and trace elements) can potentially act as bioactive substances. Food proteins as such or as precursors constitute the major source of health-enhancing components that may be incorporated in functional food preparations (Korhonen and Pihlanto, 2006). The role of proteins and protein-derived bioactive

117

peptides as health-enhancing components is being increasingly studied and discussed (see Korhonen and Pihlanto, 2006, for review). Many of the proteins that occur naturally in raw food materials (plants and animals) exert their biological action either directly or upon enzymatic digestion *in vitro* or *in vivo* to generate bioactive peptide sequences. This review focuses on the recently reported works on phosphopeptides as potential carriers for mineral bioavailability.

5.2 Mineral-Binding Proteins in Food

Food proteins, especially those from milk and eggs interact specifically or chelate strongly with minerals and trace elements such as iron, calcium, zinc, manganese, and copper. There are two different groups of proteins according to the type of interaction with metals. The first group includes proteins with specific binding site(s) for minerals such as lactoferrin and α-lactalbumin from milk and ovotransferrin from egg white. The other group includes proteins rich in acidic clusters, that is, phosphoproteins such as phosvitin from egg yolk and caseins from milk. These proteins are able to sequester a high amount of divalent cations, in particular, through electrostatic interactions between highly negative clusters and positive charges of metals.

Lactoferrin and ovotransferrin are two transferring proteins with iron-binding property, showing 49% homology (González-Chávez et al., 2008). Transferrins are a superfamily of single-chain, glycosylated proteins that transport iron from plasma to cells and contribute to regulating iron levels in biological fluids. Lactoferrin and ovotransferrin are proteins with a molecular weight of 80,000 Da. Their tertiary structure consists of two globular lobes having a similar amino-acid sequence and one iron-binding site per lobe. Iron in its ferric form (Fe^{3+}) is the most common metal ion associated with the native form of these proteins. The concentration of lactoferrin in milk varies widely among species, with a high content in human milk (Lonnerdal, 2003). Lactoferrin possesses a greater iron-binding affinity with the ability to retain this metal over a wide pH range including an extremely acidic pH (González-Chávez et al., 2008). However, the protein can also bind other metals such as copper, manganese, and aluminum (Lonnerdal, 2003).

Lactoferrin has been reported to be a multifunctional agent: inhibition toward bacterial adhesion, a role in iron uptake by the intestinal mucosa, bacteriostatic, involvement in phagocytic killing and immune response, and anticancer activity. As a result, lactoferrin from bovine milk has been commercialized in various areas of functional foods (Steijns and van Hooijdonk, 2000). Although several studies exist on the physiological role of these proteins, in particular regarding their involvement in iron absorption, firm evidence *in vivo* is still lacking, particularly in humans. Ovotransferrin (formerly conalbumin) is present in birds as a major protein of egg white (up to 13%; Stevens, 1991). Ovotransferrin is also thought to affect calcium carbonate crystals and calcite morphology, suggesting that ovotransferrin has a role in nucleation and growth process of calcite crystals *in vivo*. As with lactoferrin, it is suggested that ovotransferrin can be used as a functional food ingredient.

α-Lactalbumin is a globular protein from milk whey with a molecular weight of 14,200 Da. α-Lactalbumin was extensively used as a model in numerous studies of biophysical properties of globular proteins (see Permyakov and Berliner, 2000, for review). Its concentration in bovine milk is about 1.5 g/l and it possesses a calcium binding site with a high affinity (dissociation constant Kd = 0.1 nM). The presence of calcium atom in its binding site is very important for the conformational stability of the protein and for its biological function (regulation of lactose synthesis). α-Lactalbumin is also able to bind other divalent cations such as zinc and manganese but with lower affinities (Veprintsev et al., 1996).

Caseins are a mixture of a least four phosphoproteins (αs1-casein, MW 23,000 Da; αs2-casein, MW 25,000 Da; β-casein, MW 24,000 Da; and κ-casein, MW 19,000 Da). They account for 80% of the total protein (30–35 g/l) in bovine milk. The common compositional factor is that caseins are phosphorylated proteins (Swaisgood, 2003). Casein molecules exist in a colloidal particle known as the casein micelle in association with calcium phosphate. In bovine milk, about two-thirds of the calcium and one-half of the inorganic phosphate are bound to various species of caseins, forming colloidal micelles with a calcium/phosphate/casein molar ratio of 30:21:1 (Holt, 1992). Hence, about 90% of the calcium content of skim milk (1 g/l) is associated in a colloidal form with the casein micelle. These phosphate groups are important to the structure of the casein micelle. Calcium binding by the individual caseins is proportional to the phosphate content. The phosphorylated sequences can be isolated throughout enzymatic digestion of a casein mixture (see Section 5.3).

Phosvitin, a highly phosphorylated glycoprotein, represents the major fraction of hen egg yolk phosphoproteins. This protein has a molecular weight of 34,000 Da and contains 220 amino acid residues, 50% of which are phosphoserine. The presence of highly acidic clusters explains the powerful sequestering ability of this protein toward divalent cations (iron, magnesium, calcium, etc.) and also its resistance to proteolytic enzymes. The potential binding capacity of phosvitin is very high with 60 iron atoms per protein molecule (Webb et al., 1973). Therefore, phosvitin is often considered nutritionally negative due to its contribution toward lowering the bioavailability of iron (Ishikawa et al., 2007).

5.3 Mineral-Binding Peptides: Analytical and Structural Features

Because of their chelating ability toward minerals, phosphorylated peptides are the most extensively studied mineral-binding peptides from food proteins, in particular those derived from caseins (Kitts, 2006). Caseinophosphopeptides (CPPs) are bioactive phosphorylated casein-derived peptides latent until they are released from α_{s1}-, α_{s2}-, and β-casein during gastrointestinal digestion of milk. Considering variations in enzyme specificity, side chemical reactions such as deamination or partial dephosphorylation, 40 to 50 or more molecular species can be present in a given CPPs preparation. CPPs contain a common cluster sequence of three phosphoseryl groups followed by two glutamic acid residues S(P)S(P)S(P)EE representing the binding sites for minerals (especially calcium, magnesium, iron, and zinc). These peptides could contribute to an increase in solubility of such minerals at intestinal pH, playing an important role in their bioavailability (FitzGerald, 1998; Meisel and FitzGerald, 2003).

Various methods have been developed to purify and characterize food phosphopeptides. Chromatographic methods (affinity, reverse phase, size exclusion, hydrophobic interactions) are the main methods used for such purposes. Affinity chromatography has been developed to prepare enriched fractions of phoshphopeptides from complex peptide mixtures. Immobilized metal affinity chromatography (IMAC), based on the affinity between negatively charged phosphate groups and positive charges of immobilized metal ions (Fe^{3+}, Fe^{2+}, Cu^{2+}, Zn^{2+}, Ni^{2+}, Co^{2+}, etc.), is widely used. The IMAC-based enrichment allows the sequencing, that is, identification of phosphorylated peptides throughout subsequent analysis by electrospray or nanoelectrospray mass spectrometry (Kocher, Allmaier, and Wilm, 2003). This coupled separation analysis procedure allows accurate identification of phosphopeptides even if other nonphosphorylated peptides are present in the original mixture (Kocher et al., 2003). Currently, improved IMAC strategies coupled with iterative mass spectrometry-based scanning techniques have been extensively reported for phosphoproteomic studies (Cirruli et al., 2008). These approaches allow sensitive detection of phosphorylated peptides in biological fluids (Cirruli et al., 2008). Another highly selective affinity chromatography for phosphorylated peptides enrichment based on TiO2 columns was developed recently (Pinkse et al., 2004; Martin et al., 2005). Using MALDI mass spectrometry analysis of enriched fractions, this procedure was found to be more selective for binding phosphopeptides than the IMAC procedure (Martin et al., 2005).

The physicochemical properties of CPPs derived from various caseins were recently reviewed in detail (Kitts, 2006). CPPs are able to chelate and then to solubilize high amounts of divalent cations, up to 250 mg calcium/g peptides, for example. The dissociation constant (affinity) of calcium ions depends on the phosphopeptide sequence but is generally reported to be in the mM range. The affinity of CPPs toward trace elements (iron, zinc, copper) is known to be higher than for calcium. We recently found, using isothermal titration calorimetry, dissociation constants in the μM range for iron and zinc (unpublished results). In another study, the binding of zinc to casein phosphopeptides has been analyzed using nanoelectrospray mass spectrometry. The study showed a direct relationship between phosphorylation degree and zinc binding, with a decrease in zinc-bound forms of peptide paralleling the decrease in phosphorylation degree of CPPs (Wang et al., 2007).

Few data are available on how mineral-binding affects the structure and conformation of phosphopeptides in solution. The effects of calcium binding on conformation of two different phosphopeptides, β-casein-(1-25) and αS1-casein-(59-79), were compared using NMR spectroscopy (Cross et al., 2001). Four structured regions have been identified in the calcium-bound β-casein-(1-25): residues Arg^1 to Glu^4 were found to be involved in a loop-type structure and residues Val^8 to Glu^{11}, $Ser(P)^{17}$ to Glu^{20}, and Glu^{21} to Thr^{24} were found to be implicated in β-turn conformations. In contrast, a loop-type structure was identified in the phosphorylated region of αS1-casein-(59-79) in the presence of calcium. Comparison of NMR patterns suggested that, despite their high degree of sequential and functional similarity, phosphopeptides β-casein-(1-25) and αS1-casein-(59-79) exhibited distinctly different conformations in the presence of calcium ions (Cross et al., 2001). More recently, using a variety of physical techniques including X-ray diffraction, scanning, and transmission electron microscopies as well as a library of synthetic analogues of casein phosphopeptides, the same research group reported on the structure of the complexes formed by the casein phosphopeptides with calcium phosphate. Interestingly, their results reveal that although the fully phosphorylated seryl-cluster motif is pivotal for the interaction with calcium and phosphate, other factors are also important for calcium binding, in particular, number of negative charge, number of amino-acids, and sequence of used phosphopeptide (Cross et al., 2005). Considering these structural and conformation differences, important differences are expected regarding the biological responses.

5.4 Health-Promoting Activity of Mineral-Binding Peptides

5.4.1 Phosphopeptides as Carriers for Mineral Absorption and Bioavailability

It has already been consistently proven that phosphorylated fragments of casein, caseinophosphopeptides (CPPs), can form organophosphate salts with minerals, such as iron, magnesium, manganese, copper, and selenium. Such sequestration prevents precipitation of minerals because it improves their solubility and stability under different physicochemical conditions in particular pH. Consequently, CPPs have been considered physiologically beneficial in the prevention of osteoporosis, dental caries, and anemia through improvement of mineral absorption and bioavailability (Bouhallab and Bouglé, 2004; Korhonen and Pihlanto, 2006). Several pieces of evidence exist today that CPPs have a positive effect on the absorption of calcium and other minerals in the intestine. From this, the use of CPPs has become widespread for various applications. However, inconsistent and conflicting results have been obtained in published studies; see the review by Korhonen and Pihlanto (2006).

The literature report conflicting results for the various minerals and a detailed conflict focusing on the role of CPPs in zinc bioavailability was published recently (Miquel and Farré, 2007). Studies performed in animals report some positive effect of CPPs on calcium absorption, whereas others have failed to find any effect in the presence of CPPs (FitzGerald, 1998). In human studies, Hansen et al. (1997) demonstrated an increase of both calcium and zinc absorption in adult volunteers administered a rice-based infant formula enriched with CPPs. In another clinical study, Aït-Oukhatar et al. (2002) showed that a pure CPP from β-casein, β-casein-(1-25) improves iron bioavailability in comparison with iron sulphate. The quantity of iron stored in the organs seven days after ingestion of a unique dose is higher in the group having received the Fe/CPP complex. These results confirmed those obtained in the rat model showing that the same pure CPP exhibited a positive effect on iron bioavailability (Aït-Oukhatar et al., 1999; Bouhallab et al., 2002).

Experiments were performed with Caco-2 cells to explain how uncharacterized peptide fractions from milk enhance iron uptake. It was found that these fractions enhanced the dialyzing ability of iron but did not affect the relative expression of gene coding for an intestinal iron transporter (Argyri et al., 2009). Other recent studies in humans failed to demonstrate a positive influence of CPPs on calcium absorption. Teucher et al. (2006) concluded from their study on 15 adults that there was no difference in terms of fractional absorption of calcium between the control group and a group receiving CPPs added in drinks (initial calcium content in test drinks ≅400–500 mg). The same trend was observed by the same research group showing that milk with 120 to 160 mg/100 ml supplemented with commercial CPPs did not significantly increase labeled calcium absorption (Lopez-Huertas et al., 2006). In these two latter studies, commercial- or laboratory-prepared crude whole CPPs fractions were used for supplementation experiments.

Another recent study performed on rats reports on the absorption and deposition of calcium from enriched milks as influenced by either bovine or caprine CPPs. The results indicated that fortification of milk with CPP has a beneficial effect in growing rats through stimulation of calcium bioavailability and deposition but no significant difference was observed between the two types of CPPs (Mora-Gutierrez et al., 2007). It was also reported in a rat model that whole CPPs can both limit the inhibitory effect of free phosphate on calcium availability and increase calcium transport across the distal small intestine. These results are considered to be of nutritional relevance when the calcium intake is lower than recommended (Erba, Ciappellano, and Testolin, 2001).

There are many factors that could explain the inconsistency of data, among them: purity and composition of CPP preparations, mineral to CPP ratio, conformation and aggregation state of CPP/mineral complexes, test meal composition, length of administration in animal or human studies, experimental approaches, and models used. Consequently, control of these factors and exploration of molecular mechanisms have been considered in more recent studies of this field (Ferraretto et al., 2003; Bouhallab et al., 2002; Kibangou et al., 2005; Gravaghi et al., 2007). These studies focused on structure–function relationships, that is, the ability of a given phosphopeptide structure to induce a positive or negative effect on mineral absorption. They highlighted some molecular elements explaining the potential role of CPPs in mineral bioavailability. One of these elements is that well-defined primary structure including the phosphorylated region as well as the N-terminal part is required for a positive effect of CPP from β-casein, β-casein-(1-25), on the uptake of calcium by HT-29 human cells in culture (Ferraretto et al., 2003). The required peptide structure includes the N-terminal portion characterized by the presence of a loop and a β-turn in addition to a specific cluster sequence. Also, distinct effects of various purified CPP on calcium and iron absorption have been demonstrated in cell culture models (Ferraretto et al., 2003; Kibangou et al., 2005). It was shown that αS1-casein-(59-79) (having five phosphate groups) exerted a lowest and modest mineral (Ca^{2+} or Fe^{3+}) uptake by cells as compared to the much more pronounced effect exerted by β-casein-(1-25), having four phosphate groups.

These functional differences were attributed to possible structural differences, aggregation state, and sensitivity to digestive phosphatases. The role of intestinal brush border phosphatase was assessed in particular for iron absorption using a rat duodenal loop model. It was shown that the uptake and absorption of iron bound to the CPP β-casein-(1-25), was enhanced in the presence of a phosphatase inhibitor. Such inhibition prevented the release of free insoluble iron from the CPP/iron complex throughout the action of intestinal phosphatase (Ani-Kibangou et al., 2005). The importance of the aggregation state of CPP in relation to their biological effect was illustrated by the work of Gravaghi et al. (2007) using various CPP preparations (commercial mixtures and purified fractions). This work provides evidence on the requirement of optimal CPP aggregation for calcium uptake by differentiated HT-29 cells *in vitro*. Together with a suitable calcium/CPP ratio, the presence of Ca^{+2}/CPP aggregates in the correct conformation was discussed by the authors as critical conditions to explain the conflicting data of human studies on the efficacy of CPPs as carriers for calcium absorption.

Although nutritional studies are abundant on CPPs from milk caseins, little work has been carried out on products and phosphopeptides from other origins. The work performed by Ishikawa et al. (2007) reported the effect of egg yolk protein on the absorption of divalent cations in rats in comparison with casein or soy proteins. The apparent absorptions of studied cations (calcium, magnesium, and iron) from a yolk protein-based diet were found to be lower than those from casein- or soy protein-based diets. The authors attributed these findings to the higher resistance of phosvitin in yolk to proteolytic action. An earlier study was performed to examine the efficiency of phosvitin tryptic hydrolysate in enhancing calcium absorption. It was found that the rate of intestinal calcium absorption and its accumulation in bones was significantly higher in groups of animals in which phosvitin peptides were added in diets. The authors suggested that phosvitin-derived peptides improved bioavailability of calcium and increased its incorporation into bones (Choi et al., 2005).

5.4.2 Phosphopeptides as Antioxidative Substances

In addition to their role as mineral carriers, another biological interest of caseinophosphopeptides is their protective ability toward oxidation. This property could be used to prevent the side effects of iron fortification, for example. Studies on the antioxidant activity of CPPs in various model systems suggest that CPPs are promising sources of natural antioxidants for foods. The antioxidant properties of casein-derived CPPs was reviewed by Kitts (2005). These properties are linked both to primary and secondary antioxidant activity toward transition ferrous ion sequestering as well as direct free radical quenching activities in both aqueous and lipid emulsion systems. Diaz and Decker (2004) reported that whole CPPs at about 1 mg/ml or more exhibited interesting antioxidant ability in a phosphatidylcholine liposome model system. In a recent study performed in our laboratory, an *in vivo* model of gut peroxidation aimed to confirm the *in vitro* protective effects of CPP from β-casein on iron-induced peroxidation. Compared to ferrous sulphate as control, the results showed an enhancing effect on iron absorption in β-casein CPP group, which was indeed associated with a protective outcome against enterocyte peroxidation (Kibangou et al., 2008). Potential cyto-protective effect of CPPs against stress induced by H_2O_2 coupled with a synergistic result with another antioxidant component from a fruit beverage was also seen using a caco-2 cell model (Laparra et al., 2008). Because CPPs are produced *in vivo* following the action of digestive enzymes on consumed milk, it is likely that their antioxidant property could have an additional health benefit through significant reduction of oxidative stress.

Phosphopeptides from hen egg yolk were also reported to exert protective activity against tissue oxidative stress in intestinal epithelial cells (Katayama et al., 2006). Interestingly, it was shown that phosphopeptides with a high content of phosphorus exhibited higher protection against H2O2-induced oxidative stess than phosphopeptides without phosphorus.

5.4.3 Phosphopeptides as Anticariogenic Substances

The ability of CPPs to stabilize minerals, in particular, calcium phosphate and thereby enhance mineral solubility was also explored to investigate their potential application in the treatment of dental caries. In fact, several groups, including the Reynolds group in Australia have focused their investigations on the protective action of CPPs against dental erosion, known to be caused by acid attack on tooth mineral (Reynolds et al., 2003). CPP-calcium phosphate complexes have been shown to be anticariogenic and then to remineralize early stages of enamel caries in animal as well as human studies. CPPs exert their anticariogenic property through inhibition of lesions from caries and recalcification of dental enamel. The available scientific evidence for the efficacy of calcium/CPPs complex as an anticarcinogenic substance has been recently reviewed by Reynolds (2008). Studies are still in progress to determine the exact mechanism behind this clear biological effect of CPPs and the physicochemical conditions (concentration, pH, etc.) for optimal protective action. A possible formation of a protective coat of CPP over the tooth surface via an adsorption process was proposed as a plausible mechanism (Kanekanian, 2008). Table 5.1 summarizes the recently reported results on the biological role of phosphorylated peptides in relation to mineral nutrition.

5.5 Production and Application of Mineral-Binding Phosphopeptides

The potential health benefits of phosphopeptides have been the subject of increasing commercial interest and development for health-promoting functional foods and ingredients. Tailored dietary formulations, with health claims, are currently being developed worldwide to optimize health through nutrition. CPPs are used as ingredients in commercial products containing casein phosphopeptides, stabilized amorphous calcium phosphate for remineralization systems, as well as for enrichment of yogurt and other milk-based drinks.

Several ion exchange chromatographic methods are available today for the production of enriched fractions of CPPs from protein hydrolysates (Ellegard, 1999), but the production costs of these methods are prohibitive for large-scale operation. Other methods for large preparation purposes of phosphopeptides fractions, such as selective precipitation or ultrafiltration

TABLE 5.1

Different Food Protein Derived Phosphopeptides in Mineral Nutrition[a]

Source	Mean Results	References
Phosvetin Phosphopeptides	Compared to casein, yolk protein decreases calcium and magnesium absorption via the resistance of phosvitin to proteolytic enzymes	Ishikawa et al., 2007
CPPs	CPPs from both bovine and caprine milks exhibit beneficial effect on the absorption of Ca in growing rats	Mora-Gutierrez et al., 2007
CPPs	Increase of zinc and calcium absorption in humans administered rice-based diet	Hansen et al., 1997
CPPs	The aggregation state of CPPs affects their biological effect and calcium uptake by differentiated HT-29 cells	Gravaghi et al., 2007
αs-Casein CPP	The structure and phosphorylation state dramatically affect zinc binding	Wang et al., 2007
CPPs	CPPs added to milk are unsuitable as ingredients to deliver improved calcium nutrition	Teucher et al., 2006; Lopez-Huertas et al., 2006
CPPs	CPPs limit the inhibitory effect of free phosphate on calcium absorption	Erba et al., 2001
Phosvitin-derived peptides	Improvement of calcium bioavailability and its incorporation into bones	Choi et al., 2005
β-casein CPP	Improvement of iron bioavailability in rat	Bouhallab et al., 2002
β-casein CPP	Improvement of iron absorption and storage in human	Aït-Oukhatar et al., 2002
β-casein CPP	Increase of calcium uptake by HT-29 human cells	Ferraretto et al., 2003
β-casein or αS-casein CPP	β-casein CPP allowed better absorption and uptake of calcium or iron than did αS-casein CPP	Ferraretto et al., 2003; Kibangou et al., 2005
CPPs	CPPs protect caco-2 cell against stress induced by H_2O_2	Laparra et al., 2008
β-casein CPP	Induction of a protective effect against enterocyte peroxidation	Kibangou et al., 2008
CPPs	CPPs are promising sources of natural antioxidants for foods	Diaz et al., 2004
CPPs	Anticarcinogenic effect in animal and human studies	Reynolds et al., 2003
CPPs	Act as anticarcinogenic compounds, CPPs form a protective coat over the tooth surface	Kanekanian et al., 2008

[a] Absorption, uptake, and protective effects.

have been described. Selective precipitation of phosphorylated peptides from peptide mixtures was obtained using an adequate mix of ethanol and divalent cations (barium, calcium). This method was optimized by Adamson and Reynolds (1995) for large-scale production of whole CPPs. The ultrafiltration technology, combined with complexation of CPPs by minerals, seems to be the more adequate technique for large isolation of CPPs inasmuch as it leads to very high yields (Brulé and Fauquant, 1982). This membrane technology has now been well developed for industrial scale operation by different international companies for the production of commercially available CPPs mixtures.

Recent studies support the need to use purified, well-characterized CPP molecules for better control and reproducibility of their biological and nutritional impact. Adaptation and development of production processes and technologies for specific isolation of pure phosphopeptide on a large scale and at low cost, still need to be performed.

5.6 Conclusion

The major part of published studies clearly supports the fact that CPPs have wide applications as nutraceuticals. Phosphopeptides as mineral-binding peptides, formed *in vitro* and *in vivo*, have emerged as multifunctional compounds with different promising biological applications. However, most of the available data concerning the bioactivity of these molecules are the results of *in vitro*, *ex vivo*, and animal studies. More clinical and human studies are necessary to demonstrate and to prove the potential of CPPs and derived peptides in (1) enhancing dietary mineral bioavailability under given specific physiological conditions; (2) protecting efficiency toward oxidation *in vivo*; and (3) modulating bone formation and inhibiting caries development. Simultaneously, it is now evident that deeper structure/activity relationships need to be established to link the role of mineral-induced conformational changes to the biological function of phosphopeptides with various amino acid sequences and degree of phosphorylation. Also, as demonstrated from recent research papers, the degree of aggregation of phosphopeptides in the presence of divalent cations must be taken into consideration as this phenomenon highly affects the biological response. Finally, low cost processes for the production of pure, well-characterized phosphopeptide from food protein have to be developed and studied in detail.

References

Adamson, N.J. and Reynolds, E.C. 1995. Characterization of tryptic casein phospho-peptdes prepared under industrially relevant conditions. *Biotechnol. Bioeng.* 45: 196–204.

Ait-Oukhatar, N., Bouhallab, S., Arhan, P., Maubois, J.L., Drosdowsky, M., and Bouglé, D. 1999. Iron tissue storage and hemoglobin levels of deficient rats repleted with iron bound to the caseinophosphopeptide 1-25 of beta-casein. *J. Agric. Food Chem.* 47: 2786–2790.

Ait-Oukhatar, N., Peres, J.M., Bouhallab, S., Neuville, D., Bureau, F., Bouvard, G. et al. 2002. Bioavailability of caseinophosphopeptide-bound iron. *J. Lab. Clin. Med.* 140: 290–294.

Ani-Kibangou, B., Bouhallab, S., Mollé, D., Henry, G., Bureau, F., Neuville, D. et al. 2005. Improved absorption of caseinophosphopeptide-bound iron: Role of alkaline phosphatise. *J. Nutr. Biochem.* 16: 398–401.

Argyri, K., Tako, E., Miller, D.D., Glahn, R.P., Komaitis, M., and Kapsokefalou, M. 2009. Milk peptides increase iron dialyzability in water but do not affect DMT-1 expression in Caco-2 Cells. *J. Agric. Food. Chem.* 57: 1538–1543.

Bouhallab, S. and Bouglé, D. 2004. Biopeptides of milk: Caseinophosphopeptides and mineral bioavailability. *Reprod. Nutr. Dev.* 44: 493–498.

Bouhallab, S., Cinga, V., Ait-Oukhatar, N., Bureau, F., Neuville, D., Arhan, P. et al. 2002. Influence of various phosphopeptides of caseins on iron absorption. *J. Agric. Food Chem.* 50: 7127–7130.

Brulé, G. and Fauquant, J. 2002. Interactions des protéines du lait et des oligoélé-ments. *Lait* 62: 323–331.

Choi, I., Jung, C., Choi, H., Kim, C., and Ha, H. (2005). Effectiveness of phosvitin peptides on enhancing bioavailability of calcium and its accumulation in bones. *Food Chem.* 93: 577–583.

Cirulli, C., Chiappetta, G., Marino, G., Mauri, P., and Amoresano, A. 2008. Identification of free phosphopeptides in different biological fluids by a mass spectrometry approach. *Anal. Bioanal. Chem.* 392: 147–159.

Cross, K.J., Huq, N.L., Bicknell, W., and Reynolds, E.C. 2001. Cation-dependent structural features of β-casein-(1–25). *Biochem. J.* 356: 277–286.

Cross, K.J., Huq, N.L., Palamara, J.E., Perich, J.W., and Reynolds, E.C. 2005. Physicochemical characterization of casein phosphopeptide-amorphous calcium phosphate nanocomplexes. *J. Biol. Chem.* 280: 15362–15369.

Diaz, M. and Decker, E.A. 2004. Antioxidant mechanisms of caseinophosphopep-tides and casein hydrolysates and their application in ground beef. *J. Agric. Food Chem.* 52: 8208–8213.

Diplock, A.T, Aggett, P.J., Ashwell, M., Bornet, F., Fern E.B., and Roberfroid, M.B. 1999. Scientific concepts of functional foods in Europe Consensus document. *Br. J. Nutr.* 81 (Suppl 1): S1–S27.

Ellegard, K.H., Gammelgad-Larsen, C., Sørensen, E.S., and Fedosov, S. 1999. Process scale chromatographic isolation, characterization and identification of tryptic bioactive casein phosphopeptides. *Int. Dairy J.* 9: 639–652.

Erba, D., Ciappellano, S., and Testolin, G. 2001. Effect of caseinphosphopeptides on inhibition of calcium intestinal absorption due to phosphate. *Nutr. Res.* 21: 649–656.

Ferraretto, A., Gravaghi, C., Fiorilli, A., and Tettamanti, G. 2003. Casein-derived bioactive phosphopeptides: Role of phosphorylation and primary structure in promoting calcium uptake by HT-29 tumor cells. *FEBS Lett.* 551: 92–98.

FitzGerald, R.J. 1998. Potential uses of caseinophosphopeptides. *Int. Dairy J.* 8: 451–457.

González-Chávez, S.A., Arévalo-Gallegos, S., and Rascón-Cruz, Q. 2008. Lactoferrin: Structure, function and applications. *Int. J. Antimicrob. Agents.* doi:10.1016 /j.ijantimicag. 2008.07.020.

Gravaghi, C., Del Favero, E., Cantu, L., Donetti, E., Bedoni, M., Fiorilli, A. et al. 2007. Casein phosphopeptide promotion of calcium uptake in HT-29 cells: Relationship between biological activity and supramolecular structure. *FEBS J.* 274: 4999–5011.

Hansen, M., Sandström, B., Jensen, M., and Sorrensen, S.S. (1997). Casein phosphopeptides improve zinc and calcium absorption from rice-based but not from whole-grain infant cereal. *J. Pediatr. Gastr. Nutr.* 24: 56–62.

Holt, C. 1992. Structure and stability of bovine casein micelles. *Adv. Protein Chem.* 43: 63–151.

Ishikawa, S.I., Tamaki, S., Arihara, K., and Itoh, M. 2007. Egg yolk protein and egg yolk phosvitin inhibit calcium, magnesium, and iron absorptions in rats. *J. Food Sci.* 72: S412–S419.

Kanekanian, A.D., Williams, R.J.H., Brownsell, V.L., and Andrews, A.T. 2008. Caseinophosphopeptides and dental protection: Concentration and pH studies. *Food Chem.* 107: 1015–1021.

Katayama, S., Xu, X., Fan, M.Z., and Mine, Y. 2006. Antioxidative stress activity of oligophosphopeptides derived from hen egg yolk phosvitin in Caco-2 Cells. *J. Agric. Food Chem.* 54: 773–778.

Kibangou, I., Bouhallab, S., Bureau, F., Allouche, S., Thouvenin, G., and Bouglé, D. 2008. Caseinophosphopeptide-bound iron: Protective effect against gut peroxidation. *Ann. Nutr. Metab.* 52: 177–180.

Kibangou, I.B., Bouhallab, S., Henry, G., Bureau, F., Allouche, S., Blais, A. et al. 2005. Milk proteins and iron absorption: Contrasting effects of different caseinophosphopeptides. *Pediatr. Res.* 58: 731–734.

Kitts, D.D. 2005. Antioxidant properties of caseinphosphopeptides. *Trends Food Sci. Technol.* 16: 549–554.

Kitts, D.D. 2006. Calcium binding peptides. In *Nutraceutical Proteins and Peptides in Health and Disease* (Y. Mine, and F. Shahidi, Eds.), Boca Raton, FL: Taylor & Francis, pp. 11–27.

Kocher, T., Allmaier, G., and Wilm, M. 2003. Nanoelectrospray-based detection and sequencing of substoichiometric amounts of phosphopeptides in complex mixtures. *J. Mass Spectrom.*, 38: 131–137.

Korhonen, H. and Pihlanto, A. 2006. Bioactive peptides: Production and functionality. *Int. Dairy J.* 16: 945–960.

Laparra, J.M., Alegria, A., Barbera, R., and Farré, R. 2008. Antioxidant effect of casein phosphopeptides compared with fruit beverages supplemented with skimmed milk against H_2O_2-induced oxidative stress in caco-2 cells. *Food Res. Int.* 41: 773–779.

Lonnerdal, B. 2003. Lactoferrin. In *Advanced Dairy Chemistry, Proteins Part A* (P.F. Fox and P.L.H. McSweeney, Eds.), New York: Kluwer Academic/Plenum, pp. 449-466.

Lopez-Huertas, E., Teucher, B., Boza, J.J., Martinez-Férez, A., Majsak-Newman, G., Baro, L. et al. 2006. Absorption of calcium from milks enriched with fructooligosaccharides, caseinophosphopeptides, tricalcium phosphate, and milk solids. *Am. J. Clin. Nutr.* 83: 310–316.

Martin, R., Larsen, M.R., Thingholm, T.E., Jensen, O.N., Roepstorff, P., and Jørgensen, T.J.D. 2005. Highly selective enrichment of phosphorylated peptides from peptide mixtures using titanium dioxide microcolumns. *Mol. Cell. Proteomics* 4: 873–886.

Meisel, H. and FitzGerald, R.J. (2003). Biofunctional peptides from milk proteins: Mineral-binding and cytomodulatory effects. *Curr. Pharm. Des.* 9: 1289–1295.

Miquel, E. and Farré, R. 2007. Effects and future trends of casein phosphopeptides on zinc bioavailability. *Trends Food Sci. Technol.* 18: 139–143.

Mora-Gutierrez, A., Farrell, H.M., Attaie, R., McWhinney, V.J., and Wang, C.Z. 2007. Influence of bovine and caprine casein phosphopeptides differing in alpha s1-casein content in determining the absorption of calcium from bovine and caprine calcium-fortified milks in rats. *J. Dairy Res.* 74: 356–366.

Permyakov, E.A. and Berliner, L.J. 2000. α-Lactalbumin: Structure and function. *FEBS Lett.* 473: 269–274.

Pinkse, M.W., Uitto, P.M., Hilhorst, M.J., Ooms, B., and Heck, A.J. 2004. Selective isolation at the femtomole level of phosphopeptides from proteolytic digests using 2D-NanoLC-ESI-MS/MS and titanium oxide precolumns. *Anal. Chem.* 76: 3935–3943.

Reynolds, E.C. 2008. Calcium phosphate-based remineralization systems: Scientific evidence? *Austral. Dent. J.* 53: 268–273.

Reynolds, E.C., Cai, F., Shen, P., and Walker, G.D. 2003. Retention in plaque and remineralisation of enamel lesions by various forms of calcium in a mouth rinse or sugar-free chewing gum. *J. Dent. Res.* 82: 206–211.

Roberfroid, M.B. 2000. Concepts and strategy of functional food science: The European perspective. *Am. J. Clin. Nutr.* 71: 1660S–1664S.

Steijns, J.M. and van Hooijdonk, A.C. 2000. Occurence, structure, biochemical properties and technological characteristics of lactoferrin. *Br. J. Nutr.* 84 (Suppl 1): S11–S17.

Stevens, L. 1991. Egg white proteins. *Comp. Biochem. Physiol. B* 100: 1–9.

Swaisgood, H.E. 2003. Chemistry of caseins. In *Advanced Dairy Chemistry, Proteins Part A* (P.F. Fox and P.L.H. McSweeney, Eds.), New York: Kluwer Academic/Plenum, pp. 139–202.

Teucher, B., Majsak-Newman, G., Dainty J.R., McDonagh, D., FitzGerald, R.J., and Fairweather-Tait, S. 2006. Calcium absorption is not increased by caseinophosphopeptides. *Am. J. Clin. Nutr.* 84: 162–166.

Veprintsev, D.B., Permyakov, E.A., Kalinichenko, L.P., and Berliner, L.J. 1996. Pb^{2+} and Hg^{2+} binding to α-lactalbumin. *Biochem. Mol. Biol. Int.* 39: 1255–1265.

Wang, J., Green, K., McGibbon, G., and McCarry, B. 2007. Analysis of effect of casein phosphopeptides on zinc binding using mass spectrometry. *Rapid Commun. Mass Spectrom.* 21: 1546–1554.

Webb, J., Multani, J.S., Saltman, P., Beach, N.A, and Gray, H.B. 1973. Spectroscopic and magnetic studies of iron (III) phosvitins. *Biochemistry* 12: 1797–1802.

6

Food Peptides as Antihypertensive Agents

Amaya Aleixandre and Marta Miguel

CONTENTS

6.1 Introduction

Some fragments of the sequence of food proteins may be released by hydrolysis, and, once released, they show biological activity. These fragments, or bioactive peptides, are usually generated *in vivo* by the action of gastrointestinal enzymes. They can also be obtained *in vitro* using specific enzymes, or be produced during the manufacture of certain foods.

Since their discovery in 1979, bioactive peptides with different biological activity have been described (Gobbeti et al., 2000). Some are able to produce a decrease in arterial tone and control hypertension. This illness is defined as a chronic increase in systolic (SBP) or diastolic blood pressure (DBP) to above normal values. Hypertension is a significant health problem in today's society. There can be no doubt that its presence has far-reaching implications, as it is so widespread, and also because it is a principal risk factor in cardiovascular disease. It is, in fact, the primary cause of death in developed countries. It is estimated that, for every decrease of 5 mm Hg in DBP, the risk of cardiovascular disease drops by approximately 16% (Collins et al., 1990; MacMahon et al., 1990). For these reasons, health experts are trying to make the treatment of hypertension more effective and to begin the treatment earlier on. Recently the figures for what constitutes normal arterial blood pressure have been revised downward. Following the current criteria, many individuals who were previously considered normotensive are now described as prehypertensive. The new guidelines (National Heart, Lung and Blood Institute, 2003) for the classification of degrees of blood pressure are as follows:

Normal: PAS <120 mm Hg/PAD <80 mm Hg

Prehypertension: PAS 120–139 mm Hg/PAD 80–90 mm Hg

Hypertension stage 1: PAS 140–159 mm Hg/PAD 90–99 mm Hg

Hypertension stage 2: PAS >160 mm Hg/PAD >100 mm Hg

Primary hypertension, also known as idiopathic or essential hypertension, which develops with no obvious organic cause, is the most common (90–95% of hypertensive patients have primary hypertension). Genetic, environmental, psychosocial, and dietary factors all play an important part in its development. Although its detection and control are relatively straightforward, in truth many hypertensive patients are unaware of their condition, and others are diagnosed but receive inappropriate treatment. In Spain, for example, only between 5 and 10% of hypertensive patients receive appropriate treatment (Wolf-Maier et al., 2004). Nevertheless, the annual cost of antihypertensive treatment is high. Taking into account both direct and indirect costs, Spain annually spends almost 3 billion euros on tackling the illness (Rodicio Díaz, 2000). The cost of drugs used in the treatment of arterial hypertension and associated pathologies in the United States is also considerable. In 2004 it amounted to more than 15 billion dollars (Fischer and Avorn, 2004). Nonpharmacological strategies for controlling arterial blood pressure are therefore highly desirable. Obviously, one such strategy which is currently proving of great interest is the research into antihypertensive peptides of food origin. Functional foods containing these may in fact represent a new strategy for the prevention or treatment of hypertension.

Antihypertensive peptides have been obtained from food protein of both animal and plant origin. The principal ones are antihypertensive peptides derived from milk proteins. This is important, as milk is a primary food. Eggs, also a fundamental product in today's society, and other frequently consumed foods such as fish, maize, soy, and various vegetables, may also be an important source of antihypertensive peptides. The present chapter relates the most important findings in current research into hydrolysates and antihypertensive peptides derived from food proteins. The possible mechanisms of action of these peptides are discussed, and the strategies used in their selection and evaluation are described. The prospects for the use of these products are also analyzed.

6.2 Mechanisms Involved in the Antihypertensive Effect of Food Peptides

6.2.1 Inhibition of Angiotensin-Converting Enzyme

The main mechanism involved in the antihypertensive effect of food-derived peptides is the inhibition of the angiotensin-converting enzyme (ACE). The activity of this enzyme is decisive for the effectiveness of a system, namely the renin–angiotensin system, which plays a crucial part in the maintenance of arterial blood pressure, and in the organic damage that occurs when this variable is increased. The change in arterial tone caused by the renin–angiotensin system may in reality be critical in some hypertensive patients. Below we set out the biochemical routes of this system, commenting on the physiological importance of the peptides that form within it. Figure 6.1 shows a diagram of these routes and peptides. In this section, we also present several concepts regarding the kinin system, as the ACE is also active within this system. In addition, we establish the structural requirements of the compounds and peptides that inhibit ACE.

Renin is a glucoprotein composed of 350 amino acids, with protease activity, that shows high substratum specificity. It is synthesized and stored in specialized muscular cells, known as the juxtaglomerular cells, situated in the wall of the efferent renal artery. Its secretion is controlled by a range of factors, the most important being the decrease in renal perfusion pressure when arterial blood pressure is lowered, and a reduction in the renal level of sodium. Figure 6.2 shows a diagram of physiological regulation for the secretion of renin in the renin–angiotensin system. Although renin is not in itself a pressor substance, it has the ability to initiate the formation of an active peptide from a proteinic substance known as angiotensinogen. Angiotensinogen is synthesized in the liver, is present in the α_2 globulinic plasma fraction, and is in reality the only substratum for renin. It is made up of glucoproteins that contain a peptide residue of 14 amino acids, of which the first 10 correspond to the angiotensin sequence. This sequence is released when renin acts on angiotensinogen at the level of Leu–Leu link, but the decapeptide thus released, known as angiotensin I, has no activity. ACE, a plasmatic glycoprotein, catalyzes the conversion of angiotensin I into angiotensin II, an octapeptide with potent vasoconstrictor activity.

ACE is a nonspecific *trans*-membrane metaloenzyme, containing Zn. Its function is to separate carboxi-terminal dipeptides from different proteinic

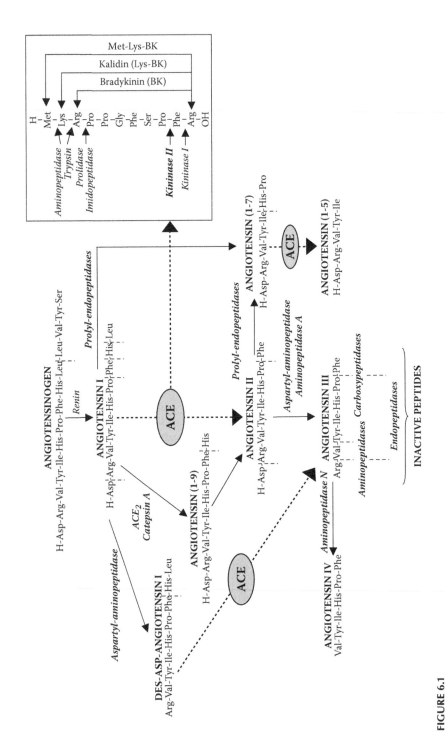

FIGURE 6.1
Biochemical routes and peptides of the rennin-angiotensin system.

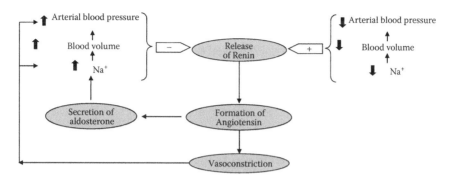

FIGURE 6.2
Physiological regulation for the secretion of rennin in the rennin–angiotensin system.

substrata, and for this reason is known as peptidildipeptidase. It promotes the separation of the angiotensin I carboxi-terminal dipeptide by acting on the Phe–His link of this compound (see Figure 6.1). Human ACE appears in two forms. One of these, somatic ACE, has a molecular weight of 150–180 kDa, and contains two homologous domains which, in accordance with their position in the chain, are known as N-terminal, or N-domain, and C-terminal, or C-domain. Both domains contain Zn at the binding point and an active center (Deddish et al., 1996). The other form of ACE has low molecular weight (90–100 kDa), is found in the testicle (germinative or testicular ACE), and only contains the C-domain (Bree, Hamon, and Tillement, 1992; Williams et al., 1992).

Angiotensin II acts by binding itself to specific receptors. Of these, the best known are the AT_1 and AT_2 receptors. The characteristic actions of angiotensin II are mediated by the AT_1 receptors, whose activation promotes, among other things, a rise in intracellular calcium concentration, with an increased contraction of cardiac muscle and arteriovenous tone. The activation of these receptors also stimulates the synthesis and release of aldosterone in the suprarenal cortex.

Angiotensin III is the first metabolite of angiotensin II. It is formed when aspartil-aminopeptidase splits the amino acid Asp from the extreme amino-terminal of angiotensin II. For this reason it is also called des-Asp-angiotensin II. This heptapeptide still maintains an important physiological activity, and also has an affinity for AT_1 and AT_2 receptors. It may be responsible for some of the effects observed when angiotensin II is administered, but its properties have not yet been completely explained, and may vary in different tissues. It is known that angiotensin III is the principal peptide effector in the cerebral renin–angiotensin system. This compound may exercise central tonal control over arterial blood pressure. In brain tissue, the main enzyme responsible for the formation of angiotensin III is aminopeptidase A. A way exists of synthesizing angiotensin III without forming angiotensin II. It involves the enzymes acting in a different order. Angiotensin I is hydrolyzed first with an aspartil-aminopeptidase, thereby forming the nonapeptide des-Asp-angio-tensin II. Next, ACE acts on des-Asp-angiotensin I, and angiotensin III is formed (see Figure 6.2; Gaynes, Szidon, and Oparil, 1978; Sexton et al., 1979; García del Río, Smellie, and Morton, 1981).

The attack on angiotensin III by the aminopeptidase N produces another active metabolite composed of six amino acids, angiotensin IV. Angiotensin IV may interact with the classic angiotensin II receptors, AT_1 and AT_2, but specific binding points have been identified for angiotensin IV, namely the AT_4 receptors (Swanson et al., 1992). These receptors are found in differ-ent tissues (brain, cardiac membrane, kidney, and human collector conduit cells). They also appear in some cell cultures. They are widely distributed in the central nervous system, where their activation by angiotensin IV seems to be linked to the processes of memory, learning, and neurone develop-ment. A functional role in regulation of blood flow in different tissues is also attributed to angiotensin IV. In the lung, this peptide would act as a vascular relaxant, as in the endothelial cells of this organ, the angiotensin IV recep-tor mediates the activation of nitric oxide (NO) synthase (Patel et al., 1998). In other vascular layers, it has been observed that angiotensin IV acts as a vasoconstrictor, via AT_1 receptors (Gardiner et al., 1993; Garrison et al., 1995; Loufrani et al., 1999). The subsequent attack on angiotensins III and IV by amino- and carboxylpeptidases produces inactive peptides (Song and Healy, 1999; Turner and Hooper, 2002).

The kinins are peptides formed from substrata known as kininogens, which are present in the plasma, lymph, and interstitial fluid of mammals. They are formed by the action on these substrata of a group of serum pro-teases similar to other well-known enzymes such as trypsin, trombine, or plasmine. The two best-known kinins are calidine, also known as Lys-bradykinin, with 10 amino acids, and bradykinin, a potent vasodilator con-taining 9 amino acids. Apparently another kinin with 11 amino acids exists, called Met-Lys-bradykinin. These kinins are inactivated by different plasma and tissue enzymes, that is to say, enzymes found in the blood or in cell membranes. The attack of the kinins in the extreme amino-terminal cannot be truly considered an inactivation, as the aminopeptidases that act in this extreme release the amino acids Met and Lys, and permit the formation of calidine from Met-Lys-bradykinin, and of bradykinin from calidine. These aminopeptidases are similar to trypsin. Both prolidases and imidopepti-dases might act on the Arg–Pro link of bradykinin, but it is doubtful whether these enzymes participate physiologically in the inactivation of bradykinin. It is the proteolytic enzymes that act on the level of the C-terminal group of kinins which are, in reality, in charge of the inactivation of these peptides. Two enzymes are known that act on this group, namely kinase I and kinase II. Both are metaloproteins. Kinase I produces active metabolites, and it is thought that kinase II, responsible for supplying biologically inactive metab-olites, is identical to ACE. ACE therefore is also responsible for the hydroly-sis and inactivation of bradykinin and other potent vasodilatory peptides (see Figure 6.1). Bradykinin exercises its vasodilatory effect via the endothelial B_2 receptors which mediate the synthesis and release of prostaglandins, NO, and endothelium-derived hyperpolarizing factor (Bernier, Haldar, and Michel, 2000; Tom et al., 2002; Landmesser and Drexler, 2006).

The inhibition of ACE presupposes the inhibition of the formation of different vasoconstrictor compounds (including angiotensin II, which is the most potent pressor substance we know of), and the inhibition and degradation of different vasodilatory substances (including bradykinin, the most potent vasodilator we know of). Nevertheless, when ACE is inhibited, the production of angiotensin II is not completely blocked. This is in part due to the conversion of angiotensin I into angiotensin II by the action of different chimases. Of particular importance is the chimase that hydrolyzes angiotensin I, isolated in mastocytes and in the endothelial cells of the human heart (Husain, 1993). In the left ventricle of the human heart, the formation of angiotensin II by the action of this chimase seems to be more significant than the formation of this compound by the action of ACE (Urata, Nishimura, and Ganten, 1996; Song and Healy, 1999; Turner and Hooper, 2002).

It should also be remembered that, in addition to the classic route, there are other enzymatic routes for angiotensin II synthesis. A bridging route has been postulated for the formation of angiotensin II avoiding ACE, in which angiotensin 1-9 is formed first (see Figure 6.1). This route is of greater importance in tissues such as the heart, blood vessels, and nervous system. Thus, in the human heart, catapsin A and ACE_2 are responsible for converting angiotensin I into angiotensin 1-9 (Donoghue et al., 2000). Other metalopeptidases (prolil-endopeptidases) produce angiotensin 1-7 from angiotensin I. Angiotensin 1-7 is a peptide that can also be generated from angiotensin II. Angiotensin 1-7 can be considered a paracrine hormone that negatively counterbalances the actions of angiotensin II in the cardiovascular system, kidney, and central nervous system. It has antiproliferative and vasodilatory effects that are mediated by the release of NO and prostaglandins (Almeida et al., 2000). ACE also has the property of processing angiotensin 1-7 into angiotensin 1-5, and it does this 10 times more quickly via the N-dominion than via the C-dominion (Deddish et al., 1998), which proves that the two ACE dominions can perform different functions. The concentration of angiotensin 1-7 increases significantly during the administration of ACE-inhibitors, and it is thought that the increase in this peptide is linked to the beneficial effect of these compounds. Angiotensin 1-5 barely appears in plasma. It does not participate in the modulation of arterial blood pressure and its function appears to be purely central.

Traditionally the renin–angiotensin system was thought to be a circulating system with endocrinal action, but today we know that most of its components express themselves in varying degrees depending on the tissue. This system performs a range of functions in these tissues. We know that the vasculature in particular can synthesize and secrete angiotensin II. This peptide has a local paracrine/autocrine/intracrine effect on vascular function. Nevertheless, it does not seem that local synthesis of angiotensin II contributes greatly to the production of this peptide in vessels. Allusion has been made to an internalization of plasmatic renin in the vascular wall. It is thought that renin and angiotensin II of systemic origin control, above all, acute vascular functions, such as vascular tone and blood pressure. The local renin–angiotensin system would, however, be responsible for tissue maintenance and repair.

ACE-inhibitors were discovered in the venom of some snakes as inhibitory peptides of chimase II which boosted bradykinin (Cushman et al., 1973; Ondetti, Rubin, and Cushman, 1977). These peptides, which had between 5 and 13 amino acids, were first isolated and then chemically synthesized. They were potent and specific, but they did not prove to be ideal compounds for the inhibition of ACE, as their molecules were relatively too large for enzymatic interaction, and they were not active when administered orally. The most active was a peptide named teprotide, with 9 amino acids. Its interaction with ACE, as well as the interaction of angiotensin I with the same enzyme, is shown in Figure 6.3.

Later came the synthesis of captopril, an octapeptide that specifically inhibited the enzyme and also had a more suitable structure and proved active when administered orally. This compound headed a pharmacological group, known properly as ACE-inhibitors, which at present is of great importance in the treatment of hypertension. The drugs in this group work by interacting with the Zn group which contains ACE in its active center. This is also the binding point of the enzyme with angiotensin I. Figure 6.3 shows the interaction of captopril and angiotensin with ACE. The inhibitors of the enzyme prevent the transformation of angiotensin I into angiotensin II and block the renin–angiotensin–aldosterone system, but they do not impede the actions of angiotensin II. Some of the members of this group contain a sulphidryl group, and are structurally related to captopril. Others present a different structure, and many are inactive prodrugs that have better bioavailability, but require an esterase to act on them *in vivo* in order to generate the active compound. It has recently been established that ACE-inhibitors are also able to stimulate NO synthesis by means of a direct activation of B_1 receptors, which are expressed particularly in pathological situations (Marceau et al., 1995, 1997; Ni, Chao, and Chao, 1998).

We do not know the exact structural requirements of the peptides of food origin that inhibit ACE. The inhibition of the enzyme seems to involve the interaction of these peptides with an anionic zone distinct from the catalytic site of the enzyme, or with subsites of the enzyme not normally occupied by other substrata (Meisel, 1997). These peptides are usually sequences of between 2 and 12 amino acids, although some contain up to 27 amino acids. The structure–activity relationship seems to be clearer with those containing less than 7 amino acids. The binding with the enzyme is specifically conditioned by the carboxi-terminal tripeptide sequence of the peptides, which may interact with three regions of the active center of ACE. This bond is

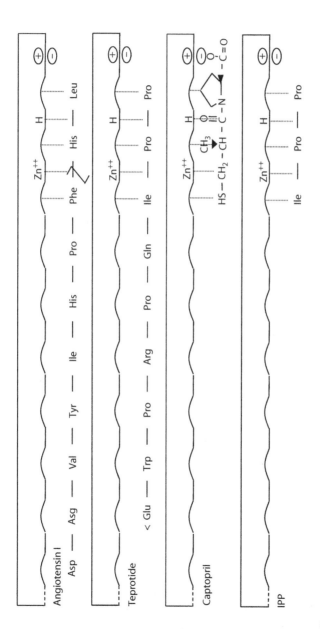

FIGURE 6.3
Interaction with angiotensin converting enzyme of angiotensin I, teprotide, captopril, and the food-derived peptide IPP (Ile-Pro-Pro).

made more favorable by the presence of hydrophobic amino acids in these last three positions. The residues Trp, Tyr, Phe, and Pro are those that in principle most favor the bond (Ondetti and Cushman, 1982).

What is most important and decisive for the bond is that the amino acid Pro is located right at the carboxylic extreme (Rohrbach, Williams, and Rolstad, 1981). It should be borne in mind that the rigid structure of the Pro ring means that the orientation of the carboxyl group is highly favorable for interaction with the enzyme (Cushman et al., 1977). The presence of the residue Pro in the penultimate position of the carboxyl extreme does not however appear to favor the bond (Cheung et al., 1980). The carboxi-terminal sequence Phe-Ala-Pro (FAP), analogous to that shown by the ACE-inhibitor first found in snake venom extracts, is one of the most favorable to bind the enzyme's catalytic site. Peptides may adopt different configurations, and this is also significant. The change in the carboxi-terminal position from the *trans-* form to the *cis-* form of the amino acid Pro may cause important alterations in activity (Gómez Ruiz, Ramos, and Recio, 2004a). It has also been suggested that the inhibitory activity of ACE increases due to the presence of the amino acid Leu in the carboxi-terminal extreme (Kim et al., 2001; Gómez-Ruiz, Recio, and Belloque, 2004b). Likewise it increases due to the presence in this extreme of positively charged amino acids such as Lys with the ε-amino group, and Arg with the guanidine group. The presence of an aromatic amino acid in the antepenultimate position also favors the bond with ACE (Cheung et al., 1980). ACE by contrast shows little affinity with the substrata which present dicarboxyl amino acids in its carboxyl extreme (Erdös, 1976; Cushman et al., 1977; Cheung et al., 1980). The amino-terminal extreme of the peptide may also influence the inhibitory activity of ACE. It is increased by the presence of Val or Ile in this position (Cheung et al., 1980). The conformation D or L is also highly significant for the inhibitory activity of ACE. Several L enantiomers present up to 100 times less activity than the corresponding D enantiomers (Cushman et al., 1977).

The inhibition of the formation of angiotensin II *in vitro*, which reflects the interaction of a compound with ACE under these conditions, is a common test for evaluating antihypertensive drugs. The potency for ACE inhibition is usually expressed by the concentration or dose of a compound necessary to inhibit 50% of the enzyme's activity (IC_{50}). Captopril, a drug considered the prototype for ACE inhibition, presents an IC_{50} value close to 0.02 μM (Suetsuna, 1998; Fujita and Yoshikawa, 1999; Matsui, Li, and Osajima, 1999). In the case of hydrolysates and peptides derived from food proteins, spectrophotometric, fluorimetric, chromatographic, and capillary electrophoresis techniques have been used to measure their ability to inhibit ACE *in vitro*. Once their IC_{50} value has been established, this also represents an approximation of the possible antihypertensive effect of these compounds (Li et al., 2005). One of the most common methods for obtaining the IC_{50} value of these peptides is that developed by Cushman and Cheung (1971) and later modified by Nakamura et al. (1995a). This method uses the compound hypuril histidil leucine instead of angiotensin I as a substrate to react with ACE. When ACE reacts with angiotensin I, the dipeptide His–Leu (HL) and angiotensin II are released. When this enzyme acts on this other substrate, it releases the same dipeptide and hypuric acid. The amount of hypuric acid released, reflecting ACE activity, is quantified after the acid has been extracted with ethyl acetate. In order to establish the ACE inhibition that produces a compound, another spectrophotometric technique has been used, based on the release of the dipeptide Gly-Gly (GG) when the enzyme reacts with the furanocriloil derivative of the tripeptide Phe-Gly-Gly (FGG) (Vermeirssen, Van Camp, and Verstraete, 2002). Attempts have also been made to improve this method by introducing certain modifications (Murray, Walsh, and FitzGerald, 2004). It should, however, be remembered that none of the procedures used is exact, and that the results they provide are not homogeneous, as each method differs in sensitivity.

It is also important to note that small peptides from food origin might also control blood pressure by antagonizing the proliferative effect of angiotensin II. In this context, Toshiro et al. (2005) suggested that the dipeptide Val-Tyr inhibited the proliferation of vascular smooth muscle cells by serving as a natural L-type Ca^{2+} channel blocker.

6.2.2 Vascular Effects

The antihypertensive effect of some peptides derived from food proteins is greater than the effect predicted by the potency presented *in vitro* by these same peptides for ACE inhibition. One such case is the peptide Leu-Lys-Pro-Asn-Met (LKPNM), obtained by the digestion with thermolysin of "Katsuobushi," a traditional Japanese dish made from dried bonito. Another such peptide is Leu-Lys-Pro (LKP), produced when ACE hydrolyzes the sequence LKPNM. The IC_{50} values of these peptides and their antihypertensive effect in hypertensive rats have been compared with those of captopril (Fujita and Yoshikawa, 1999). Captopril had an IC_{50} value of 0.022 µM, and its minimum effective dose was 1.25 mg/kg. The peptide LKPNM, with a value of $IC_{50} = 2.4$ µM, significantly higher than that of captopril, was considered an ACE inhibitor similar to the prodrugs that inhibit the enzyme. Its minimum effective dose was 8 mg/kg, and its maximum effect, as with captopril, was observed four hours after oral administration. The peptide LKP, with a value of $IC_{50} = 0.32$ µM, also greater than that of captopril, showed its maximum effect two hours after oral administration, and presented a minimum effective dose of 2.25 mg/kg. Using all these measurements, it was established that both peptides, LKPNM and LKP, showed antihypertensive effects clearly greater than what their ACE-inhibitory potency *in vitro* had led the researchers to expect. This fact was initially linked to these peptides' tissular affinity and their slow elimination. It was later suggested that additional mechanisms other than ACE-inhibition might collaborate in the antihypertensive effect of these and other food-derived peptides.

Today we know that, in reality, some peptides derived from food proteins, including dried bonito derivatives, present direct effects in vascular smooth muscle (Kuono et al., 2005). In many other cases, the antihypertensive effect observed when food-derived peptides are administered can be explained by the fact that these compounds are able to modulate the release from the endothelium of factors that relax or contract vascular smooth muscle. These peptides therefore produce an endothelium-dependent vascular effect. In a tuna hydrolysate, for example, peptides were found that inhibit the endothelin-converting enzyme (Okitsu et al., 1995). This enzyme is necessary for the synthesis of endothelin, a potent endothelial vasoconstrictor (Yanagisawa et al., 1988). Similarly, lactokinin, a peptide corresponding to the sequence Ala-Leu-Met-Pro-His-Ile-Arg (ALMPHIR), inhibits the release of endothelin in the endothelium (Maes et al., 2004).

When interpreting the effect of lactokinin it should, however, be borne in mind that proteinases may reduce the quality of this peptide when administered orally, and that when they act, these enzymes also generate products with ACE-inhibitory activity (Murakami et al., 2004; Walsh et al., 2004). Usually, however, the peptides whose antihypertensive effect cannot be explained by their ACE-inhibitory potency facilitate the endothelial release of factors that relax the vascular smooth muscle, such as prostacyclin or NO, and they may do this by activating different endothelial receptors. We know, for example, that the vasodilatory effect of ovokinin, an antihypertensive peptide derived from egg proteins (described in more detail in a later section of this chapter) may be linked to the activation of B_1 endothelial receptors which stimulate the release of prostacyclin in this tissue (Fujita et al., 1995a). In the same way, an ovokinin derivative, ovokinin (2-7), presents vascular and antihypertensive relaxing effects, and these effects have been linked to the activation of B_2 receptors in the endothelium and the release of NO in this tissue (Matoba et al., 1999, 2001; Scruggs et al., 2004).

We also know that α-lactorphin, fragment f(50-53) of the α-lactoglobulin which corresponds to the sequence Tyr-Gly-Leu-Phe (YGLF), produces endothelium-dependent relaxation in the mesenteric arteries of hypertensive rats. This relaxation is inhibited by an NO synthase inhibitor, and it is therefore possible that the antihypertensive effect of this peptide may be linked to the endothelial release of NO (Sipola et al., 2002). Nalaxone antagonizes the effect of α-lactorphin when this peptide is administered subcutaneously, but the antihypertensive dose of α-lactorphin causes neither antinociception nor sedation, nor does it provoke other effects mediated by opioid receptors in the central nervous system (Ijäs et al., 2004). It is of course unlikely that antihypertensive peptides derived from food proteins could have an effect on arterial blood pressure by means of central action. It has been suggested that the antihypertensive effect of α-lactorphin, and the release of NO provoked by this peptide, may be mediated by the stimulus of peripheral opioid receptors (Nurmeinin et al., 2000; Sipola et al., 2002). Opioid receptors do in fact exist that are linked to the regulation of arterial blood pressure in different peripheral tissues such as the vascular endothelium (Saeed et al., 2000), some sympathetic nerves (Hughes, Kosterlitz, and Smith, 1977), and adrenal glands (Viveros et al., 1979). Some endogenous opioid peptides may modulate this variable by acting on the aforementioned receptors (Sirén and Feuerstein, 1992).

It is worth discussing in a little more depth the possibility that food-derived peptides activate opioid receptors. The relationship between this activation and the regulation of arterial blood pressure has been widely discussed. We have known for some time that lactic peptides in particular can behave as agonists of these receptors, and can control arterial blood pressure. The study of opioid activity in a hydrolysate of milk proteins even gave rise to the term "exorphin." This term is applied to an opioid peptide of food origin, as opposed to the term "endorphin," used for endogenous opioid peptides (Fazel, 1998). Studies have been made of the structural characteristics which determine that food-derived peptides activate opioid receptors. Except for the case of peptides derived from α-casein, we can say that the common structural characteristic of exogenous and endogenous peptide opioid antagonists is the presence of Tyr in the amino-terminal extreme. The negative charge located in the Tyr phenolic group seems to be essential for opioid activity (Chang et al., 1981). The presence of another aromatic amino acid, Phe or Tyr, in the third or fourth position of the aminic extreme also favors the binding of the peptide to the opioid receptor (Meisel, 1998). Similarly, the presence of Pro in the second position is crucial for biological activity, as this amino acid maintains the orientation of the Tyr and Phe chains (Mierke et al., 1990). Peptides of food origin that behave as do opioid receptor antagonists have also been described. These peptides suppress the antagonist activity of encephalins, and produce the same effect as nalaxone (Fiat et al., 1993). Included in this group of peptides that antagonize opioid receptors are the casoxins, derived from the κ-casein and the αs_1-casein found in milk (Chiba, Tani, and Yoshikawa, 1989; Yoshikawa and Chiba, 1990). Also included are several μ receptor ligands with moderate antagonist activity (Yoshikawa, Tani, and Chiba, 1988).

6.2.3 Antioxidant Effects

A normal human metabolism produces oxygen-reactive species that originate in the mitochondria as part of the cellular metabolism. These oxidant species, which may be the cause of cell damage, are produced specifically by phagocytes and other cells as a response to external agents. They react chemically to the cells' components, modifying or suppressing their biological function. So, for example, the oxidative damage to DNA plays a crucial role in the initiation, promotion, and propagation of certain types of cancer. Similarly, lipid peroxidization contributes significantly to the onset of arteriosclerosis, and the oxidative damage to proteins is associated with chronic conditions connected with aging, such as inflammatory illnesses, cataracts, and so on (Ames, Shigenaga, and Hagen, 1993). Some studies suggest that diets rich in antioxidant components may lessen the incidence of these health problems. The relationship between the antioxidant and antihypertensive activity of many compounds seems obvious today. By contrast, we know that antioxidant deficiency is clearly involved in the onset of hypertension. One of the principal harmful effects of angiotensin II is an increase in the formation of oxygen-reactive species, because this compound activates a potent membrane oxidase, and produces superoxide anion and hydrogen peroxide (Laursen et al., 1997; Dzau, 2001). Captopril, by contrast, has antioxidant properties (Gurer et al., 1999; Baykal, Shigenaga, and Hagen, 2003), and different compounds that present antioxidant activity have the ability to inhibit ACE *in vitro*. These compounds could reduce arterial blood pressure by means of a combination of vasodilatory and antioxidant actions (Baykal et al., 2003).

Few studies have been made of the antioxidant activity of compounds of a proteinic nature, but it has been established that the antioxidant activity of proteins increases (Yee and Shipe, 1981), or at least is maintained (Rival et al., 2001a), when the proteins are hydrolyzed with different enzymes. Several casein-derived peptides (Suetsuna, Ukeda, and Ochi, 2000; Rival et al., 2001b), and some obtained from other sources (Srinivas, Shalini, and Shylaja, 1992; Kudoh et al., 2003; Sun, He, and Xie, 2004), have antioxidant properties. Antioxidant activity obtained for these peptides varies depending on the method used to determine it, which makes it difficult to establish a structure–activity relationship. Our research group has evaluated the antioxidant activity of a hydrolysate of egg albumin with pepsin.

We have also evaluated the antioxidant activity of the peptides isolated in this hydrolysate, and that of some of its amino acids. Specifically we obtained the ORAC-FL ("oxygen radical absorbance capacity-fluoresceine") value of these products, which allowed us to measure their radical-neutralizing activity (Dávalos, Gómez-Cordovés, and Bartolomé, 2004a). This experiment uses peroxyl radicals, the most widely occurring in biological systems. The peptides Tyr-Ala-Glu-Glu-Arg-Tyr-Pro-Ile-Leu (YAEERYPIL) and Ser-Ala-Leu-Ala-Met (SALAM) showed high antioxidant activity, and the peptides Tyr-Gln-Ile-Gly-Leu (YQIGL) and Tyr-Arg-Gly-Gly-Leu-Glu-Pro-Ile-Asn-Phe (YRGGLEPINF) showed intermediate antioxidant activity. The remaining peptides presented little or no antioxidant activity. The ORAC-FL value of the peptide YAEERYPIL was almost six times higher than that of α-tocopherol, but was approximately two times lower than that of caffeic acid, and three times lower than that of the flavonoid quercetin (Appel et al., 1997; Dávalos et al., 2004b). We should emphasize that it was also established that the peptide YAEERYPIL was a very potent ACE inhibitor, with a value of $IC_{50} = 4.7$ μM. Of all the amino acids studies, Tyr showed the highest ORAC-FL value. The antioxidant activity of the peptides YAEERYPIL, YQIGL, and YRGLEPINF could be due to the presence of this amino acid, which occupies the amino-terminal position in them.

Some peptide hormones in fact lose their antioxidant activity when the Tyr residue is removed (Moosmann and Behl, 2002). As is the case with phenolic compounds, the amino acid Tyr can transfer the hydrogen atom of its hydroxil group, thereby breaking the electron transfer chain. It is likely that the oxygen radical captures the Tyr hydrogen atom, which would imply the formation of a more stable radical. The activity to neutralize radicals of meat protein hydrolysates, is attributed to the presence within them of Tyr, His, and Met (Saiga, Tanabe, and Nishimura, 2003). Met could be principally responsible for the scavenger activity of free radicals of the peptide SALAM. The explanation for the antioxidant activity of this amino acid is that its sulphidryl group is easily oxidized to form sulphoxide (Stadman et al., 1993; Vogt, 1995). For antioxidant activity, it is not only the presence of certain amino acids in a peptide that is important: the position of these amino acids in the peptide chain may also be decisive (Tsuge et al., 1991). Chen et al., in 1996, confirmed this idea when they studied the antioxidant activity of 28 peptides derived from the peptide Leu-Leu-Pro-His-His (LLPHH). This peptide had been isolated in a soy protein digested with different enzymes, and it was established that the antioxidant activity of the peptide Pro-His-His (PHH) was much greater than that of the peptides His-Pro-His (HPH) or His-His-Pro (HHP; Chen et al., 1996).

The oxidization of lipids favors the onset of arteriosclerosis. More specifically, the oxidation of low-density lipoproteins (LDL) clearly contributes to its onset. In many cases, this pathology accompanies and favors the hypertensive condition. For this reason, laboratory methods have also been developed to detect and quantify the effect of peptides on lipid oxidization. These methods make it possible to predict whether food-derived peptides will be able to antagonize the effect of LDL *in vivo*. For example, the ability of peptides to inhibit LDL oxidization induced by Cu^{2+} can be determined in the laboratory. The peptide YAEERYPIL, a potent ACE inhibitor which also shows a high degree of free radical neutralizing activity (3.8 times greater than that of Trolox, a vitamin E analogue), in addition delays the LDL oxidization induced by Cu^{2+} (Dávalos et al., 2004b).

To close this section, it is worth pointing out that some experiments have shown that milk caseins are able to fix iron *in vitro* and inhibit lipid peroxidization (Cervato, Cazzola, and Cestaro, 1999). Several recent studies have also described the antioxidant activity of human milk (VanderJagt et al., 2001) and of whey proteins (Tong et al., 2000), but further studies are necessary in order to corroborate the antioxidant action of milk proteins in laboratory animals and in humans.

6.3 Strategies to Obtain Antihypertensive Food Peptides

In order to obtain antihypertensive peptides from food proteins, the usual practice is to provoke hydrolysis of these proteins by exposing them to different enzymes. Enzymatic protein hydrolysis has various advantages, including the speed of the process, its high specificity, its moderate cost, and the possibility of obtaining high-quality products that may be marketed on a large scale (Clemente et al., 2000). Table 6.1 shows the principal proteins of animal origin that have provided antihypertensive hydrolysates and peptides.

It is assumed that in most cases antihypertensive peptides derived from food proteins will be ACE inhibitors. Enzymatic hydrolysis of the protein of origin produces a hydrolysate that should therefore be characterized as an inhibitor of this enzyme. This hydrolysate should also be characterized by its possible antihypertensive effect, because, as we explain below, compounds that show ACE-inhibitory activity *in vitro* do not always present effects on arterial blood pressure when administered. This hydrolysate is then subjected to ultrafiltration via membranes, making it possible to obtain a fraction or permeate of the hydrolysate that usually has a value lower than the IC_{50} of the hydrolysate itself (Visser et al., 1989; Turgeon and Gauthier, 1990; Vreeman, Both, and Slangen, 1994; D'Alvise et al., 2000). This could be easily explained as this fraction is enriched with small peptides which are those with the best ACE-inhibitory potency (Mullally, Meisel, and FitzGerald, 1997a; Fujita, Yamagami, and Ohshima, 2001; Miguel et al., 2004).

Most of the peptides that inhibit ACE in reality correspond to short sequences containing between three and seven amino acids (Pihlanto-Leppälä, Rokka, and Korhonen, 1998). The permeate rich in peptide sequences that inhibit ACE is split up by means of various chromatographic techniques into subfractions, and the ability of these subfractions to inhibit ACE is once again determined. Peptide subfractions rich in amino acids of a hydrophobic nature usually escape at the base of the column, and it is these subfractions that frequently show lower IC_{50} values. This is not surprising, as the hydrophobic character of the amino acids at the carboxi-terminal extreme is decisive for the bind of the peptides at the active center of ACE (Cheung et al., 1980). The next step is to analyze the subfractions that present low IC_{50} values. Different techniques, including high-pressure liquid chromatography (HPLC) together with mass spectrometry (ESI-MS), and tandem mass spectrometry (ESI-MS/MS), make it possible to discover the approximate molecular mass of the peptides present in these subfractions. The final identification of these peptides is obtained by contrasting different sequences that correspond to this molecular mass in a protein database. Following this, the potency of ACE inhibition of the identified peptides should be established, and a study should be made of the *in vivo* effect of those showing low IC_{50} values. The objective is therefore to obtain peptide concentrates with both ACE-inhibitory and antihypertensive activities, with an ever-decreasing number of peptides, until concrete sequences presenting these activities are achieved. Purifying the identified peptides, and obtaining them from hydrolysate in sufficient quantities, is a complex and costly process. For this reason, for studies evaluating peptide activity, it is usual to chemically synthesize the sequences of interest. Figure 6.4 presents a diagram of the steps required to obtain antihypertensive peptides derived from food proteins.

It would be preferable if more studies were made of the structural requirements of the food-derived antihypertensive peptides that interact with ACE, but it should always be borne in mind that determining the inhibitory activity of this enzyme is only a starting point in the selection of these peptides. Many of them present ACE-inhibitory activity *in vitro*, but in reality they do not show this activity *in vivo*, nor do they exert antihypertensive effects. The following explanations help to understand why this should be so. Peptides and hydrolysates from food proteins showing ACE-inhibitory activity have greater difficulty in reaching their action points in the organism, and in producing the physiological effect, than the drugs used for this purpose in clinical practice. It is preferable that all these compounds be administered orally, and the peptides should therefore be resistant to digestion processes. In order to produce the effect, the peptides should also pass through the intestinal epithelium and be distributed intact until they reach the target organs.

The hydrolysis of peptides by pepsin in the stomach and by some pancreatic enzymes, such as trypsin and chymotrypsin, usually generates smaller peptides which present a different ACE-inhibitory activity from that of the administered peptide. It might be that the ACE-inhibitory activity of the products obtained by hydrolysis is nonexistent, or smaller than that of the administered peptide, but what might also happen is that peptides with poor ACE-inhibitory activity are administered, and that these peptides hydrolyze and generate other products with more ability to inhibit the enzyme. In this last case, an antihypertensive effect might be observed that is contrary to what was expected. Thus, for example, the peptide Tyr-Lys-Val-Pro-Gln-Leu (YKVPQL), which figures in the sequences identified in a casein hydrolysate, showed high ACE-inhibitory activity *in vitro* (IC_{50} = 22 µM), but presented no antihypertensive effect. In the same hydrolysate, another peptide sequence, Lys-Val-Leu-Pro-Val-Pro-Gln (KVLPVPQ), was identified, which showed antihypertensive activity, in spite of its low ACE-inhibitory activity *in vitro*. It was discovered that, following pancreatic digestion of this sequence, the extreme C-terminal was released and the peptide Lys-Val-Leu-Pro-Val (KVLPV) was formed. This peptide showed potent ACE-inhibitory activity *in vitro* (IC_{50} = 5 µM), and was probably responsible for the antihypertensive activity observed with the larger peptide (Maeno, Yamamoto, and Takano, 1996). In reality, studies that simulate gastrointestinal digestion of the peptides identified in protein hydrolysates are always necessary, and these studies should be carried out with the sequences which show ACE-inhibitory activity in particular. At present there exist diverse procedures that simulate *in vitro* the enzymatic conditions

TABLE 6.1

Antihypertensive Peptides Derived from Food Proteins

Sequence	Origin	Enzyme	Activity	Reference
VPP, IPP	β-casein	Proteinase of *Lactobacillus helveticus*	IACE/Antihypertensive	Nakamura et al., 1995a,b, 1996; Sipola et al., 2001
VYP, VYPFPG	β-casein	Proteinase K	Antihypertensive	Abubakar et al., 1998
KVLPVP, KVLPVPQ	β-casein	Proteinase of *Lactobacillus helveticus*	Antihypertensive	Maeno et al., 1996
LHLPLP	β-casein	Proteinasa de *Enterococcus faecalis*	IACE/Antihypertensive	Muguerza et al., 2006; Quirós et al., 2007
FFVAPFPEVFGK	αs_1-casein	Trypsin	IACE/Antihypertensive	Karaki et al., 1990; Townsend et al., 2002
ALPMHIR	β-lactoglobulin	Trypsin	IACE/Antihypertensive	Mullally et al., 1997a,b
LLF, LKQW	β-lactoglobulin	Thermolysin	IACE/Antihypertensive	Hernández-Ledesma et al., 2002, 2007
IPA	β-lactoglobulin	Several enzymes	IACE/Antihypertensive	Abubakar et al., 1998
YGLF	α-lactoglobulin	Several enzymes	Antihypertensive	Mullally et al., 1996; Nurmeinin et al., 2000
FRADHPFL	Ovalbumin	Pepsin	Vasodilator/Antihypertensive	Fujita et al., 1995a,b
RADHPF	Ovalbumin	Chymotrypsin	Vasodilator/Antihypertensive	Matoba et al., 1999; Scruggs et al., 2004
KVREGTTY	Ovotransferrina	Not mentioned	IACE/Antihypertensive	Lee et al., 2006a,b
YAEERYPIL RADHPFL IVF	Egg white proteins	Pepsin	IACE/Antihypertensive	Miguel et al., 2004, 2005b
LW	Ovalbumin	Pepsin	IACE/Antihypertensive	Fujita et al., 2000
RADHP	Egg white proteins	Pepsin/Corolase PP	IACE/Antihypertensive	Miguel et al., 2006a
Oligopeptides	Egg yolk proteins	Several enzymes	IACE/Antihypertensive	Yoshii et al., 2001

Peptide	Source	Enzyme	Activity	References
LKPNM, LKP	Bonito proteins	Thermolysin	IACE/Antihypertensive	Yokohama et al., 1992; Fujita et al., 1993; Karaki et al., 1993; Fujita and Yoshikawa, 1999
YRPY, GHF, VRP, ILP, IRP, LRP	Bonito proteins	Not mentioned	IACE/Antihypertensive	Fujita et al., 1993; Matsumura et al., 1996
THILTGD	Tuna proteins	Not mentioned	IACE/Inhibition of endothelin production	Kohama et al., 1989; 1994
VY	Sardine proteins	Proteinase of *Bacillus licheniformis*	IACE/Antihypertensive	Matsui et al., 1993; Matsufuji et al., 1994, 1995

IACE: Inhibition of angiotensin converting enzyme.

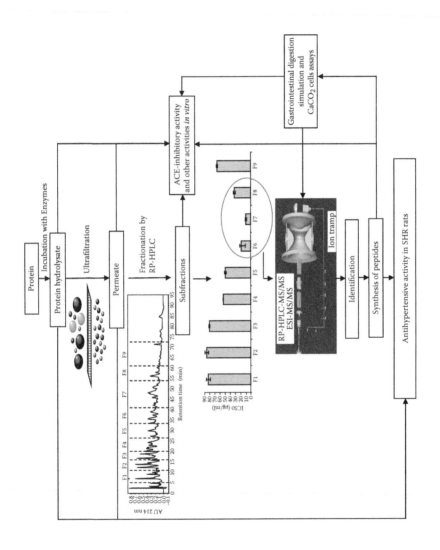

FIGURE 6.4
Diagram of the steps required to obtain antihypertensive peptides derived from food proteins.

peculiar to the human digestive system (Astwood, Leach, and Fuschs, 1996; Alting, Meijer, and Van Beresteijn, 1997; Zarate et al., 2000; Vermeirssen et al., 2003). These methods provide a fairly accurate idea of the peptides generated *in vivo* that are resistant to digestive enzymes. Our research group uses one method in particular that allows us to determine fairly precisely which sequences resist these processes (Miguel et al., 2006a). If we wish to find out which specific sequences are responsible for the antihypertensive effect, we also need to subsequently evaluate the ACE-inhibitory activity of these products of hydrolysis.

The results from different studies indicate that peptides usually require specific transport mechanisms and transporters to pass through the intestinal epithelium. Peptides are actively carried through the apical enterocite membrane (Ganapathy and Leibach, 1985; Leibach and Ganapathy, 1986), but in principle the transport systems would only be valid for di- and tripeptides, as peptides with more than four amino acids would be difficult to recognize (Daniel, Morse, and Adibi, 1992). Nevertheless, some studies have shown that larger peptides can be absorbed (Gardner, 1994). After being ingested, large peptides, sometimes composed of up to 20 amino acids, have been detected in plasma (Chabance et al., 1998). It is possible that these peptides are absorbed by a transcytolitic process and that they use other macromolecule transport systems (Heyman and Desjeux, 1992).

A passive means of transport has also been postulated for these peptides, as they could be absorbed principally by a paracellular process, by crossing intercellular junctions (Burton et al., 1992; Conradi et al., 1992, 1993; Kim, Burton, and Borchardt, 1993; Adson et al., 1994; Pappenheimer et al., 1994; Goodwin et al., 1999). Most of the studies that have determined the mechanisms by which peptides pass through the intestinal epithelium are *in vitro* studies using CaCo-2 cells. These are epithelial cells, originally isolated in human colon carcinoma, which form single layers in culture and present morphological and biochemical characteristics similar to those of differentiated enterocytes. These cells express many of the peptide transport systems (Hidalgo, Raub, and Borchardt, 1989; Hilgers, Conradi, and Burton, 1990). Results from studies in these cells correlate well with results of *in vivo* absorption studies. For this reason, the U.S. Food and Drug Administration (FDA) has officially accepted this cellular model for determining the permeability of a drug (FDA, 1999). The model appears in the guide for requesting exception from bioequivalency studies (CDER/FDA, 2000). CaCo-2 cells also express many intestinal enzymes, and make it possible to study the processes of hydrolysis of peptides by the peptidases located in the cells of the intestinal epithelium (Satake et al., 2002). Peptides may be hydrolyzed when these peptidases act, and the speed of intestinal transportation of long sequences depends fundamentally on their susceptibility to these peptidases (Shimizu, Tsunogai, and Arai, 1997; Vermeirssen et al., 2002, 2004).

Orally absorbed peptides may also deteriorate when they reach systemic circulation because plasmatic peptidases act on them. The possible metabolic modifications to orally administered peptides are therefore multiple. For this reason animal experiments are required nowadays to evaluate the antihypertensive effect of products that present ACE-inhibitory activity *in vitro*. We discuss these experiments later on. These studies attempt to establish the precise sequences that are able to inhibit ACE *in vivo*, or control arterial blood pressure by means of some other mechanism when administered. Nevertheless, we should be aware that these strategies are only an approximation of what might happen when these products are used in humans. In many cases, human metabolic and physiological processes differ substantially from those of animals. It should also be borne in mind that the mechanisms by which antihypertensive products decrease arterial tone may have a different relevance in different species. It is therefore obvious that, before routinely using food protein-derived products and peptides in prehypertensive or hypertensive subjects, studies are always necessary to guarantee their safety and effectiveness in these patients.

We have indicated that, as part of the action strategy necessary for obtaining food protein-derived antihypertensive products, experiments on animals are unavoidable and of particular importance. For these experiments, spontaneously hypertensive rats (SHR) are normally used. This animal model was developed from Wistar rats bred at Kyoto University (Okamoto and Aoki, 1963). They are the most widely used experimental model for hypertension at the present time. Different researchers agree that the basic principles associated with the onset of hypertension in these animals and in humans are surprisingly similar (Trippodo and Frohlich, 1981; Zicha and Kunes, 1999). Wistar Kyoto rats (WKY) are nowadays considered the normotensive control of SHR rats. These animals are used in experiments with food protein-derived hydrolysates and peptides in order to establish how these products act when arterial tone is not altered. In all these experiments, the rat's arterial blood pressure is usually measured by a modification of the tail cuff method, originally described by Buñag (1973; Buñag and Butterfield, 1982). With the original technique it was only possible to obtain SBP measurements. Several of the studies carried out with food protein-derived antihypertensive peptides continue to provide information about SBP modification only. It should, however, be mentioned that nowadays several types of equipment exist which make it possible to differentiate between the animals' SBP and DBP, and reliable measurements of these two variables can be obtained. It is recommended that data be provided on the modification of both. Before fitting the cuff and the transductor, the rats are exposed to a temperature of around 30°C to facilitate dilation of the caudal artery. The animals should become accustomed to the procedure, as SHR rats are particularly nervous. It is also recommended that at least three consecutive SBP and DBP readings are made and that the average reading is used to establish the definitive value of these variables. Figure 6.5 shows the measurement of arterial blood pressure in a rat using the tail cuff method.

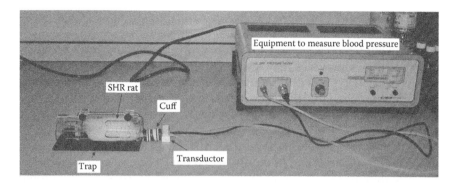

FIGURE 6.5
Measurement of arterial blood pressure in a rat using the tail cuff method.

In experimental studies of chronic administration of peptides and hydrolysates derived from food proteins, arterial blood pressure in rats is always measured at the same time of day (usually in the morning) and it is usual to take weekly SBP and DBP readings. In these studies the products are usually administered in the animals' drinking water, and treatment begins at weaning. In this way the development of hypertension in the rats is studied, as well as the part played by the products administered during this development. After sufficient time has passed (usually at about 20 weeks of life) the treatment can be withdrawn. This allows us to evaluate the possible reversal of the effect. Nevertheless, the most common studies are those evaluating the acute effect of a hydrolysate or peptide following the intragastric administration of a single dose of the product. The rats' arterial blood pressure should be measured before this administration, and then afterwards at various times. Normally four to five measurements are made following administration, establishing intervals of 1 hour 30 minutes or 2 hours between measurements. Arterial blood pressure is again measured 24 hours after administration. For these acute administration experiments, rats with stable arterial blood pressure values should be used. Arterial blood pressure begins to stabilize in SHR when the rats are 11–12 weeks old, but is not completely stable until they reach 20 weeks of life. It is therefore recommended that rats close to this age be used.

6.4 Antihypertensive Food Peptides

6.4.1 Antihypertensive Milk Peptides

Milk proteins are those most commonly studied as a source of antihypertensive peptides (see reviews by Jauhiainen and Korpela, 2007; Saito et al., 2008).

These peptides can be produced by hydrolyzing milk proteins with different enzymes, and also by fermenting milk with different bacteria. These methods have provided excellent results for obtaining antihypertensive products. Antihypertensive peptides may also be produced as a result of the proteolytic process which takes place at the same time as cheese ripens. It has been known for some time that both bovine caseins (Maruyama and Suzuki, 1982; Maruyama et al., 1987a) and human caseins (Kohmura et al., 1989) behave as do ACE-inhibitors, and at the end of the twentieth century attempts were already being made to market several products with antihypertensive effect derived from milk protein. The study carried out by Sekiya et al. (1992) made it possible to market a casein hydrolysate obtained by treating this protein with trypsin; consumption of 20 g per day of this hydrolysate for four weeks caused a drop in both SBP and DBP in hypertensive patients. The product was marketed in Japan under the name "Casein DP Peptio Drink" (Sugai, 1998). Subsequently, in Holland, another tryptic hydrolysate of casein went on sale which also lowered arterial blood pressure in both animals and hypertensive patients (Karaki et al., 1990; Townsend et al., 2004; Cadee et al., 2007). The commercial name of this new hydrolysate was "C12 Peption."

At the end of the twentieth and the beginning of the twenty-first century, whey protein began to be studied as a source of antihypertensive peptides with ACE-inhibitory activity (Mullally, Meisel, and FitzGerald, 1997a,b; Pihlanto-Leppälä et al., 2000). Some of the whey protein derivatives that showed this activity also had a beneficial effect on plasma cholesterol. It is worth mentioning a product supplemented with whey protein obtained by fermenting milk with *Lactobacillus casei* TMC0409 and *Streptococcus thermophilus* TMC 1543. The product in question lowers total cholesterol in rats and humans (Kawase et al., 2000). Also of interest are the studies carried out by Pins and Keenan (2002, 2003, 2006). These researchers showed that a daily consumption of 20 g of an hydrolysate obtained from whey proteins for six weeks significantly reduced SBP and DBP in prehypertensive patients receiving no medication. Five weeks after hydrolysate intake had finished, its effect could still be appreciated. The patients in this study presented hypercholesterolemia, and the hydrolysate lowered LDL cholesterol levels. This product is currently marketed in the United States under the name "Biozate." In this context it is also important to mention that Lee et al. in 2007 described the blood pressure lowering effect of a milk drink supplemented with whey peptides in patients with mild hypertension (Lee et al., 2007).

Several ACE-inhibitory peptides with antihypertensive activity have also been isolated from hydrolysates obtained by treating milk α-lactoglobulin and β-lactoglobulin with digestive enzymes (Pihlanto-Leppälä et al., 2000; Chobert et al., 2005; Nurmeinin et al., 2000; Mullaly et al., 1996). Recently two peptides with powerful ACE-inhibitory activity have been identified, derived from caprine β-lactoglobulin (Hernández-Ledesma et al., 2002). These peptides, which correspond to the sequences Leu-Gln-Lys-Trp (LQKW) and Leu-Leu-Phe (LLF), are released when whey is incubated with thermolysin for 24 hours at 37°C. Both sequences showed antihypertensive effects in SHR when administered orally in a single dose (Hernández-Ledesma et al., 2007).

The production of antihypertensive peptides by milk fermentation is also of great importance. For thousands of years, natural or controlled fermentation has been exploited for the preservation of different foods, and in order to maintain, or alter, their nutritional or sensorial properties. Lactobaciles and bifidobacteria are the probiotics most widely used in the manufacture of different fermented milk products, such as yogurt and cheese. These bacteria contain many proteolitic enzymes, capable of hydrolyzing milk proteins. In this way they release peptides and amino acids, which serve as a source of nitrogen, essential for their growth. These peptides which are released during the fermentation process can perform different biological activities that probably contribute to the beneficial properties of fermented milk products and of the lactic bacteria themselves (Sanders, 1993; Lee and Salminen, 1995). Several of these peptides may behave like ACE-inhibitors, and some of these fermented products may therefore be used in the prevention or treatment of hypertension. *Lactobacillus helveticus* strains have greater proteolitic activity on milk proteins than other strains of lactic bacteria, and the peptide content when milk is fermented with this microorganism is higher (Yamamoto, Akino, and Takano, 1994). Many products rich in antihypertensive peptides have in fact been obtained by fermenting milk with this microorganism. ACE-inhibitory peptides have also been isolated by fermenting milk with other lactic bacteria, such as *Lactobacillus casei* subsp. *rhamnosus* (*Lactobacillus* GC; Rokka et al., 1997), or *Lactobacillus delbrueckii* subsp. *Bulgaricus* and *Lactococcus lactis* subsp. *cremoris* (Gobbetti et al., 2000). In the processes of isolating and characterizing ACE-inhibitory peptides, and for the production of antihypertensive products, some strains of yeast are also being used that are able to hydrolyze milk proteins. The combination of *Lactobacillus helveticus* and *Saccharomyces cerevisiae* has produced good results. An example of this is the milk product marketed in Japan by Calpis (Calpis Co. Ltd., Japan) which is prepared by fermenting skimmed milk with *Lactobacillus helveticus* and *Saccharomyces cerevisiae*. In 1995, Nakamura et al. established that Calpis milk showed antihypertensive effects in SHR following acute oral administration in these animals (Nakamura et al., 1995b). The Japanese group showed that the peptides Val-Pro-Pro (VPP) and Ile-Pro-Pro (IPP) were mainly responsible for this milk's antihypertensive activity (Nakamura et al., 1995a,b; Nakamura, Masuda, and Takano, 1996; Masuda, Nakamura, and Takano, 1996). These peptides also exerted an antihypertensive effect in humans (Hata et al., 1996; Mizushima et al., 2004; Aihara et al., 2005; Mizuno et al., 2005). A double-blind study carried out shortly afterwards by the same group demonstrated that consuming 95 ml of Calpis milk per day for eight weeks, significantly reduced SBP and DBP in hypertensive patients who continued to receive their antihypertensive medication during the study (Hata et al., 1996). The effect of this milk has also recently been shown in hypertensive subjects receiving no antihypertensive medication. In the study, patients consumed 160 g of milk per day for four weeks (Mizushima et al., 2004). The product, marketed by Calpis under the name Ameal Peptide®, has been added to a new milk drink launched by Unilever and sold in Spain as Flora Proactive®.

We know that in most cases ACE-inhibition explains the effect of functional products with antihypertensive activity. The sequences VPP and IPP, which are responsible for the antihypertensive effect of Calpis milk, are characterized as potent ACE-inhibitors. It is not surprising that these sequences inhibit the enzyme. Both contain the amino acid Pro in the carboxi-terminal position, and have the amino acids Val and Ile, respectively, which also favor enzymatic inhibition in the extreme amino-terminal. These sequences appear in various antihypertensive milk products, in addition to Calpis, and are responsible for their effect. In some other cases, these tripeptides are released *in vivo* when antihypertensive milk products with longer sequences are administered. Studies carried out by Sipola et al. (2001) made clear the antihypertensive effect of the peptides VPP and IPP in SHR. This group likewise showed the antihypertensive effect of a milk fermented with *Lactobacillus helveticus* LBK–16H, which contained both tripeptides (Sipola et al., 2002). Consuming 150 ml of this milk a day for several weeks caused a slight, but indisputable, decrease in arterial blood pressure in hypertensive subjects receiving no antihypertensive medication (Seppo et al., 2002, 2003; Tuomilehto et al., 2004; Jauhianien et al., 2005). This dairy product was marketed in Finland by Valio Ltd. as Evolus. It is now sold in Iceland by Mjòlkursam salan as LH, in Spain by Kaiku-Iparlat as Kaiku Vita and in Switzerland, Portugal, Malta, and Italy by Emmi as Emmi-evolus. Hirota et al. (2007) have demonstrated that the sequences VPP and IPP also improve vascular endothelial function, but it is important to note that in a recent study carried out by Engberink et al. these lactotripeptides showed no effect on human blood pressure (Engberink et al., 2008).

Robert et al., from the Nestlé group, have identified several ACE-inhibitory peptides in a milk product obtained by fermenting milk with *Lactobacillus helveticus* NCC 2765, but *in vivo* experiments have yet to be carried out that confirm its possible antihypertensive effect (Robert et al., 2004).

Our group of researchers, in collaboration with the Leche Pascual Group, has shown that some selected strains of *Enterococcus faecalis* are capable of producing ACE-inhibitory peptides other than the peptides VPP and IPP. Milk fermented by these strains of *Enterococcus faecalis* showed antihypertensive activity in SHR when administered acutely and orally (Muguerza et al., 2006). The milk also reduced the onset of hypertension in these animals when administered orally and continuously from weaning. The antihypertensive properties improved when the fermented product was enriched with calcium (Miguel et al., 2005a). The ACE-inhibitory peptides identified in the milk, of which the sequence Leu-His-Leu-Pro-Leu-Pro (LHLPLP), corresponding to the fragment f(133-138) of β-casein was of particular interest, also caused a decrease in the arterial blood pressure of SHR when administered orally. By contrast, these peptides did not modify arterial blood pressure in the normotensive WKY rats (Miguel et al., 2006b).

Peptides with biological activity may also be released during the cheese-making process. As a consequence of the proteolitic process that takes place as cheese matures, specifically peptides with ACE-inhibitory activity are formed (Abubakar et al., 1998; Gómez-Ruiz, Ramos, and Recio, 2003). The original milk, the initiating culture, and the conditions of the ripening process may all influence the inhibitory activity of the end product. Thus in 1995, Okamoto et al. established that ACE-inhibitory activity was present in ripe Camembert-type cheeses, in blue cheeses, and in red Cheddar cheese (Okamoto et al., 1995). These researchers saw that, by contrast, soft white cheeses of the cottage cheese-type, with low proteolysis levels, did not show this activity. These facts were corroborated by Meisel, when studying enzymatic inhibition produced by different cheeses (Meisel, 1997, 1998, 2001). The inhibition percentages for ripe cheeses obtained by these researchers were close to 70%, but those obtained for cottage cheese- and quark-type cheeses were less than 13% and 27%, respectively.

It is also true that when the cheese-ripening process is too prolonged, at a certain point ACE-inhibitory activity decreases. This can be easily explained because the peptides that are released during the cheese-ripening process by the action of proteolitic enzymes from lactic bacteria later disintegrate and generate inactive fragments. Thus, for example, in Parmesan cheese ripened for 6 months, ACE-inhibitory peptides derived from αs_1-casein were isolated, but these peptides were not detectable in the same cheese after 15 months of ripening (Addeo et al., 1994). Similarly, the ACE-inhibitory activity of Gouda cheese after 8 months of ripening was 70%, but the ACE-inhibitory activity of this cheese after 24 months of ripening was only 34.6% (Meisel, 1998, 2001). Similarly, it has more recently been established that there is a decrease in ACE-inhibitory activity in a semi-mature cheese manufactured using conventional fermenting methods and strains of *Lactobacillus* and *Bifidobacterium* during its long storage period (Ryhänen, Pilhlanto-Leppälä, and Pahkala, 2001).

Peptide fragments with ACE-inhibitory activity have been isolated in other types of cheese. These include the fragment f(58-72) of β-casein isolated in Crescenza cheese (Smacchi and Gobbetti, 1998) and in cheddar cheese (Stepaniak et al., 1995), and the fragment f(43-52) of β-casein, isolated in the latter cheese (Jiang et al., 1998). Similarly, a fraction with casein-derived peptide fragments, and with antihypertensive peptide precursors, has been extracted in an enzymatically modified cheese produced by using *Lactobacillus casei* and different commercial enzymes (Haileselassie, Lee, and Gibbs, 1999). In 2000, Saito et al. also isolated several peptide fragments in Gouda cheese ripened for 8 months which showed high ACE-inhibitory activity. These fragments were released from αs_1-casein and from β-casein (Saito et al., 2000). The fragment f(60-68) of β-casein isolated in feta cheese by the same researchers also showed ACE-inhibitory activity (Saito et al.,

2000). The angiotensin-converting enzyme inhibitory tripeptides Val-Pro-Pro and Ile-Pro-Pro have been recently identified in different cheese varieties of Swiss origin (Bütikofer et al., 2008). Moreover, Tonouchi et al. (2008) have demonstrated the antihypertensive effect of an angiotensin converting enzyme inhibitory peptide from enzyme-modified cheese.

6.4.2 Antihypertensive Egg Peptides

In the middle of the twentieth century, the first studies were carried out into the possible physiological effect of egg proteins (Mine et al., 1995), but until now, very few bioactive peptides derived from egg proteins have been described (Miguel and Aleixandre, 2006). The first were two antihypertensive peptides with direct activity in vessels. Initially, an antihypertensive and vasorelaxing octapeptide was isolated with the aminoacidic sequence Phe-Arg-Ala-Asp-His-Pro-Phe-Leu (FRADHPFL; Fujita et al., 1995a). This sequence corresponds to fragment 358-365 of egg albumin, which is the chief protein present in egg white. The peptide in question showed partially endothelium-dependent vasodilatory activity in canine mesenteric arteries, and was given the name ovokinin. Its effect was partly mediated by B_1 receptors which stimulated the release of prostacyclin. Ovokinin displayed antihypertensive effects when administered in high doses to SHR, an effect that increased when the peptide was administered orally, in the form of an emulsion in egg yolk. It was postulated that the phospholipids in the egg yolk increased the oral availability of ovokinin because they improved its intestinal absorption and protected the peptide from digestion by intestinal peptidases (Fujita et al., 1995b).

The second egg white-derived peptide with vasorelaxing properties was a hexapeptide, characterized as fragment 2-7 of ovokinin. It was given the name ovokinin (2-7) and its sequence was: Arg-Ala-Asp-His-Phe-Leu (RADHPF). This sequence was purified from an egg albumin hydrolysate using chymotrypsin, and it was observed that it corresponded to residues 359-364 of this protein. The peptide caused endothelium-dependent relaxation in mesenteric arteries of SHR, and this relaxation was mediated principally by nitric oxide. It did not, however, produce relaxation in the arteries of normotensive WKY rats. The arterial blood pressure of SHR decreased when doses of ovokinin (2-7) ten times smaller than effective doses of ovokinin were administered orally, but the blood pressure of WKY rats was not affected when identical doses of this peptide were administered in the same way. It was observed that intravenous administration of ovokinin (2-7) in the aforementioned doses produced no significant change in the arterial blood pressure of SHR. Paradoxically, administering very high concentrations of the peptide intravenously caused only a slight decrease in this variable in SHR (Matoba et al., 2001). Scruggs et al. (2004) in their studies of vascular reactivity in isolated arteries showed that ovokinin (2-7) produces its effects by activation of bradykinin B_2 vascular receptors.

Attempts have been made to improve the oral activity of antihypertensive peptides derived from egg proteins by means of structural modifications. For example, several products similar to ovokinin (2-7) have been synthesized in order to improve its oral antihypertensive activity. Of these, of particular interest are the peptides Arg-Pro-Phe-His-Pro-Phe (RPFHPF) and Arg-Pro-Leu-Lys-Pro-Trp (RPLKPW), which showed, respectively, 10 and 100 times more activity than ovokinin (2-7) when administered orally to SHR. The substitution of amino acids would probably make these sequences more resistant to digestive tract proteases (Matoba et al., 2001; Yamada et al., 2002). The sequence RPLKPW has been named novokinin and it has been recently shown that this peptide induces relaxation by AT2- and IP-receptor-dependent mechanism in the mesenteric artery from SHRs (Yamada et al., 2008).

We have previously emphasized that ACE-inhibition is the main mechanism involved in the antihypertensive effect of peptides of food origin. Within this context, we should emphasize that Fujita, Yokoyama, and Yoshikawa (2000) established that hydrolysates obtained from ovalbumin using pepsin and thermolysin exhibited ACE-inhibitory activity. The IC_{50} values for these hydrolysates were 45.0 μg/ml and 83.0 μg/ml, respectively. Six ACE-inhibitory peptides were isolated in the ovalbumin hydrolysate obtained by treatment with pepsin. These peptides had IC_{50} values of between 0.4 μM and 15 μM, but only one of them, the dipeptide Leu-Trp (LW), showed antihypertensive activity in SHR. Fujita et al. (2000) did not succeed in producing active hydrolysates when they treated egg albumin with trypsin or chymotrypsin. The IC_{50} values for these hydrolysates obtained by these researchers were greater than 1,000 μg/ml.

Several studies carried out by our research group have also shown that hydrolysis of egg white proteins with different enzymes of digestive origin produces hydrolysates with a high degree of ACE-inhibitory activity. The most potent hydrolysates were obtained following hydrolysis of egg white with pepsin. In this case, the length of time of hydrolysis was significant for the potency of the hydrolysate. After more than 30 min incubation, active hydrolysates with relatively low IC_{50} values were obtained. Hydrolysis of egg white with pepsin over a three-hour period produced a hydrolysate with potent ACE-inhibitory activity and an IC_{50} value of 55 μg/ml. Ultrafiltration of this hydrolysate made it possible to obtain a fraction with a molecular mass of less than 3,000 Da, which exhibited much more ACE-inhibitory activity than the hydrolysate itself. The IC_{50} value of this fraction was 34 μg/ml, and it contained several peptides with ACE-inhibitory activity. The most potent peptides corresponded to the sequences Tyr-Arg-Glu-Glu-Arg-Tyr-Pro-Ile-Leu (YAEERYPIL), Arg-Ala-Asp-His-Pro-Phe-Leu (RADHPFL), and Ile-Val-Phe (IVF). These sequences presented IC_{50} values of 4.7, 6.2, and 33.1 μM, respectively. In contrast to the results reported by other researchers, in our laboratory we were also able to obtain active hydrolysates when egg white protein hydrolysis was performed using trypsin or chymiotrypsin, but in this case a minimum incubation time of 24 hours was required (Miguel et al., 2004).

The hydrolysate obtained in our laboratory from egg white by treatment with pepsin for three hours, its <3,000 Da fraction, and the peptide sequences YAEERYPIL, RADHPFL, and IVF, exhibited clear antihypertensive effects. These products produced a significant decrease in both SBP and DBP when administered orally in one single dose to SHR. By contrast, this administration did not modify the arterial blood pressure of normotensive WKY rats (Miguel et al., 2005b), but it was established that the hydrolysate also attenuated the onset of arterial hypertension in SHR when administered orally from weaning (Miguel et al., 2006c). Parallel studies simulating gastrointestinal digestion indicate that the sequences YAEERYPIL and RADHPFL hydrolyze when administered orally (Miguel et al., 2006a). It is therefore highly likely that the products resulting from this hydrolysis are responsible for the effect observed when these sequences are administered, and also, in part at least, responsible for the antihypertensive effect observed when the hydrolysate is administered (Miguel et al., 2006a).

Some peptides with antihypertensive activity have also been produced by enzymatic hydrolysis of egg yolk. ACE-inhibitory oligopeptides may be produced when egg yolk is hydrolyzed with different enzymes. Oral administration of different doses of these oligopeptides to SHR produced a significant drop in both their SBP and DBP (Yoshii et al., 2001). Lastly we should mention that recently several ovotransferrin-derived peptides with both ACE-inhibitory and antihypertensive effects have been described (Lee et al., 2006a,b).

6.4.3 Antihypertensive Fish Peptides

Bioactive peptides have been obtained from biological materials other than milk and egg. In particular, many studies have been made of fish proteins, and their biological activity has also been linked to specific sequences that may be released by enzymatic hydrolysis. In 1986 Suetsuna and Osajima were the first to obtain fish-derived peptide products with ACE-inhibitory activity. These researchers obtained protein hydrolysates showing ACE-inhibitory activity from sardines, by treating these proteins with a protease derived from *Aspergillus oryzae* (Suetsuna and Osajima, 1986). Some time later, Kohama et al. (1988, 1989) purified an octapeptide derived from a tuna hydrolysate that showed high ACE-inhibitory activity.

Similarly, four ACE-inhibitory peptides were isolated in bonito intestine (Matsumura et al., 1993). Oral administration of these peptides lowered arterial blood pressure in SHR (Karaki et al., 1993). In addition, a hydrolysate obtained from the proteins of this fish by treatment with thermolysin exhibited potent ACE-inhibitory activity (Yokoyama, Chiba, and Yoshikawa, 1992), and also showed antihypertensive activity in SHR (Fujita et al., 1995c) and in hypertensive patients (Fujita et al., 1997a,b). One of the sequences identified in this hydrolysate, the sequence Leu-Lys-Pro-Asn-Met (LKPNM), and the end product of the proteolysis of this sequence, the peptide Leu-Lys-Pro (LKP), also showed antihypertensive activity when administered intravenously to SHR. Oral administration of both peptides, LKPNM and LKP, also lowered blood pressure in these animals. The greatest decrease was observed six hours after administering the sequence LKPNM, and four hours after administering the sequence LKP. The hydrolysate also exhibited antihypertensive activity in hypertensive and prehypertensive subjects (Fujita and Yoshikawa, 1999). This product was officially approved in Japan as a "food of specific benefit to health," and Nippon Supplement Inc. has marketed it there as a "peptide soup." Metagenics has also marketed this product in the United States in tablet form under the name Vasotensin™.

Hydrolysis of dried bonito protein with alkalase derived from *Bacillus licheniformis* also resulted in a product that showed considerable ACE-inhibitory activity *in vitro*. A protein fraction was obtained from this hydrolysate which, after treatment with gastrointestinal enzymes, maintained this high ACE-inhibitory activity (Matsui et al., 1993). Of the peptides identified in this protein fraction, the sequence Val-Tyr (VY) is of particular interest. This sequence lowered arterial blood pressure when administered orally to SHR (Matsufuji et al., 1995). This dipeptide also inhibited ACE (Matsufuji et al., 1995), and showed antihypertensive effects in hypertensive patients (Kawasaki et al., 2000, 2002). It has recently been suggested that antihypertensive peptides derived from dried bonito might lower arterial blood pressure by means of mechanisms other than ACE-inhibition. It has been postulated that, as well as inhibiting this enzyme, they might specifically act directly on vascular smooth muscle (Kouno et al., 2005).

Proteolitic digestion of gelatin extracts obtained from the skin of Alaska pollock, a by-product of the industrial processing of fish, has also produced several peptides with potent ACE-inhibitory activity (Byun and Kim, 2001). When these proteins were hydrolyzed with pepsin, five fractions were obtained. In the fraction that showed the greatest ACE-inhibitory activity, an octapeptide with a value of $IC_{50} = 14.7$ µM was discovered (Je et al., 2004). Studies have also been carried out with several hydrolysates of salmon protein. These hydrolysates behaved as potent ACE inhibitors, and showed antihypertensive effects when administered orally to SHR (Ono et al., 2003, 2006). Very recently, peptides with high ACE-inhibitory activity and antihypertensive effects derived from shrimp proteins have been discovered (He et al., 2006; Hai-Lun et al., 2006; Nii et al., 2008).

6.4.4 Antihypertensive Vegetal Peptides

Of the antihypertensive peptides that have been isolated from proteins of vegetable origin, those derived from soy protein are of particular interest. Before describing in detail the studies carried out in order to obtain these peptides, it is worth stating some facts about this food product. Soy and its derivatives form an important part of the diet in eastern countries. Because of its nutritional value and its beneficial effect on health, its consumption is at present also on the increase in western countries. Soy has an important role as a functional food. Diverse properties are attributed to it, including its anticarcinogenic activity, its ability to lower cholesterol, and its effectiveness in preventing cardiovascular diseases, diabetes, and obesity (Messina, 1995, 1999; Anderson, Smith, and Washnock, 1999). Soy proteins also have physico-chemical properties which can prove very useful, such as their ability to gel, emulsify, and foam, as well as their elasticity, viscosity, solubility, cohesion-adhesion, and their capacity for water and fat absorption. These properties may be modified by the conditions under which the different products are prepared. The essential amino acid content of soy is different from that of cows' milk. Soy contains a smaller quantity of Met and Phe, but it is also considered a substance with complete protein value. Soy's reserve proteins are albumins and globulins, with a majority of the latter. Its two important globulins, glycine and β-conglycin, have different structures, and also have some different properties, but it is very important for both proteins to have a gel-forming capacity, because this makes it possible to manufacture tofu (soy protein concentrate), one of the most traditional products in the eastern diet.

Bioactive peptides have been obtained from tofu, from soy seeds (*Glycine max*), from soy by-products (soy whey) and from products obtained in the manufacture of a soy drink (soy milk). Research into antihypertensive peptides derived from soy proteins is, in spite of everything, relatively recent. The first studies of this type of peptide were carried out in 1995, and in these studies, hydrolysates of soy proteins with ACE-inhibitory activity (Shin et al., 1995) and with antihypertensive effect (Yu et al., 1995) were obtained. In 2001, Wu and Ding also obtained peptides with ACE-inhibitory activity, by treating soy proteins with alkalase. These peptides also showed antihypertensive activity when administered orally to SHR (Wu and Ding, 2001).

Later several soy derivatives were obtained by sequential digestion of these proteins with pepsin and pancreatin. Using this procedure many peptide fractions with ACE-inhibitory activity were obtained (Lo and Li-Chan, 2005). Cha and Park have also obtained hydrolysates of soy proteins with a high level of ACE-inhibitory activity by treating these proteins with proteases derived from *alkalophilic Bacillus* sp. (Cha and Park, 2005). In addition, it has been established that different enzymatic hydrolysates of glycinin, the most important storage proteins to be found in soy seeds, present ACE-inhibitory properties (Mallikarjun Gouda et al., 2006). More recently, Zhu et al. (2008) identified ACE-inhibitory peptides in salt-free soy sauce that are transportable across caco-2 cell monolayers. Few studies have, however, been made to evaluate the antihypertensive activity of soy derivatives in animal models (Chen et al., 2004; Yang, H.Y. et al., 2004; Kodera et al., 2006).

Apart from soy, other proteins of vegetable origin are a potential source of peptides with antihypertensive activity. These include maize (Maruyama et al., 1989; Miyoshi et al., 1991), wheat (Matsui et al., 1999; Li et al., 2002; Motoi and Kodama, 2003), rice (He et al., 2005; Li et al., 2007), "wakame" (Suetsuna and Nakano, 2000; Sato et al., 2002a,b; Suetsuna et al., 2004), spinach (Yang, Y. et al., 2003, 2004), sunflower (Megias et al., 2004), bean (Li et al., 2006), sesame (Nakano et al., 2006), broccoli (Lee et al., 2006), and mushroom protein (Hyoung Lee et al., 2004).

Table 6.1 shows the principal antihypertensive peptides derived from food proteins that have thus far been obtained. Their origin, the enzymatic treatment used to obtain them, and the activity they show are all indicated.

6.5 Conclusion

Hypertension is an important problem in our society given its high prevalence and its critical role as a cardiovascular risk factor. The new strategies for treating hypertension based on natural products could greatly benefit this pathology. In this context, it is important to point out that some peptide sequences released by hydrolysis of food proteins can produce a decrease in arterial tone. The idea of using them for the prevention or treatment of hypertension is very attractive, but experiments on animals are unavoidable and of particular importance in the search of food peptides as antihypertensive agents. For these experiments, SHR are normally used. Above all, milk proteins have proved to be important as a source of antihypertensive peptides, and the main mechanism involved in the antihypertensive effect of food-derived peptides is the inhibition of ACE. Nevertheless, antihypertensive peptides derived from egg, fish, and vegetable proteins have also been described, and some of them present direct effects in vascular smooth muscle or can modulate the release of endothelial factors.

The usefulness of antihypertensive hydrolysates and peptides derived from food proteins is particularly clear in prehypertensive subjects who do not yet need pharmacological medication and who control their blood pressure by dietary means. It is highly likely that these products may also be of use in hypertensive patients who do not respond well to pharmacological treatment. Several of them have already proved their safety and effectiveness in hypertensive patients, and have been marketed in functional foods that are at present being used for this purpose. Table 6.2 is a summary of the functional products containing antihypertensive peptides that have already been marketed. The commercial name of each product and the peptide sequences responsible for its effect are shown.

TABLE 6.2
Functional Products with Antihypertensive Peptides

Produce	Commercial Name	Active Component	Company/Country	Clinical Studies
Fermented milk	"Ameal Peptide"[a] "Flora Proactive"[b]	IPP, VPP	Calpis Co./Japan[a] Unilever/Spain[b]	Hata et al., 1996; Mizushima et al., 2004; Aihara et al., 2005; Mizuno et al., 2005
Fermented milk	"Evolus"[a] "LH"[b] "KaikuVita"[c] "Emmi-Evolus"[d]	IPP, VPP	Valio Ltd./Finland[a] Mjölkursamsalan/Iceland[b] Kaiku-Iparlat/Spain[c] Emmi/Switzerland, Portugal, Malta and Italy[d]	Seppo et al., 2002, 2003; Tuomilehto et al., 2004; Jauhiainen et al., 2005
Casein hydrolysate	"Casein DP peptio drink"	FFVAPFEVFGK	Kanebo, Ltd/Japan	Sugai, 1998; Sekiya et al., 1992
Casein hydrolysate	"C12 Peption"	FFVAPFEVFGK	DMV International/Holland	Townsend et al., 2004; Cadee et al., 2007
Whey proteins hydrolysate	"Biozate"	Whey peptides	Davisco/USA	Pins and Keenan, 2002, 2003, 2006
Bonito proteins hydrolysate	"Peptide soup"	LKPNM	NIPPON/Japan	Fujita et al., 1997a,b; Fujita et al., 2001
Sardine proteins hydrolysate	—	VY	Approved by Japanese goverment as FOSHU*	Kawasaki et al., 2000, 2002

Note: Superscript a–d link commercial name to company/country.
* FOSHU: Foods for specified health use

References

Abubakar, A., Saito, T., Kitazawa, H., Kawai, Y., and Itoh, T. 1998. Structural analysis of new antihypertensive peptides derived from cheese whey protein by proteinase K digestion. *J. Dairy Sci.* 81: 3131–3138.

Addeo, F., Chianese, L., Sacchi, R., Musso, S.S., Ferranti, P., and Malorni, A. 1994. Characterization of the oligopeptides of Parmigiano-Reggiano cheese soluble in 120 g trichloroacetic acid/l. *J. Dairy Res.* 61: 365–374.

Adson, A., Raub, T.J., Burton, P.S. et al. 1994. Quantitative approaches to delineate paracellular diffusion in cultured epithelial cell monolayers. *J. Pharm. Sci.* 83: 1529–1536.

Aihara, K., Kajimoto, O., Hirata, H., Takahashi, R., and Nakamura, Y. 2005. Effect of powdered fermented milk with Lactobacillus helveticus on subjects with high-normal blood pressure or mild hypertension. *J. Am. Coll. Nutr.* 24: 257–265.

Almeida, A.P., Frabregas, B.C., Madureira, M.M., Santos, R.J., Campagnole-Santos, M.J., and Santos, R.A. 2000. Angiotensin-(1-7) potentiates the coronary vasodilatatory effect of bradykinin in the isolated rat heart. *Braz. J. Med. Biol. Res.* 33: 709–713.

Alting, A.C., Meijer, R J.G.M., and Van Beresteijn, E.C.H. 1997. Incomplete elimination of the ABBOS epitope of bovine serum albumin under simulated gastrointestinal conditions of infants. *Diabetes Care* 20: 875–880.

Ames, B.N., Shigenaga, M.K., and Hagen, T.M. 1993. Oxidants, antioxidants, and the degenerative diseases of aging. *Proc. Natl. Acad. Sci.* 90: 7915–7922.

Anderson, J.W., Smith, B.M., and Washnock, C.S. 1999. Cardiovascular and renal benefits of dry bean and soybean intake. *Am. J. Clin. Nutr.* 70: 464S–474S.

Appel, L.J., Moore, T.J., Obarzanek, E. et al. 1997. A clinical trial of the effects of dietary patterns on blood pressure. *New Engl. J. Med.* 336: 1117–1124.

Astwood, L.D., Leach, J.N., and Fuschs, R.L. 1996. Stability of food allergens to digestion *in vitro*. *Nat. Biotechnol.* 14: 1269–1273.

Baykal, Y., Yilmaz, M.I., Celik, T. et al. 2003. Effects of antihypertensive agents, alpha receptor blockers, beta blockers, angiotensin-converting enzyme inhibitors, angiotensin receptor blockers and calcium channel blockers, on oxidative stress. *J. Hypertens.* 21: 1207–1211.

Bernier, S.G., Haldar, S., and Michel, T. 2000. Bradykinin-regulated interactions of the mitogen-activated protein kinase pathway with the endothelial nitric-oxide synthase. *J. Biol. Chem.* 275: 30707–30715.

Bree, F., Hamon, G., and Tillement, J.P. 1992. Evidence for two binding sites on membrane-bound angiotensin-converting enzymes (ACE) for exogenous inhibitors except in testis. *Life Sci.* 51: 787–794.

Buñag, R.D. 1973. Validation in awake rats of a tail-cuff method for measuring systolic pressure. *J. Appl. Physiol.* 34: 279–282.

Buñag, R.D. and Butterfield, J. 1982. Tail-cuff blood-pressure measurement without external preheating in awake rats. *Hypertension* 4: 898–903.

Burton, P.S., Conradi, R.A., Hilgers, A.R., Ho, N.F.H., and Maggiora, L.L. 1992. The relationship between peptide structure and transport across epithelial cell monolayers. *J. Control Release* 19: 87–98.

Bütikofer, U., Meyer, J., Sieber, R., Walther, B., and Wechsler, D. 2008. Occurrence of the angiotensin-converting enzyme inhibiting tripeptides Val-Pro-Pro and Ile-Pro-Pro in different cheese varieties of Swiss origin. *J. Dairy Sci.* 91: 29–38.

Byun, H. and Kim, S. 2001. Purification and characterization of angiotensin I converting enzyme (ACE) inhibitory peptides from Alaska Pollack (Theragra chalcogramma) skin. *Process Biochem.* 36: 1155–1162.

Cadee, J.A., Chang, C.Y., Chen, C.W., Huang, C.N., Chen, S.L., and Wang, C.K. 2007. Bovine casein hydrolysate (c12 Peptide) reduces blood pressure in prehypertensive subjects. *Am. J. Hypertens.* 20: 1–5.

CDER/FDA. 2000. Guidance for industry: Waiver of *in vivo* bioavailability and bioequivalence studies for immediate-release solid oral dosage form based on a biopharmaceutical classification system (August).

Cervato, G., Cazzola, R., and Cestaro, B. 1999. Studies on the antioxidant activity of milk caseins. *Int. J. Food Sci. Nutr.* 50: 291–296.

Cha, M. and Park, J.R. 2005. Production and characterization of a soy protein-derived angiotensin I-converting enzyme inhibitory hydrolysate. *J. Med. Food* 8: 305–310.

Chabance, B., Marterua, P., Rambaud, J.C. et al. 1998. Casein peptide release and passage to the blood in humans during digestion of milk or yogurt. *Biochimie* 80: 155–165.

Chang, K.J., Lillian, A., Hazum, E., and Cuatrecasas, P. 1981. Morphicetin: A potent and specific agonist for morphine (μ) receptors. *Science* 212: 75–77.

Chen, H.M., Muramoto, K., Yamauchi, F., and Nokihara, K. 1996. Antioxidant activity of peptides designed based on the antioxidative peptide isolated from digests of a soybean protein. *J. Agric. Food Chem.* 44: 2619–2623.

Chen, J.R., Yang, S.C., Suetsuna, K., and Chao, J.C.J. 2004. Soybean protein-derived hydrolysate affects blood pressure in spontaneously hypertensive rats. *J. Food Biochem.* 28: 61–73.

Cheung, H.S., Wang, F.L., Ondetti, M.A., Sabo, E.F., and Cushman, D.W. 1980. Binding of peptides substrates and inhibitors of angiotensin-converting enzyme. *J. Biol. Chem.* 25: 401–407.

Chiba, H., Tani, F., and Yoshikawa, M. 1989. Opioid antagonist peptides derived from κ-casein. *J. Dairy Res.* 56: 363–366.

Chobert, J.M., El-Zahar, K., Sitohy, M. et al. 2005. Angiotensin I-converting-enzyme (ACE)-inhibitory activity of tryptic peptides of ovine beta-lactoglobulin and of milk yoghurts obtained by using different starters. *Lait* 85: 141–152.

Clemente, A. 2000. Enzymatic protein hydrolysates in human nutrition. *Trends Food Sci.* 11: 254–262.

Collins, R., Peto, R., MacMahon, S. et al. 1990. Blood pressure, stroke, and coronary heart disease. Part 2, Short-term reductions in blood pressure: Overview of randomised drug trials in their epidemiological context. *Lancet* 335: 827–838.

Conradi, R.A., Hilgers, A.R., Ho, N.F.H., and Burton, P.S. 1992. The influence of peptide structure on transport across Caco-2 cells. II. Peptide bond modification which results in improved permeability. *Pharm. Res.* 9: 435–439.

Conradi, R.A., Wilkinson, K.F., Rush, B.D., Hilgers, A.R., Ruwart , M.J., and Burton, P.S. 1993. *In vitro/in vivo* models for peptide oral absorption: Comparison of Caco-2 cell permeability with rat intestinal absorption of renin inhibitory peptides. *Pharm. Res.* 10: 1790–1792.

Cushman, D.W. and Cheung, H.S. 1971. Spectrophotometric assay and properties of the angiotensin-converting enzyme of rabbit lung. *Biochem. Pharmacol.* 20: 1637–1648.

Cushman, D.W., Cheung, H.S., Sabo, E.F., and Ondetti, M.A. 1977. Design of potent competitive inhibitors of angiotensin-converting enzyme. Carboxyalkanoyl and mercaptoalkanoyl amino acids. *Biochemistry* 16: 5484–5491.

Cushman, D.W., Pluscec, J., Williams, N.J. et al. 1973. Inhibition of angiotensin-converting enzyme by analogs of peptides from Bothrops jaraca venom. *Experientia* 29: 1032–1035.

D'Alvise, N., Lesueur-Lambert, C., Ferin, B., Dhulster, P., and Guillochon, D. 2000. Hydrolysis and large scale ultrafiltration study of alfalfa protein concentrate enzymatic hydrolysate. *Enzyme Microb. Tech.* 27: 286–294.

Daniel, H., Morse, E.L., and Adibi, S.A. 1992. Determinants of substrate affinity for the oligopeptide/H^+ symporter in the renal brush border membrane. *J. Biol. Chem.* 267: 9565–9573.

Dávalos, A., Gómez-Cordovés, C., and Bartolomé, B. 2004a. Extending applicability of the oxygen radical absorbance capacity (ORAC-fluorescein) assay. *J. Food Prot.* 67: 1939–1944.

Dávalos, A., Miguel, M., Bartolomé, B., and López-Fandiño, R. 2004b. Antioxidant activity of peptides derived from egg white proteins by enzymatic hydrolysis. *J. Food Prot.* 67: 1939–1944.

Deddish, P.A., Marcic, B., Jackman, H.L., Wang, H.Z., Skidgel, R.A., and Erdös, E.G. 1998. N-domain-specific substrate and C-domain inhibitors of angiotensin-converting enzyme: angiotensin-(1-7) and keto-ACE. *Hypertension* 31: 912–917.

Deddish, P.A., Wang, L.X., Jackman, H.L. et al. 1996. Single-domain angiotensin I converting enzyme (kininase II): Characterization and properties. *J. Pharmacol. Exp. Ther.* 279(3): 1582–1589.

Donoghue, M., Hsieh, F., Baronas, E. et al. 2000. A novel angiotensin-converting enzyme-related carboxypeptidase (ACE2) converts angiotensin I to angiotensin 1-9. *Circ. Res.* 87: E1–E9.

Dzau, V.J. 2001. Theodore Cooper lecture: Tissue angiotensin and pathobiology of vascular disease: a unifying hypothesis. *Hypertension* 37: 1047–1052.

Engberink, M.F., Schouten, E.G., Kok, F.J., van Mierlo, L.A., Brouwer, I.A., and Geleijnse, J.M. 2008. Lactotripeptides show no effect on human blood pressure: Results from a double-blind randomized controlled trial. *Hypertension* 51: 399–405.

Erdös, E.G. 1976. Conversion of angiotensin I to angiotensin II. *Am. J. Med.* 60: 749–759.

Fazel, A. 1998. Nutritional and health benefits of yogurt and fermented milks. Functional peptides. *Danone World News.*

FDA. 1999. Guidance for industry waiver of *in vivo* bioavailability and bioequivalence studies base don Biopharmaceutics Classification system. CDER draft – January.

Fiat, A.M., Migliore-Samour, D., Jollès, P., Drouet, L., Bal dit Sollier, C., and Caen, J. 1993. Biologically active peptides from milk proteins with emphasis on two examples concerning antithrombotic and immunomodulating activities. *J. Dairy Sci.* 76: 301–310.

Fischer, M.A. and Avorn, J. 2004. Economic implications of evidence-based prescribing for hypertension: Can better care cost less? *JAMA* 291: 1850–1856.

Fitzgerald, R.J., Murria, B.A., and Walsh, G.J. 2004. Hypotensive peptides from milk proteins. *J. Nutr.* 134: 980S–988S.

Fujita, H., Sasaki, R., and Yoshikawa, M. 1995b. Potentiation of the antihypertensive activity of orally administered Ovokinin, a vasorelaxing peptide derived from ovalbumin, by emulsification in egg phosphatidyl-choline. *Biosci. Biotech. Biochem.* 59: 2344–2345.

Fujita, H., Usui, H., Kurahashi, K., and Yoshikawa, M. 1995a. Isolation and characterization of Ovokinin, a bradykinin B1 agonist peptide derived from ovalbumin. *Peptides* 16: 785–790.

Fujita, H., Yamagami, T., and Ohshima, K. 2001. Effects of an ACE-inhibitory agent, katsuobushi oligopeptide, in the spontaneously hypertensive rat and in borderline and mildly hypertensive subjects. *Nutr. Res.* 21: 1149–1158.

Fujita, H., Yasumoto, R., Hasegawa, M., and Ohshima, K. 1997a. Human volunteers study on antihypertensive effect of "Katsuobushi Oligopeptide" (I). *Jpn. Pharmacol. Ther.* 25: 147–151.

Fujita, H., Yasumoto, R., Hasegawa, M., and Ohshima, K. 1997b. Human volunteers study on antihypertensive effect of "Katsuobushi Oligopeptide" (II) – A placebo-controlled study on the effect of "peptide soup" on blood pressure in borderline and hypertensive subjects. *Jpn. Pharmacol. Ther.* 25: 153–157.

Fujita, H., Yokoyama, K., and Yoshikawa, M. 2000. Classification and antihypertensive activity of angiotensin I-converting enzyme inhibitory peptides derived from food proteins. *J. Food Sci.* 65: 564–569.

Fujita, H., Yokoyama, K., Yaumoto, R., and Yoshikawa, M. 1995c. Antihypertensive effect of thermolysin digest of dried bonito in spontaneously hypertensive rat. *Clin. Exp. Pharmacol. Physiol.* Suppl 22: S304–S305.

Fujita, H. and Yoshikawa, M. 1999. LKPNM: a prodrug-type ACE-inhibitory peptide derived from fish protein. *Immunopharmacology* 44: 123–127.

Ganapathy, V. and Leibach, F.H. 1985. Is intestinal peptide transport energized by a proton gradient? *Am. J. Physiol.* 249: G153–G160.

García Del Río, C., Smellie, W.S., and Morton, J.J. 1981. Des-Asp-angiotensin I: Its identification in rat blood and confirmation as a substrate for converting enzyme. *Endocrinology* 108: 406–412.

Gardiner, S.M., Kemp, P.A., March, J.E., and Bennett, T. 1993. Regional haemodynamic effects of angiotensin II (3-8) in conscious rats. *Br. J. Pharmacol.* 110: 159–162.

Gardner, M.I.G. 1994. Absorption of intact proteins and peptides. In *Physiology of the Gastrointestinal Tract* (L.R. Johnson, Ed.), New York: Raven Press, pp. 1795–1820.

Garrison, E.A., Santiago, J.A., Osei, S.Y., and Kadowitz, P.J. 1995. Analysis of responses to angiotensin peptides in the hindquarters vascular bed of the cat. *Am. J. Physiol.* 268: H2418–H2425.

Gaynes, R.P., Szidon, J.P., and Oparil, S. 1978. *In vivo* and *in vitro* conversion of des-1-Asp angiotensin I to angiotensin III. *Biochem. Pharmacol.* 27: 2871–2877.

Gobbetti, M., Ferranti, P., Smacchi, E., and Goffredi, F. 2000. Production of angiotensin I-converting peptides in fermented milks started by Lactobacillus delbruckii subsp bulgaricus SS1 and Lactococcus lactis subsp cremoris FT5. *Appl. Environ. Microbiol.* 66: 3898–3904.

Gómez-Ruiz, J.A., Ramos, M., and Recio, I. 2003. Angiotensin-converting enzyme-inhibitory peptides in Manchego cheeses manufactured with different starter cultures. *Int. Dairy J.* 12: 697–706.

Gómez-Ruiz, J.A., Ramos, M., and Recio, I. 2004a. Angiotensin-converting enzyme-inhibitory peptides isolated from Manchego cheeses. Stability under simulated gastrointestinal digestion. *Int. Dairy J.* 14: 1075–1080.

Gómez-Ruiz, J.A., Recio, I., and Belloque, J. 2004b. ACE-Inhibitory activity and structural properties of peptide Asp-Lys-Ile-His-Pro [β-CN f(47-51)]. Study of the peptide forms synthesized by different methods. *J. Agric. Food Chem.* 52: 6315–6319.

Goodwin, J.T., Mao, B., Vidmar, T.J., Conradi, R.A., and Burton, P.S. 1999. Strategies toward predicting peptide cellular permeability from computed molecular descriptors. *J. Pept. Res.* 53: 355–369.

Gurer, H., Neal, R., Yang, P., Oztezcan, S., and Ercal, N. 1999. Captopril as an antioxidant in lead-exposed Fisher 344 rats. *Hum. Exp. Toxicol.* 18: 27–32.

Haileselassie, S.S., Lee, B.H., and Gibbs, B.F. 1999. Purification and identification of potentially bioactive peptides from enzyme-modified cheese. *J. Dairy Sci.* 82: 1612–1617.

Hai-Lun, H., Xiu-Lan, C., Cai-Yun, S., Yu-Zhong, Z., and Bai-Cheng, Z. 2006. Analysis of novel angiotensin-I-converting enzyme inhibitory peptides from protease-hydrolyzed marine shrimp Acetes chinensis. *J. Pept. Sci.* 12: 726–733.

Hata, Y., Yamamoto, M., Ohni, M., Nakajima, K., Nakamura, Y., and Takano, T. 1996. A placebo-controlled study of the effect of sour milk on blood pressure in hypertensive subjects. *Am. J. Clin. Nutr.* 64: 767–771.

He, G.Q., Xuan, G.D., Ruan, H., Chen, Q.H., and Xu, Y. 2005. Optimization of angiotensin I-converting enzyme (ACE) inhibition by rice dregs hydrolysates using response surface methodology. *J. Zhejiang Univ. Sci. B* 6: 508–513.

He, H., Chen, X., Sun, C., Zhang, Y., and Gao, P. 2006. Preparation and functional evaluation of oligopeptide-enriched hydrolysate from shrimp (Acetes chinensis) treated with crude protease from Bacillus sp. SM98011. *Bioresour. Technol.* 97: 385–390.

Hernández-Ledesma, B., Miguel, M., Amigo, L., Aleixandre, M.A., and Recio, I. 2007. Effect of simulated gastrointestinal digestion on the antihypertensive properties of β-lactoglobulin peptides. *J. Dairy Res.* 74: 336–339.

Hernández-Ledesma, B., Recio, I., Ramos, M., and Amigo, L. 2002. Preparation of ovine and caprine β-lactoglobulin hydrolysates with ACE-inhibitory activity. Identification of active peptides from caprine β-lactoglobulin hydrolysed with thermolysin. *Int. Dairy J.* 12: 805–812.

Heyman, M. and Desjeux, J.F. 1992. Significance of intestinal food protein transport. *J. Pediat. Gastroent. Nutr.* 15: 48–57.

Hidalgo, I.J., Raub, T.J., and Borchardt, R.T. 1989. Characterization of the human coloncarcinoma cell line (Caco-2) as a model system form intestinal epithelial permeability. *Gastroenterology* 96: 736–749.

Hilgers, A.R., Conradi, R.A., and Burton, P.S. 1990. Caco-2 cell monolayers as a model for drug transport across the intestinal mucosa. *Pharm. Res.* 7: 902–910.

Hirota, T., Ohki, K., Kawagishi, R., Kajimoto, Y., Mizuno, S., Nakamura, Y., and Kitakaze, M. 2007. Casein hydrolysate containing the antihypertensive tripeptides Val-Pro-Pro and Ile-Pro-Pro improves vascular endothelial function independent of blood pressure-lowering effects: contribution of the inhibitory action of angiotensin-converting enzyme. *Hypertens. Res.* 30: 489–496.

Hughes, J., Kosterlitz, H.W., and Smith, T.W. 1977. The distribution of methionine-enkephalin and leucine-enkephalin in the brain and peripheral tissues. *Br. J. Pharmacol.* 61: 639–647.

Husain, A. 1993. The chymase-angiotensin system in humans. *J. Hypertens.* 11: 1155–1159.

Hyoung Lee, D., Ho Kim, J., Sik Park, J., Jun Choi, Y., and Soo Lee, J. 2004. Isolation and characterization of a novel angiotensin I-converting enzyme inhibitory peptide derived from the edible mushroom Tricholoma giganteum. *Peptides* 25: 621–627.

Ijäs, H., Collin, M., Finchenberg, P. et al. 2004. Antihypertensive opioid-like milk peptide alpha-lactorphin: Lack of effect on behavioural test in mice. *Int. Dairy J.* 14: 201–205.

Jauhiainen, T. and Korpela, R. 2007. Milk peptides and blood pressure. *J. Nutr.* 137: 825S–829S.

Jauhiainen, T., Vapaatalo, H., Poussa, T., Kyronpalo, S., Rasmussen, M., and Korpela, R. 2005. Lactobacillus helveticus fermented milk lowers blood pressure in hypertensive subjects in 24-h ambulatory blood pressure measurement. *Am. J. Hypertens.* 18: 1600–1605.

Je, J.Y., Park, P.J., Kwon, J.Y., and Kim, S.K. 2004. A novel angiotensin I converting enzyme inhibitory peptide from Alaska pollack (Theragra chalcogramma) frame protein hydrolysate. *J. Agric. Food Chem.* 52: 7842–7845.

Jiang, H., Grieve, P.A., Marschke, R.J., Wood, A.F., Dionysius, D.A., and Alewood, P.F. 1998. Application of tandem mass spectrometry in the characterisation of flavour and bioactive peptides. *Aust. J. Dairy Technol.* 53: 119.

Karaki, H., Doi, K., Sugano, S. et al. 1990. Antihypertensive effect of tryptic hydrolysate of milk casein in spontaneously hypertensive rats. *Comp. Biochem. Physiol.* 96C: 367–371.

Karaki, H., Kuwahara, M., Sugano, S. et al. 1993. Oral administration of peptides derived from bonito bowels decreases blood pressure in spontaneously hypertensive rats by inhibiting angiotensin converting enzyme. *Comp. Biochem. Physiol. C* 104: 351–353.

Kawasaki, T., Jun, C.J., Fukushina, Y. et al. 2002. Antihypertensive effect and safety evaluation of vegetable drink with peptides derived from sardine protein hydrolysates on mild hypertensive, high-normal, and normal blood pressure subjects. *Fukuoka Igaku Zasshi* 93: 208–218.

Kawasaki, T., Seki, E., Osajima, K. et al. 2000. Antihypertensive effect of Valyl-tyrosine, a short chain peptide derived of sardine muscle hydrolyzate, on mild hypertensive subjects. *J. Hum. Hypertens.* 14: 519–523.

Kawase, M., Hashimoto, H., Hosada, M., Morita, H., and Hosono, A. 2000. Effect of administration of fermented milk containing whey proteins concentrate to rats and healthy men on serum lipids and blood pressure. *J. Dairy Sci.* 83: 255–263.

Kim, D.C., Burton, P.S., and Borchardt, R.T.A. 1993. A correlation between the permeability characteristics of a series of peptides using an *in vitro* cell culture model (Caco-2) and those using an in situ perfused rat ileum model of the intestinal mucosa. *Pharm. Res.* 10: 1701–1714.

Kim, S.K., Byun, H.G., Park, P.J, and Shahidi, F. 2001. Angiotensin I converting enzyme inhibitory peptides purified from bovine skin gelatin hydrolysate. *J. Agric. Food Chem.* 49: 2992–2997.

Kodera, T. and Nio, N. 2006. Identification of an angiotensin I-converting enzyme inhibitory peptides from protein hydrolysates by a soybean protease and the antihypertensive effects of hydrolysates in spontaneously hypertensive rats models. *J. Food Sci.* 71: C164–C173.

Kohama, Y., Matsumoto, S., Oka, H., Teramoto, T., Okabe, M., and Mimura, Y. 1988. Isolation of angiotensin-converting enzyme inhibitor from tuna muscle. *Biochem. Biophys. Res. Commun.* 155: 332–337.

Kohama, Y., Oka, H., Matsumoto, S. et al. 1989. Biological properties of angiotensin converting enzyme inhibitor derived from tuna muscle. *J. Pharmacobio.-Dynan* 12: 566–571.

Kohmura, M., Nio, N., Kubo, K., Minishima, Y., Munekata, E., and Ariyoshi, Y. 1989. Inhibition of angiotensin-converting enzyme by synthetic peptides of human β-casein. *Agric. Biol. Chem.* 53: 2107–2114.

Kudoh, K., Matsumoto, M., Onodera, S., Takeda, Y., Ando, K., and Shiomi, N. 2003. Antioxidative activity and protective effect against ethanol-induced gastric mucosal damage of a potato protein hydrolysate. *J. Nutr. Sci. Vitaminol.* 49: 451–455.

Kuono, K., Hirano, S., Kuboki, H., Kasai, M., and Hatae, K. 2005. Effects of dried bonito (Katsuobushi) and captopril, an angiotensin I-converting enzyme inhibitor, on rat isolated: A possible mechanism of antihypertensive action. *Biosci. Biotech. Biochem.* 69: 911–915.

Landmesser, U. and Drexler, H. 2006. Effect of angiotensin II type 1 receptor antagonism on endothelial function: role of bradykinin and nitric oxide. *J. Hypertens.* Suppl. (1): S39–S43.

Laursen, J.B., Rajagopalan, S., Galis, Z., Tarpey, M., Freeman, B.A., and Harrison, D.G. 1997. Role of superoxide in angiotensin II-induced but catecholamine-induced hypertension. *Circulation* 95: 588–593.

Lee, J.E., Bae, I.Y., Lee, H.G., and Yang, C.B. 2006. Tyr-Pro-Lys, an angiotensin I-converting enzyme inhibitory peptide derived from broccoli (Brassica oleracea Italica). *Food Chem.* 99: 143–148.

Lee, N.Y., Cheng, J.T., Enomoto, T., and Nakano, Y. 2006a. One peptide derived from hen ovotransferrin as prodrug to inhibit angiotensin converting enzyme. *J. Food Drug. Anal.* 14: 31–35.

Lee, N.Y., Cheng, J.T., Enomoto, T., and Nakano, Y. 2006b. Antihypertensive effect of angiotensin converting enzyme inhibitory peptide obtained from hen ovotransferrin. *J. Chinese Chem. Soc.* 53: 495–501.

Lee, Y.K. and Salminen, S. 1995. The coming of age of probiotics. *Trends Food Sci. Technol.* 6: 241–245.

Lee, Y.M., Skurk, T., Hennig, M., and Hauner, H. 2007. Effect of a milk drink supplemented with whey peptides on blood pressure in patients with mild hypertension. *Eur. J. Nutr.* 46: 21–27.

Leibach, F.H. and Ganapathy, V. 1986. Peptide transporters in the intestine and the kidney. *Ann. Rev. Nutr.* 16: 99–119.

Li, C. H., Matsui, T., Matsumoto, K., Yamasaki, R., and Kawasaki, T. 2002. Latent production of angiotensin I-converting enzyme inhibitors from buckwheat protein. *J. Pept. Sci.* 8: 267–274.

Li, G.H., Liu, H., Shi, Y.H., and Le, G.W. 2005. Direct spectrophotometric measurement of Angiotensin I-converting ezyme inhibitory activity for screening bioactive peptides. *J. Pharm. Biomed. Anal.* 37: 219–224.

Li, G.H., Qu, M.R., Wan, J.Z., and You, J.M. 2007. Antihypertensive effect of rice protein hydrolysate with *in vitro* angiotensin I-converting enzyme inhibitory activity in spontaneously hypertensive rats. *Asia Pac. J. Clin. Nutr.* 16 Suppl: 275–280.

Li, G.H., Wan, J.Z., Le, G.W., and Shi, Y.H. 2006. Novel angiotensin I-converting enzyme inhibitory peptides isolated from Alcalase hydrolysate of mung bean protein. *J. Pept. Sci.* 12: 509–514.

Lo, W.M. and Li-Chan, E.C. 2005. Angiotensin I converting enzyme inhibitory peptides from *in vitro* pepsin-pancreatin digestion of soy protein. *J. Agric. Food Chem.* 53: 3369–3376.

Loufrani, L., Henrion, D., Chansel, D., Ardaillou, R., and Levy, B.I. 1999. Functional evidence for an angiotensin IV receptor in rat resistance arteries. *J. Pharmacol. Exp. Ther.* 291: 583–588.

MacMahon, S., Peto, R., Cutler, J. et al. 1990. Blood pressure, stroke, and coronary heart disease. Part 1, Prolonged differences in blood pressure: Prospective observational studies corrected for the regression dilution bias. *Lancet* 335: 765–774.

Maeno, M., Yamamoto, N., and Takano, T. 1996. Identification of an antihypertensive peptide from casein hydrolysate produced by a proteinase from Lactobacillus helveticus CP790. *J. Dairy Sci.* 79: 1316–1321.

Maes, W., Van Camp, J., Vermeirssen, V. et al. 2004. Influence of the lactokinin Ala-Leu-Pro-Met-His-Ile-Arg (ALPMHIR) on the release of endothelial cells. *Regul. Pept.* 18: 105–109.

Mallikarjun Gouda, K.G., Gowda, L.R., Rao, A.G., and Prakash, V. 2006. Angiotensin I-converting enzyme inhibitory peptide derived from glycinin, the 11S globulin of soybean (Glycine max). *J. Agric. Food Chem.* 54: 4568–4573.

Marceau, F. 1995. Kinin B1 receptors: A review. *Immunopharmacology* 30: 1–26.

Marceau, F., Larrivee, J.F., Saint-Jacques, E., and Bachvarov, D.R. 1997. The kinin B1 receptor: An inducible G protein coupled receptor. *Can. J. Physiol. Pharmacol.* 75: 725–730.

Maruyama, S., Mitachi, H., Awaya, J., Kurono, M., Tomizuka, N., and Suzuki, H. 1987a. Angiotensin I-converting enzyme inhibitory activity of the C-terminal hexapeptide of α s1-casein. *Agric. Biol. Chem.* 51: 2557–2561.

Maruyama, S., Mitachi, S., Tanaka, H., Tomizuka, N., and Suzuki, H. 1987b. Studies on the active site antihypertensive activity of angiotensin I-converting inhibitors derived from casein. *Agric. Biol. Chem.* 51: 1581–1586.

Maruyama, S., Miyoshi, S., Kaneko, T., and Tanaka, H. 1989. Angiotensin I-converting enzyme inhibitory activities of synthetic peptides related to the tandem repeated sequence of a maize endosperm protein. *Agric. Biol. Chem.* 53: 1077–1081.

Maruyama, S. and Suzuki, H. 1982. A peptide inhibitor of angiotensin-I converting enzyme in the tryptic hydrolyzate of casein. *Agric. Biol. Chem.* 46: 1393–1394.

Masuda, O., Nakamura, Y., and Takano, T. 1996. Antihypertensive peptides are present in aorta after oral administration of sour milk containing these peptides to spontaneously hypertensive rats. *J. Nutr.* 126: 3063–3068.

Matoba, N., Usui, H., Fujita, H., and Yoshikawa, M. 1999. A novel antihypertensive peptide derived from ovalbumin induces nitric oxide-mediated vasorelaxation in an isolated SHR mesenteric artery. *FEBS Lett.* 452: 181–184.

Matoba, N., Yamada, Y., Usui, H., Nakagiri, R., and Yoshikawa, M. 2001. Designing potent derivatives of Ovokinin (2-7), an antihypertensive peptide derived from ovalbumin. *Biosci. Biotech. Biochem.* 65: 636–639.

Matsufuji, H., Matsui, T., Ohshige, S., Kawasaki, T., Osajima, K., and Osajima, T. 1995. Antihypertensive effects of angiotensin fragments in SHR. *Biosci. Biotechnol. Biochem.* 59: 1398–1401.

Matsufuji, H., Matsui, T., Seki, E., Osajima, K., Nakashima, M., and Osajima, Y. 1994. Angiotenin I-converting enzyme inhibitory peptides in an alkaline protease hydrolyzate derived from sardine muscle. *Biosci. Biotechnol. Biochem.* 58: 2244–2245.

Matsui, T., Li, C.H., and Osajima, Y. 1999. Preparation and characterization of novel bioactive peptides responsible for angiotensin I-converting enzyme inhibition from wheat germ. *J. Peptide Sci.* 5: 289–297.

Matsui, T., Matsufuji, H., Seki, E., Osajima, K., Nakashima, M., and Osajima, Y. 1993. Inhibition of angiotensin I-converting enzyme by Bacillus licheniformis alkaline protease hydrolyzates derived from sardine muscle. *Biosci. Biotechnol. Biochem.* 57: 922–925.

Matsumura, N., Fujii, M., Takeda, Y., Sugita, K., and Shimizu, T. 1993. Angiotensin I-converting enzyme inhibitory peptides derived from bonito bowels autolysate. *Biosci. Biotech. Biochem.* 57: 695–697.

Megias, C., del Mar Yust, M., Pedroche, J. et al. 2004. Purification of an ACE inhibitory peptide after hydrolysis of sunflower (Helianthus annuus L.) protein isolates. *J. Agric. Food Chem.* 52: 1928–1932.

Meisel, H. 1997. Biochemical properties of bioactive peptides derived from milk proteins: Potential nutraceuticals for food and pharmaceutical applications. *Liv. Prod Sci.* 50: 125–138.

Meisel, H. 1998. Overview on milk protein-derived peptides. *Int. Dairy J.* 8: 363–373.

Meisel, H. 2001. Bioactive peptides from milk proteins: A perspective for consumers and producers. *Aust. J. Dairy Technol.* 56: 83–92.

Messina, M. 1995. Modern applications for an ancient bean: Soybeans and the prevention and treatment of chronic disease. *J. Nutr.* 125: 567S–569S.

Messina, M. 1999. Legumes and soybeans: Overview of their nutritional profiles and health effects. *Am. J. Clin. Nutr.* 70: 439S–450S.

Mierke, D.F., Nöbner, G., Schiller, P.W., and Goodman, M. 1990. Morphicetin analogs containing 2-aminocyclopentane carboxylic acid as a peptidomimetic for proline. *Int. J. Pep. Res.* 35: 34–45.

Miguel, M. and Aleixandre, A. 2006. Antihypertensive peptides derived from egg proteins. *J. Nutr.* 136: 1457–1460.

Miguel, M., López-Fandiño, R., Ramos, M., and Aleixandre, M.A. 2005b. Short-term effect of egg white hydrolysate products on the arterial blood pressure of hypertensive rats. *Br. J. Nutr.* 94: 731–737.

Miguel, M., López-Fandiño, R., Ramos, M., and Aleixandre, M.A. 2006c. Long-term antihypertensive effect of egg white treated with pepsin in hypertensive rats. *Life Sci.* 78: 2960–2966.

Miguel, M., Muguerza, B., Sánchez, E. et al. 2005a. Changes in arterial blood pressure of milk fermented by Enterococcus faecalis CECT 5728 in spontaneously hypertensive rats. *Br. J. Nutr.* 93: 1–9.

Miguel, M., Ramos, M., Aleixandre, M.A., and López-Fandiño, R. 2006a. Antihypertensive peptides obtained from egg white proteins by enzymatic hydrolysis. Stability under simulated gastrointestinal digestion. *J. Agric. Food Chem.* 54: 726–731.

Miguel, M., Recio, I., Gómez-Ruiz, J.A., Ramos, M., and López-Fandiño, R. 2004. Angiotensin I-converting enzyme inhibitory activity of peptides derived from egg white proteins by enzymatic hydrolysis. *J. Food Prot.* 67: 1914–1920.

Miguel, M., Recio, I., Ramos, M., Delgado, M.A., and Aleixandre, M.A. 2006b. Effect of ACE-inhibitory peptides obtained from Enterococcus faecalis fermented milk in hypertensive rats. *J. Dairy Sci.* 89: 3352–3359.

Mine, Y. 1995. Recent advances in the understanding of egg white protein functionality. *Trends Food Sci. Technol.* 6: 225–232.

Miyoshi, S., Ishikawa, H., Kaneko, T., Fukui, F., Tanaka, H., and Maruyama, S. 1991. Structures and activity of angiotensin-converting enzyme inhibitors in an α-zein hydrolysate. *Agric. Biol. Chem.* 55: 1313–1771.

Mizuno, S., Matsura, K., Gotou, T. et al. 2005. Antihypertensive effect of casein hydrolysate in a placebo-controlled study in subjects with high-normal blood pressure and mild hypertension. *Br. J. Nutr.* 94: 84–91.

Mizushima, S., Ohshige, K., Watanabe, J. et al. 2004. Randomized controlled trial of sour milk on blood pressure in borderline hypertensive men. *Am. J. Hypertens.* 17: 701–706.

Moosmann, B. and Behl, C. 2002. Secretory peptide hormones are biochemical antioxidants: Structure-activity relationship. *Mol. Pharmacol.* 61: 260–268.

Motoi, H. and Kodama, T. 2003. Isolation and characterization of Angiotensin I-coverting enzyme inhibitory peptides from wheat gliadin hydrolysate. *Nahrung/Food* 47: 354–358.

Muguerza, B., Ramos, M., Sánchez, E. et al. 2006. Antihypertensive activity of milks fermented by Enterococcus faecalis strains isolated from raw milk. *Int. Dairy J.* 16: 61–69.

Mullally, M.M., Meisel, H., and FitzGerald, R. J. 1996. Synthetic peptides corresponding to α-lactalbumin and α-lactoglobulin sequences with angiotensin-I-converting enzyme inhibitory activity. *Biol. Chem.* 377: 259–260.

Mullally, M.M., Meisel, H., and FitzGerald, R.J. 1997a. Angiotensin-I-converting enzyme inhibitory activities of gastric and pancreatic proteinase digest of whey proteins. *Int. Dairy J.* 7: 299–303.

Mullally, M.M., Meisel, H., and FitzGerald, R.J. 1997b. Identification of a novel angiotensin-I-converting enzyme inhibitory peptide α-lactoglobulin corresponding to a tryptic fragment of bovine. *FEBS Lett.* 402: 99–101.

Murakami, M., Tonouchi, H., Takahashi, R. et al. 2004. Structural analysis of a new antihypertensive peptide (beta-lactosin B) isolated from a commercial whey product. *J. Dairy Sci.* 87: 1967–1974.

Murray, B.A., Walsh, D.J., and FitzGerald, R.J. 2004. Modification of the furanacryloyl-L-phenylalanylglycylglycine assay for determination of angiotensin-I-converting enzyme inhibitory activity. *J. Biochem. Biophys. Methods* 59: 127–137.

Nakamura, Y., Masuda, O., and Takano, T. 1996. Decrease of tissue angiotensin I-converting enzyme activity upon feeding sour milk in spontaneously hypertensive rats. *Biosci. Biotech. Biochem.* 60: 488–489.

Nakamura, Y., Yamamoto, N., Sakai, K., Okubo, A., Yamazaki, S., and Takano, T.J. 1995b. Antihypertensive effect of sour milk and peptides isolated from it that are inhibitors to angiotensin I-converting enzyme. *J. Dairy Sci.* 78: 1253–1257.

Nakamura, Y., Yamamoto, N., Sakai, K., Takano, T., Okubo, A., and Yamazaki, S. 1995a. Purification and characterization of angiotensin-converting enzyme inhibitors from sour milk. *J. Dairy Sci.* 78: 777–783.

Nakano, D., Ogura, K., Miyakoshi, M. et al. 2006. Antihypertensive effect of angiotensin I-converting enzyme inhibitory peptides from a sesame protein hydrolysate in spontaneously hypertensive rats. *Biosci. Biotechnol. Biochem.* 70: 1118–1126.

National Heart, Lung and Blood Institute. 2003. The seventh report of the joint national committee on prevention, detection, evaluation, and treatment of high blood pressure. 03-5233. Bethesda, MD: National Institutes of Health.

Ni, A., Chao, L., and Chao, J. 1998. Transcription factor nuclear factor kappaB regulates the inducible expression of the human B1 receptor gene in inflammation. *J. Biol. Chem.* 273: 2784–2791.

Nii, Y., Fukuta, K., Yoshimoto, R., Sakai, K., and Ogawa, T. 2008. Determination of antihypertensive peptides from an izumi shrimp hydrolysate. *Biosci. Biotechnol. Biochem.* 72: 861–864.

Nurmeinin, M.L., Sipola, M., Kaarto, H. et al. 2000. α-Lactorphin lowers blood pressure measured by radiotelemetry in normotensive and spontaneously hypertensive rats. *Life Sci.* 66: 535–543.

Okamoto, A., Hanagata, H., Matsumoto, E., Kawamura, Y., Koizumi, Y., and Yanadiga, F. 1995. Angiotensin I converting enzyme inhibitory activities of various fermented milks. *Biosci. Biotechnol. Biochem.* 59: 1147–1149.

Okamoto, K. and Aoki, K. 1963. Development of a strain of spontaneously hypertensive rats. *Japan Circ. J.* 27: 282–293.

Okitsu, M., Morita, A., Kakitani, M., Okada, M., and Yokogoshi, H. 1995. Inhibition of the endothelin-converting enzyme by pepsin digests of food proteins. *Biosci. Biotechnol. Biochem.* 59: 325–326.

Ondetti, M.A. and Cushman, D.W. 1982. Enzymes of the renin-angiotensin system and their inhibitors. *Ann. Rev. Biochem.* 51: 283–308.

Ondetti, M.A., Rubin, B., and Cushman, D.W. 1977. Design of specific inhibitors of angiotensin-converting enzyme: new class of orally active antihypertensive agents. *Science* 196: 441–444.

Ono, S., Hosokawa, M., Miyashita, K., and Takahashi, K. 2003. Isolation of peptides with angiotensin I-converting enzyme inhibitory effect derived from hydrolysate of upstream chum salmon muscle. *J. Food Sci.* 68: 1611–1614.

Ono, S., Hosokawa, M., Miyashita, K., and Takahashi, K. 2006. Inhibition properties of dipeptides from salmon muscle hydrolysate on angiotensin I-converting enzyme. *Int. J. Food Sci. Tech.* 41: 4383–386.

Pappenheimer, J.R., Dahl, C.E., Karnovsky, M.L., and Maggio, J.E. 1994. Intestinal absorption and excretion of octapeptides composed of D-aminoacides. *Proc. Natl. Acad. Sci. USA* 91: 1942–1945.

Patel, J.M., Martens, J.R., Li, Y.D., Gelband, C.H., Raizada, M.K., and Block, E.R. 1998. Angiotensin IV receptor-mediated activation of lung endothelial NOS is associated with vasorelaxation. *Am. J. Physiol.* 275: L1061–L1068.

Pihlanto-Leppälä, A., Koskinen, P., Piilola, K., Tupasela, T., and Korhonen, H. 2000. Angiotensin I-converting enzyme inhibitory properties of whey protein digests: Concentration and characterization of active peptides. *J. Dairy Res.* 1: 53–64.

Pihlanto-Leppälä, A., Rokka, T., and Korhonen, H. 1998. Angiotensin-I-converting enzyme inhibitory peptides derived from bovine milk proteins. *Int. Dairy J.* 8: 325–331.

Pins, J.J. and Keenan, J.M. 2002. The antihypertensive effects of a hydrolysated whey protein isolate supplement (Biozate® 1). *Cardiovasc. Drugs Ther.* 16: 68.

Pins, J.J. and Keenan, J.M. 2003. The antihypertensive effects of a hydrolysated whey protein isolate supplement (Biozate® 1): A pilot study. *FASEB J.* 17:A1110.

Pins, J.J. and Keenan, J.M. 2006. Effects of whey peptides on cardiovascular risk factors. *J. Clin. Hypertens.* 8: 775–782.

Quirós, A., Ramos, M., Muguerza, B. et al. 2007. Identification of novel antihypertensive peptides in milk fermented with Enterococus faecalis. *Int. Dairy J.* 17: 33–41.

Rival, S.G., Boeriu, C.G., and Wichers, H. J. 2001b. Casein and casein hydrolysates. 2. Antioxidant properties and relevance to lipoxygenase inhibition. *J. Agric. Food Chem.* 49: 295–302.

Rival, S.G., Fornaroli, S., Boeriu, C.G., and Wichers, H.J. 2001a. Caseins and casein hydrolysates. 1. Lipoxygenase inhibitory properties. *J. Agric. Food Chem.* 49: 287–294.

Robert, M.C., Razaname, A., Mutter, M., and Juillerat, M.A. 2004. Identification of angiotensin-I-converting enzyme inhibitory peptides derived from sodium caseinate hydrolysates produced by Lactobacillus helveticus NCC 2765. *J. Agric. Food Chem.* 52: 6923–6931.

Rodicio Díaz, J. L. 2000. Comentario sobre el coste/beneficio de la hipertensión arterial. *Gestión y evaluación de costes sanitarios* 1: 6–10.

Rohrbach, M.S., Williams, E.B., and Rolstad, R.A. 1981. Purification and substrate specificity of bovine angiotensin-converting enzyme. *J. Biol. Chem.* 256: 225–230.

Rokka, T., Syvaoja, E.L., Tuominen, J., and Korhonen, H. 1997. Release of bioactive peptides by enzymatic proteolysis of Lactobacillus GC fermented UHT-milk. *Milchwissenchaft* 52: 675–678.

Ryhänen, E.L., Pilhlanto-Leppälä, A., and Pahkala, K. 2001. A new type of ripened, low-fat cheese with bioactive properties. *Int. Dairy J.* 11: 441–447.

Saeed, R.W., Stefano, G.B., Murga, J.D. et al. 2000. Expression of functional delta opioid receptors in vascular smooth muscle. *Int. J. Mol. Med.* 6: 673–677.

Saiga, A.I., Tanabe, S., and Nishimura, T. 2003. Antioxidant activity of peptides obtained from porcine myofibrillar proteins by protease treatment. *J. Agric. Food Chem.* 51: 3661–3667.

Saito, T. 2008. Antihypertensive peptides derived from bovine casein and whey proteins. *Adv. Exp. Med. Biol.* 606: 295–317.

Saito, T., Nakamura, T., Kitazawa, H., Kawai, Y., and Itoh, T. 2000. Isolation and structural analysis of antihypertensive peptides that exist naturally in Gouda cheese. *J. Dairy Sci.* 83: 1434–1440.

Sanders, M.E. 1993. Effect of consumption of lactic cultures on human health. *Adv. Food Nutr. Res.* 37: 67–130.

Satake, M., Enjho, M., Nakamura, Y., Takano, T., Kawamura, Y., and Arai, S. 2002. Transepithelial transport of the bioactive tripeptide, Val-Pro-Pro, in human intestinal Caco-2 cell monolayers. *Biosci. Biotech. Biochem.* 66: 378–384.

Sato, M., Hosokawa, T., Yamaguchi, T. et al. 2002a. Angiotensin I-converting enzyme inhibitory peptides derived from wakame (Undaria pinnatifida) and their antihypertensive effect in spontaneously hypertensive rats. *J. Agric. Food Chem.* 50: 6245–6252.

Sato, M., Oba, T., Yamaguchi, T. et al. 2002b. Antihypertensive effects of hydrolysates of wakame (Undaria pinnatifida) and their angiotensin-I-converting enzyme inhibitory activity. *Ann. Nutr. Metab.* 46: 259–267.

Scruggs, P., Filipeanu, C.M., Yang, J., Kang Chang, J., and Dun, N.J. 2004. Interaction of ovokinin (2-7) with vascular bradykinin 2 receptors. *Reg. Pep.* 120: 85–91.

Sekiya, S., Kobayashi, Y., Kita, E., Imamura, Y., and Toyama, S. 1992. Antihypertensive effects of tryptic hydrolysate of casein on normotensive and hypertensive volunteers. *J. Nutr. Food Sci.* 45: 513–517.

Seppo, L., Jauhiainen, T., Poussa, T., and Korpela, R. 2003. A fermented milk high in bioactive peptides has a blood pressure-lowering effect in hypertensive subjects. *Am. J. Clin. Nutr.* 77: 326–330.

Seppo, L., Kerojoki, O., Suomalainen, T., and Korpela, R. 2002. The effect of a Lactobacillus helveticus LBK-16 H fermented milk on hypertension: A pilot study on humans. *Milchwissenschaft* 57: 124–127.

Sexton, J.M., Britton, S.L., Beierwaltes, W.H., Fiksen-Olsen, M.J., and Romero, J.C. 1979. Formation of angiotensin III from [des-Asp1] angiotensin I in the mesenteric vasculature. *Am. J. Physiol.* 237: H218–H223.

Shimizu, M., Tsunogai, M., and Arai, S. 1997. Transepithelial transport of oligopeptides in the human intestinal cell, Caco-2. *Peptides* 18: 681–687.

Shin, Z.I., Ahn, C.W., Nam, H.S., Lee, H.J., Lee, H.J., and Moon, T.H. 1995. Fraction of angiotensin converting enzyme (ACE) inhibitory peptide from soybean paste. *Kor. J. Food Sci. Tech.* 27: 230–234.

Sipola, M., Finckenberg, P., Korpela, R., Vapaatalo, H., and Nurminen, M.A. 2002. Effect of long-term intake of milk products on blood pressure in hypertensive rats. *J. Dairy Res.* 69: 103–111.

Sipola, M., Finckenberg, P., Santisteban, J., Korpela, R., Vapaatalo, H., and Nurminen, M.L. 2001. Long-term intake of milk peptides attenuates development of hypertension in spontaneously hypertensive rats. *J. Phys. Pharm.* 52: 745–754.

Sirén, A.L. and Feuerstein, G. 1992. The opioid system in circulatory control. *NIPS* 7: 26–30.

Smacchi, E. and Gobbetti, M. 1998. Peptides from several Italian cheeses inhibitory to proteolytic enzymes of acid lactic bacteria, Pseudomonas fluorescens ATCC 948 and to the angiotensin I-converting enzyme. *Enzyme Microb. Technol.* 22: 687–694.

Song, L. and Healy, D.P. 1999. Kidney aminopeptidase A and hypertension, part II: Effects of angiotensin II. *Hypertension* 33: 746–752.

Srinivas, L., Shalini, V.K., and Shylaja, M. 1992. Turmerin: A water soluble antioxidant peptide from tumeric (Curcuma longa). *Arch. Biochem. Biophys.* 292: 617–623.

Stadman, E.R. 1993. Oxidation of free amino acids and amino acid residues in proteins by radiolysis and by metal-catalyzed reactions. *Ann. Rev. Biochem.* 62: 797–821.

Stepaniak, L., Fox, P.F., Sorhaug, T., and Grabska, J. 1995. Effect of peptides from the sequence 58-72 of β-casein on the activity of endopeptidase, aminopeptidase and x-prolyldipeptidyl aminopeptidase from Lactococcus lactis ssp. lactis MG1363. *J. Agric. Food Chem.* 43: 849–853.

Suetsuna, K. 1998. Isolation and characterization of angiotensin I-converting enzyme inhibitor dipeptides derived from sardine muscle. *J. Nutr. Biochem.* 9: 415–419.

Suetsuna, K., Maekawa, K., and Chen, J.R. 2004. Antihypertensive effects of Undaria pinnatifida (wakame) peptide on blood pressure in spontaneously hypertensive rats. *J. Nutr. Biochem.* 15: 267–272.

Suetsuna, K. and Nakano, T. 2000. Identification of an antihypertensive peptide from peptic digests of wakame (Undaria pinnatifida). *J. Nutr. Biochem.* 11: 450–454.

Suetsuna, K. and Osajima, K. 1986. The inhibitory activities against angiotensin I-converting enzyme of basic peptides originating from sardine and hair tail meat. *Bull. Jpn. Soc. Sci. Fish* 5: 1981–1984.

Suetsuna, K. and Osajima, K. 1989. Blood pressure reduction and vasodilatory effects *in vivo* of peptides originating from sardine muscle. *Nippon Eiyou, Shokuryou Gakkaishi* 52:47–51.

Suetsuna, K., Ukeda, H., and Ochi, H. 2000. Isolation and characterization of free radical scavenging activities of peptides derived from casein. *J. Nutr. Biochem.* 11: 128–131.

Sugai, R. 1998. ACE inhibitors and functional foods. *Bull. IDF* 336: 17–20.

Suh, H.J. and Whang, J.H. 1999. A peptide from corn gluten hydroysate that is inhibitory toward angiotensin-I converting enzyme. *Biotechnol. Lett.* 21: 1055–1058.

Sun, J., He, H., and Xie, B. J. 2004. Novel antioxidant peptides from fermented mushroom Ganoderma lucidum. *J. Agric. Food. Chem.* 52: 6646–6652.

Swanson, G.N., Hanesworth, J.M., Sardinia, M.F. et al. 1992. Discovery of a distinct binding site for angiotensin II (3-8), a putative angiotensin IV receptor. *Regul. Pept.* 40: 409–419.

Tom, B., Dendorfer, A., de Vries, R., Saxena, P.R., and Jan Danser, A.H. 2002. Bradykinin potentiation by ACE inhibitors: A matter of metabolism. *Br. J. Pharmacol.* 137: 276–284.

Tong, L.M., Sasaki, S., McClements, D.J., and Decker, E.A. 2000. Mechanisms of the antioxidant activity of a high molecular weight fraction of whey. *J. Agric. Food Chem.* 48: 1473–1478.

Tonouchi, H., Suzuki, M., Uchida, M., and Oda, M. 2008. Antihypertensive effect of an angiotensin converting enzyme inhibitory peptide from enzyme modified cheese. *J. Dairy Res.* 75: 284–290.

Toshiro, M., Ueno, T., Tanaka, M., Oka, H., Miyamoto, T., Osajima, K., and Matsumoto, K. 2005. Antiproliferative action of an angiotensin I-converting enzyme inhibitory peptide, Val-Tyr, via an L-type Ca2+ channel inhibition in cultured vascular smooth muscle cells. *Hypertens. Res.* 28: 545–552.

Townsend, R.R., McFadden, C.B., Ford, V., and Cadeé, J.A. 2004. A randomized, doubled-blind, placebo controlled trial of casein protein hydrolysate (C12 peptide) in human essential hypertension. *Am. J. Hyperten.* 17: 1056–1058.

Trippodo, N.C. and Frohlich, E.D. 1981. Similarities of genetic (spontaneous) hypertension in man and rat. *Circ. Res.* 48: 09–319.

Tsuge, N., Eikawa, Y., Nomura, Y., Yamamoto, M., and Sugisawa, K. 1991. Antioxidative activity of peptides prepared by enzymic hydrolysis of egg-white albumin. *J. Agric. Chem. Soc. Japan* 65: 1635–1641.

Tuomilehto, J., Lindstrom, J., Hyyrynen, J. et al. 2004. Effect of ingesting sour milk fermented using Lactobacillus helveticus bacteria producing tripeptides on blood pressure in subjects with mild hypertension. *J. Hum. Hypertens.* 18: 705–802.

Turgeon, S.L. and Gauthier, S.F. 1990. Whey peptide fractions obtained with a 2-step ultrafiltration process—Production and characterization. *J. Food Sci.* 55: 106–108.

Turner, A.J. and Hooper, N.M. 2002. The angiotensin-converting enzyme gene family: Genomics and pharmacology. *Trends Pharmacol Sci* 23: 177–183.

Urata, H., Nishimura, H., and Ganten, D. 1996. Chymase-dependent angiotensin II forming systems in humans. *Am. J. Hypertens.* 9: 277–284.

VanderJagt, D.J., Okolo, S.N., Costanza, A., Blackwell, W., and Glew, R.H. 2001. Antioxidant content of the milk of Nigerian women and the sera of their exclusively breastfed infants. *Nutr. Res.* 21: 121–128.

Vermeirssen, V., Deplancke, B., Tappenden, K.A., Van Camp, J., Gaskins, H.R., and Verstraete, W. 2002. Intestinal transport of the lactokinin Ala-Leu-Pro-Met-His-Ile-Arg through a Caco-2 monolayer. *J. Pept. Sci.* 8(3): 95–100.

Vermeirssen, V., Van Camp, J., Decroos, K., Van Wijmelbeke, L., and Verstraete, W. 2003. The impact of fermentation and *in vitro* digestion on the formation of angiotensin-I-converting enzyme inhibitory activity from pea and whey protein. *J. Dairy Sci.* 86: 429–438.

Vermeirssen, V., Van Camp, J., and Verstraete, W. 2002. Optimisation and validation of an angiotensin-converting enzyme inhibition assay for the screening of bioactive peptides. *J Biochem. Biophys. Methods* 51:75–87.

Vermeirssen, V., Van Camp, J., and Verstraete, W. 2004. Bioavailability of angiotensin I converting enzyme inhibitory peptides. *Br. J. Nutr.* 92: 357–366.

Visser, S., Noorman, H.J., Slangeen, C.J., and Rollema, H.S. 1989. Action of plasmin on bovine β-casein in a membrane reactor. *J. Dairy Res.* 56: 323–333.

Viveros, O.H., Diliberto, E.J., Hazum, E.L., and Chang, K.J. 1979. Opiate-like material in the adrenal medulla: evidence for storage and secretion with catecholamines. *Mol. Pharmacol.* 16: 1101–1108.

Vogt, W. 1995. Oxidation of methionyl residues in proteins: Tools, targets, and reversal. *Free Radic. Biol. Med.* 18: 93–105.

Vreeman, H.J., Both, P., and Slangen, C.J. 1994. Rapid procedure for isolating the bitter carboxyl-terminal fragment 193-209 of β-casein on a preparative-scale. *Neth. Milk Dairy J.* 48: 63–70.

Walsh, D.J., Bernard, H., and Murray, B.A. et al. 2004. *In vitro* generation and stability of the lactokinin beta-lactoglobulin fragment (142-148). *J. Dairy Sci.* 87: 3845–3857.

Williams, T.A., Barnes, K., Kenny, A.J., Turner, A.J., and Hooper, N.M. 1992. A comparison of the zinc contents and substrate specificities of the endothelial and testicular forms of porcine angiotensin converting enzyme and the preparation of isoenzyme-specific antisera. *Biochem. J.* 288: 875–881.

Wolf-Maier, K., Cooper, R.S., Kramer, H. et al. 2004. Hypertension treatment and control in five European countries, Canada, and the United States. *Hypertension* 43: 10–17.

Wu, J. and Ding, X. 2001. Hypotensive and physiological effect of angiotensin converting enzyme inhibitory peptides derived from soy protein on spontaneously hypertensive rats. *J. Agric. Food Chem.* 49: 501–506.

Yamada, Y., Matoba, N., Usiu, H., Onishi, K., and Yoshikawa, M. 2002. Design of a highly potent antihypertensive peptide based on Ovokinin (2-7). *Biosci. Biotech. Biochem.* 66: 1213–1217.

Yamada, Y., Yamauchi, D., Yokoo, M., Ohinata, K., Usui, H., and Yoshikawa, M. 2008. A potent hypotensive peptide, novokinin, induces relaxation by AT2- and IP-receptor-dependent mechanism in the mesenteric artery from SHRs. *Biosci. Biotechnol. Biochem.* 72: 257–259.

Yamamoto, N., Akino, A., and Takano, T. 1994. Antihypertensive effects of different kinds of fermented milk in spontaneoulsy hypertensive rats. *Biosci. Biotechnol. Biochem.* 58: 776–778.

Yanagisawa, M., Kurihara, H., Kimura, S., Goto, K., and Masaki, T. 1988. A novel peptide vasoconstrictor, endothelin, is produced by vascular endothelium and modulates smooth muscle Ca^{2+} channels. *J. Hypertens.* Suppl 6: S188–S191.

Yang, H.Y., Yang, S.C., Chen, J.R., Tzeng, Y.H., and Han, B.C. 2004. Soyabean protein hydrolysate prevents the development of hypertension in spontaneously hypertensive rats. *Br. J. Nutr.* 92: 507–512.

Yang, Y., Marczak, E.D., Usui, H., Kawamura, Y., and Yoshikawa, M. 2004. Antihypertensive properties of spinach leaf protein digests. *J. Agric. Food Chem.* 52: 2223–2225.

Yang, Y., Marczak, E.D., Yokoo, M., Usui, H., and Yoshikawa, M. 2003. Isolation and antihypertensive effect of ACE-inhibitory peptides from spinach rubisco. *J. Agric. Food Chem.* 51: 4897–4902.

Yee, J.J. and Shipe, W.F. 1981. Using enzymatic proteolysis to reduce copper-protein catalysis of lipid oxidation. *J. Food Sci.* 46: 966–969.

Yokoyama, K., Chiba, H., and Yoshikawa, M. 1992. Peptide inhibitors for angiotensin I-converting enzyme from thermolysin digest of dried bonito. *Biosci. Biotech. Biochem.* 56: 1541–1545.

Yoshii, H., Tachi, N., Ohba, R., Sakamura, O., Takemaya, H., and Itani, T. 2001. Antihypertensive effect of ACE inhibitory oligopeptides from chicken egg yolks. *Com. Biochem. Physiol. Part C* 128: 27–33.

Yoshikawa, M. and Chiba, H. 1990. Biologically active peptides derived from milk proteins. *Japan J. Dairy Food Sci.* 39: A315–A321.

Yoshikawa, M., Tani, F., and Chiba, H. 1988. Structure-activity relationship of opioid antagonist peptides derived from milk proteins. In *Peptide Chemistry* (T. Shiba, Ed.), Osaka: Protein Research Foundation, pp. 473–476.

Yu, R., Park, S.A., Chung, D.K., Nam, H., and Shin, Z.I. 1996. Effect of soybean hydrolysate on hypertension in spontaneously hypertensive rats. *J. Korean Soc. Food Sci. Nutr.* 25: 1031–1036.

Zárate, G., Chaia, A.P., González, S., and Oliver, G. 2000. Viability and β-galactosidase activity of dairy propionibacteria subjected to digestion by artificial gastric and intestinal fluids. *J. Food Prot.* 63: 1214–1221.

Zhu, X.L., Watanabe, K., Shiraishi, K., Ueki, T., Noda, Y., Matsui, T., and Matsumoto, K. 2008. Identification of ACE-inhibitory peptides in salt-free soy sauce that are transportable across caco-2 cell monolayers. *Peptides* 29: 338–344.

Zicha, J. and Kunes, J. 1999. Ontogenetic aspects of hypertension development: Analysis in the rat. *Physiol. Rev.* 79: 1227–1282.

7

Therapeutic Peptides as Amino Acid Source

Hironori Yamamoto and Masashi Kuwahata

CONTENTS

7.1 Introduction

The role of peptides from dietary proteins as physiologically active components has been increasingly acknowledged (Marshall, 1994). Nutritionally, dietary peptides are a source of energy and amino acids, which are essential for growth and maintenance. In addition, it is well documented that a number of amino acids possess specific physiological properties both beneficial and detrimental; for example, they participate in many biochemical pathways and are precursors of active metabolites (Wu, 2009). The amino acids, for example, arginine, glutamine, histidine, lysine, taurine, tyrosine, and tryptophan, are considered to be physiologically beneficial and the best sources of these amino acids are meat, eggs, fish, soybeans, and dairy products. On the other hand, a few amino acid derivatives, which are formed during food processing, such as lysinoalanine, D-amino acids, and biogenic amines, may cause undesirable metabolic or even toxic events in the body.

In recent years, it has been reported that the biologically active peptides from foods might have beneficial effects on human health, especially as immunomodulating, antihypertensive, and osteoprotective substances. For example, the tripeptides IPP (isoleucyl-prolyl-proline) and VPP (valyl-prolyl-proline) from milk products with *Lactobacillus helveticus* inhibit angiotensin converting enzyme, and milk basic protein (MBP) indicates the osteoprotective effects (Möller et al., 2008).

Hyperlipidemia and hypercholesterolemia, risk factors for cardiovascular diseases, have been prevalent in many countries. To control blood lipid and cholesterol levels, many people must continually take medicine. As described in other chapters, some peptides have been demonstrated to moderate hyperlipidemia and hypercholesterolemia. Peptide-based products have then been used aiming to control mild hyperlipidemia and hypercholesterolemia, although therapeutic claims have not had governmental approval.

The observation that amino acids are absorbed more rapidly when introduced into a gut lumen as dipeptides and tripeptides rather than as a mixture of free amino acids has served as a basis of nutritional therapy for individuals with digestive or absorptive disorders (Adibi and Kim, 1981). This uptake is mediated by a transport system that is specific for peptides consisting of two or three amino acids (Ganapathy et al., 1994). The physiological role of the peptide transport system is to mediate the absorption of small peptides generated from the digestion of dietary proteins. It has been thought that peptide transporter 1 (PepT1) might play a role in the uptake of a large number of therapeutic peptides as amino acid sources in the small intestine (Daniel, 2004).

This chapter focuses on therapeutic peptides as amino acid sources. Potential therapeutic effects of peptides on mild hyperlipidemia and hypercholesterolemia are also discussed.

7.2 Supplementation of Unstable and Low Soluble Amino Acids as Peptides

Stability and solubility of some amino acids may be a limiting factor in making an adequate amino acid solution from a nutritional point of view. One approach to overcome the physicochemical limitations of amino acids is to supply the amino acid as a peptide by conjugation with other amino acids.

Glutamine (Gln) is not an essential amino acid for humans, but it is considered to be conditionally essential because of its unique role (Lacey and Wilmore, 1990). Some of the important roles for glutamine are as a preferred fuel for enterocytes and lymphocytes, a regulator of acid-base balance through the production of urinary ammonia, and a precursor for glutathione. During stress the body's requirements for Gln appear to exceed the individual's ability to produce sufficient amounts of this amino acid. Despite the nutritional effectiveness of Gln, it has not been utilized as a component of water-soluble applications due to its instability in solution (Suzuki, Motoi, and Sato, 1999). The amino and amide groups of Gln easily condense to form pyroglutamic acid under aqueous conditions. On the other hand, alanyl-glutamine, a di-peptide of alanine and glutamine, can be used as a stable source of Gln in solution. There have been a number of reports on effectiveness of this peptide in recent clinical trials (Dechelotte et al., 2006; Lima et al., 2007; Luo et al., 2008; Estivariz et al., 2008). On the basis of those findings, Gln-rich peptides have been prepared from Gln-rich protein. Glutamyl residues in wheat gluten account for about 40% of the amino acids (Kasarda et al., 1984). Due to its amino acid composition, wheat gluten has been considered a natural Gln source. Now, some beverages and clinical formulas containing Gln-rich peptides prepared from wheat gluten are commercially available.

Generally, humans do not have a requirement for a dietary source of tyrosine (Tyr), as the endogenous supply of this amino acid arises from the hydroxylation of phenylalanine (Phe) in the liver (Moss and Schoenheimer, 1940). However, Tyr may be a conditionally essential amino acid in patients with renal failure (Young and Parsons, 1973) and low birth weight infants (Heird et al., 1987), due to the reduced synthesis of Tyr from Phe. Although supplementation of Tyr is required for these subjects, poor solubility of Tyr has limited its inclusion in parenteral amino acid solutions to less than 1% of the total amino acids. On the other hand, di-peptides of Tyr, such as alanyl-tyrosine and glycyl-tyrosine have a higher solubility (Furst, Albers, and Stehle, 1990). Furthermore, the di-peptides of Tyr have been shown to be biologically effective as parenteral Tyr sources in humans (Druml et al., 1991; Roberts et al., 2001).

Very low solubility and instability of cystine (Cys) in aqueous solutions also prevents addition of this amino acid to parenteral solutions in adequate amounts. A previous study suggested that highly soluble and stable Cys-containing synthetic peptides, bis-L-alanyl-L-cystine and bis-glycyl-L-cystine, may represent efficient sources of free cystine in parenteral nutrition (Stehle et al., 1988).

7.3 Peptides for Specific Disease

The prevalence of food allergy is greatest in the first two years of life and decreases with age. The most common food allergens causing reactions in children include milk, eggs, wheat, soy, peanuts, tree nuts, fish, and shellfish. Although any food can cause anaphylaxis, the most commonly implicated foods for severe allergic reactions are peanuts, tree nuts, fish, and shellfish. Important allergens include cow's milk proteins (casein, whey), eggs (ovalbumin, ovomucoid), peanuts (vicillin, conglutin, glycinin), shellfish (tropomysin), and fish (parvalbumin; Ramesh, 2008).

Sensitization to food can occur either primarily through the GI tract (class I allergens) or secondarily through the respiratory tract via inhalation (class II allergens; Breiteneder and Ebner, 2000). The majority of the class I food allergens are heat stable, resistant to acid degradation, and resistant to proteolysis. Class I allergy is seen mainly in children and is rare in adults. Sensitization to the class II allergens, which is mainly seen in adults, occurs initially by inhaling plant and tree pollens. IgE-mediated reactions may subsequently occur when foods containing cross-reacting epitopes are eaten as seen in the oral allergy syndrome. Class II allergens are usually labile proteins, which are easily degradable.

At present, the mechanism underlying a milk allergy is not completely understood. In general, milk allergies are classified as IgE-mediated and non-IgE-mediated disorders (Sampson and Anderson, 2000). Whereas a non-IgE-mediated milk allergy is generally not considered life threatening, an IgE-mediated milk allergy is potentially fatal (Host, 1994). The IgE-mediated milk allergy involves production of IgE antibodies upon first exposure to milk protein (e.g., β-LG, α-LA, caseins) leading to sensitization of mast cells. Second and subsequent exposures to the same milk protein result in cross-linking of mast-cell-bound IgE, leading to activation and release of inflammatory mediators such as histamine. This results in clinical signs of disease such as hives, rashes, and in rare cases, potentially fatal systemic anaphylaxis (Sicherer and Sampson, 2008).

Development of novel milk products containing hypo- or nonallergenic milk proteins is an area of significant interest in the dairy science and functional food fields (Businco et al., 1993; Giampietro et al., 2001). Many industrially utilized dairy starter cultures are highly proteolytic. Peptides can, thus, be generated from milk proteins by the starter and nonstarter bacteria used in the manufacture of fermented dairy products. The proteolytic system of lactic acid bacteria (LAB; e.g., *Lactococcus lactis*, *Lactobacillus helveticus*, and *Lb. delbrueckii* ssp. *bulgaricus*) is already well characterized. This system consists of a cell wall-bound proteinase and a number of distinct intracellular peptidases, including endopeptidases, aminopeptidases, tripeptidases, and dipeptidases (Christensen et al., 1999).

A number of studies have demonstrated that hydrolysis of milk proteins by digestive or microbial enzymes may produce peptides with immunomodulatory activities (Gill et al., 2000). Sütas, Hurme, and Isolauri (1996) demonstrated that digestion of casein fractions with both pepsin and trypsin produced peptides that provoked immunomodulatory effects on human blood lymphocytes *in vitro*. Peptides derived from total casein and α_{s1}-casein mainly suppressed the proliferation of lymphocytes whereas those derived from β- and κ-casein primarily stimulated the proliferation rate. When the caseins were hydrolyzed by enzymes isolated from a probiotic strain of *Lactobacillus GG* var. *casei* prior to pepsin–trypsin treatment, all hydrolysate fractions were immunosuppressive and the highest activity was again found in α_{s1}-casein. The same hydrolysates also downregulated *in vitro* the generation of interleukin-4 by lymphocytes (Sütas et al., 1996). Matar et al. (2001) fed milk fermented with a *Lb. helveticus* strain to mice for three days and detected significantly higher numbers of IgA secreting cells in their intestinal mucosa, compared with control mice fed with similar milk incubated with a nonproteolytic variant of the same strain. The immunostimulatory effect of fermented milk was attributed to peptides released from the casein fraction. Recently it was demonstrated that commercial whey protein isolates contain immunomodulating peptides released by enzymatic digestion (Wang et al., 1995). This information is highly relevant when developing infant formulas with optimized immunomodulatory properties. These results suggest that the proteolytic system of lactic acid bacteria may modulate the immunological properties of milk proteins prior to or after oral ingestion of the product. Such modulation may be beneficial, for example, in the downregulation of hypersensitivity reactions to ingested proteins in patients with food protein allergies (Möller et al., 2008).

7.4 Peptide-Based Products and Approved Health Claims in Japan

As discussed in other chapters, it has been demonstrated that some food-derived peptides can moderate hypercholesterolemia and hyperlipidemia. In Japan, some of them are classified as food for specified health uses (FOSHU) and officially approved to present health-promoting claims on their labels. Although these products are not officially approved to use for hypercholesterolemia and hyperlipidemia therapy, these products have been used by patients suffering from mild forms of hypercholesterolemia and hyperlipidemia.

For example, the cholesterol-lowering food function contains the soy protein hydrolysate with bound phospholipids (SPHP), as a principal ingredient. In a human trial, it has been reported that SPHP significantly suppresses the increase in serum cholesterol levels when healthy males consumed excessive dietary cholesterol (Hori et al., 2000). Furthermore, SPHP has shown remarkable improving action on serum cholesterol levels in hypercholesterolemic subjects (Hori et al., 2001). These cholesterol-lowering effects of SPHP might have resulted from the inhibition of steroid absorption from the intestine. Furthermore, the bitter taste of soy protein hydrolysate prevents its use as a food ingredient in many products. However, this adverse property is improved in SPHP due to the binding of peptides and phospholipids.

The globin peptide is an oligopeptide mixture produced from enzymatic hydrolysis of red blood cells. A cross-over study in six men demonstrated that 4 g of globin peptide significantly decreased the area under the curve for serum triglycerides and chylomicrons in men after ingestion of a high fat diet (Kagawa et al., 1998). It was postulated that absorption of triglycerides and chylomicrons had been inhibited by ingestion of globin peptide. A peptide, valyl-valyl-tyroyl-proline (VVYP), which is present in globin peptide, has been identified as the active factor (Kagawa et al., 1996).

7.5 Current Application and Production of Therapeutic Peptide

One of the mechanisms for intestinal transport of oligopeptides is the transporter-mediated transport system. Irrespective of the amino acid sequence, di- and tripeptides can be actively transported by proton-coupled oligopeptide transporter 1 (PepT1; Fei et al., 1994; Liang et al., 1995). The peptides absorbed in intestinal cells are hydrolyzed into amino acids by intracellular peptidases, and the amino acids produced are either used by the absorbing cells or are released into the portal circulation via the amino acid transporters located on the basolateral membrane of these cells (Adibi, 1997). A recent study has shown that changes in serum concentrations of various amino acids in healthy adult males are greater after ingestion of soy peptide, compared with those after ingestion of nondigested soy protein or an amino acid mixture of equivalent composition (Maebuchi et al., 2007). This study suggests that the ingestion of soy peptide results in faster and more efficient absorption than the consumption of protein or amino acid mixtures in healthy humans. In addition to high solubility, low viscosity, and low allergenicity associated with low molecular weight peptides, oligopeptide mixtures as a source of nitrogen for enteral feeding might be a more effective ingredient than intact protein and amino acid mixtures.

In order to exert some physiological effects *in vivo*, peptides must be absorbed from the intestine and reach the target cells in the blood vessels in substantial concentrations, except for peptides that may act either directly in the intestinal tract or via receptors and cell signaling in the gut. It has been shown that angiotensin-converting enzyme-inhibitory peptide, valyl-prolyl-proline, could be transported across the Caco-2 cell monolayer in an intact form (Satake et al., 2002). The mechanism for this transport was mainly via paracellular diffusion, and transcytosis and transporter-mediated transport were not likely to have played major roles in the transepithelial transport of the intact tripeptide. The transepithelial flux of intact peptides would be strongly dependent on their structure and properties. If a peptide has a sequence that is highly susceptible to surface peptidases, transepithelial transport of the intact peptide would be negligible. The susceptibility to intestinal peptidases may be the most important factor determining the bioavailability of oligopeptides.

There is a potential for the application of bioactive peptides as ingredients of functional food or medical products. Enzymatic hydrolysis by pancreatic enzymes, especially trypsin, has been widely used for yielding bioactive peptides *in vitro*. However, when the amino acid sequence of a peptide is identified, the peptide can be synthesized by chemical synthesis or recombinant DNA technology. Chemical synthesis will be very useful for the production of short peptides, and recombinant DNA technology will be very useful for the generation of larger peptides. The bioactivity of the peptides may be reduced through molecular alteration during a process of purification or interaction with other ingredients in products. It will be important to develop functional foods and medical products without undesired side effects of added peptides or reduction of bioactivity.

7.6 Conclusion

A number of physiologically active peptides have been discovered in the hydrolysates of various food proteins. Furthermore, oligopeptide mixtures might be a more effective ingredient as a source of nitrogen for enteral feeding. However, the mechanism for transepithelial oligopeptide transport in the intestinal tract is not yet fully understood. In the pathophysiological state and malnutritional condition, this absorption mechanism will be modulated compared with a healthy condition. It will be very important to clarify the absorption mechanism of peptides in various conditions.

References

Adibi, S.A. 1997. The oligopeptide transporter (Pept-1) in human intestine: Biology and function. *Gastroenterology* 113: 332–340.

Adibi, S.A. and Kim, Y.S. 1981. Peptide absorption and hydrolysis. In *Physiology of the Gastrointestinal Tract*, Volume 2 (L.R. Johnson, Ed.), New York: Raven, pp. 1073–1095.

Breiteneder, H. and Ebner, C. 2000. Molecular and biochemical classification of plant-derived food allergens. *J. Allergy Clin. Immunol.* 106: 27–36.

Businco, L., Dreborg, S., and Einarsson, R. et al. 1993. Hydrolysed cow's milk formulae. Allergenicity and use in treatment and prevention. An ESPACI position paper. European Society of Pediatric Allergy and Clinical Immunology. *Pediatr. Allergy Immunol.* 4: 101–111.

Christensen, J.E., Dudley, E.G., Pederson, J.A. and Steele, J.L. 1999. Peptidases and amino acid catabolism in lactic acid bacteria. *Antonie van Leeuwenhoek*, 76: 217–246.

Daniel, H. 2004. Molecular and integrative physiology of intestinal peptide transport. *Annu. Rev. Physiol.* 66: 361–384.

Dechelotte, P., Hasselmann, M., Cynober, L. et al. 2006. L-alanyl-L-glutamine dipeptide-supplemented total parenteral nutrition reduces infectious complications and glucose intolerance in critically ill patients: The French controlled, randomized, double-blind, multicenter study. *Crit. Care Med.* 34: 598–604.

Druml, W., Lochs, H., Roth, E. et al. 1991. Utilization of tyrosine dipeptides and acetyltyrosine in normal and uremic human. *Am. J. Physiol.* 260: E280–285.

Estivariz, C.F., Griffith, D.P., and Luo, M. et al. 2008. Efficacy of parenteral nutrition supplemented with glutamine dipeptide to decrease hospital infections in critically ill surgical patients. *JPEN J. Parenter. Enteral Nutr.* 32: 389–402.

Fei, Y.J., Kanai, Y., Nussberger, S. et al. 1994. Expression cloning of a mammalian proton-coupled oligopeptide transporter. *Nature* 368: 563–566.

Furst, P., Albers, S., and Stehle, P. 1990. Dipeptides in clinical nutrition. *Proc. Nutr. Soc.* 49: 343–359.

Ganapathy, V., Brandsch, M., and Leibach, F.H. 1994. Intestinal transport of amino acids and peptides. In *Physiology of the Gastrointestinal Tract*, Volume 2, 3rd edition (L.R. Johnson, Ed.), New York: Raven, pp. 1773–1794.

Giampietro, P.G., Kjellman, N.I., Oldaeus, G., Wouters-Wesseling, W., and Businco, L. 2001. Hypoallergenicity of an extensively hydrolyzed whey formula. *Pediatr. Allergy Immunol.* 12: 83–86.

Gill, H.S., Doull, F., Rutherfurd, K.J. and Cross, M.L. 2000. Immunoregulatory peptides in bovine milk. *Br. J. Nutr.* 84(Suppl.1): S111–S117.

Heird, W.C., Dell, R.B., and Helms, R.A. et al. 1987. Amino acid mixture designed to maintain normal plasma amino acid patterns in infants and children requiring parenteral nutrition. *Pediatrics* 80: 401–408.

Hori, G., Kamiya, T., Hara, T. et al. 2000. The effect of soybean protein hydrolyzate with bound phospholipids on serum cholesterol levels in adult male subjects receiving high cholesterol diet. *Nippon Rinsyo Eiyo Gakkaishi* (in Japanese) 22: 21–27.

Hori, G., Wang, M.F., Chan, Y.C. et al. 2001. Soy protein hydrolyzate with bound phospholipids reduces serum cholesterol levels in hypercholesterolemic adult male volunteers. *Biosci. Biotechnol. Biochem.* 65: 72–78.

Host, A. 1994. Cow's milk protein allergy and intolerance in infancy. Some clinical, epidemiological and immunological aspects. *Pediatr. Allergy Immunol.* 5(Suppl.): 1–36.

Kagawa, K., Matsutaka, H., Fukuhama, C., Fujino, H., and Okuda, H. 1998. Suppressive effect of globin digest on postprandial hyperlipidemia in male volunteers. *J. Nutr.* 128: 56–60.

Kagawa, K., Matsutaka, H., Fukuhama, C., Watanabe, Y., and Fujino, H. 1996. Globin digest, acidic protease hydrolysate, inhibits dietary hypertriglyceridemia and Val-Val-Tyr-Pro, one of its constituents, possesses most superior effect. *Life Sci.* 58: 1745–1755.

Kasarda, D.D., Okita, T.W., Bernardin, J.E. et al. 1984. Nucleic acid (cDNA) and amino acid sequences of alpha-type gliadins from wheat (Triticum aestivum). *Proc. Natl. Acad. Sci. USA* 81: 4712–4716.

Lacey, J.M. and Wilmore, D.W. 1990. Is glutamine a conditionally essential amino acid? *Nutr. Rev.* 48: 297–309.

Liang, R., Fei, Y.J., Prasad, P.D. et al. 1995. Human intestinal H+/peptide cotransporter. *J. Biol. Chem.* 270: 6456–6463.

Lima, N.L., Soares, A.M., Mota, R.M. et al. 2007. Wasting and intestinal barrier function in children taking alanyl-glutamine-supplemented enteral formula. *J. Pediatr. Gastroenterol. Nutr.* 44: 365–374.

Luo, M., Fernandez-Estivariz, C., Jones, D.P. et al. 2008. Depletion of plasma antioxidants in surgical intensive care unit patients requiring parenteral feeding: Effects of parenteral nutrition with or without alanyl-glutamine dipeptide supplementation. *Nutrition* 24: 37–44.

Maebuchi, M., Samoto, M., Kohno, M. et al. 2007. Improvement in the intestinal absorption of soy protein by enzymatic digestion to oligopeptide in healthy adult men. *Food Sci. Technol. Res.* 13: 45–53.

Marshall, W.E. 1994. Amino acids, peptides, and proteins. In *Functional Foods* (I. Goldberg, Ed.), New York: Chapman & Hall, pp. 242–260.

Matar, C., Valdez, J.C., Medina, M., Rachid, M., and Perdigón, G. 2001. Immunomodulating effects of milks fermented by Lactobacillus helveticus and its non-proteolytic variant. *J. Dairy Res.* 68: 601–609.

Möller, N.P., Scholz-Ahrens, K.E., Roos, N., and Schrezenmeir, J. 2008. Bioactive peptides and proteins from foods: Indication for health effects. *Eur. J. Nutr.* 47: 171–182.

Moss, A.R. and Schoenheimer, R. 1940. The conversion of phenylalanine to tyrosine in normal rats. *J. Biol. Chem.* 135: 415–429.

Ramesh, S. 2008. Food allergy overview in children. *Clinic Rev. Allerg. Immunol.* 34: 217–230.

Roberts, S.A., Ball, R.O., Moore, A.M., Filler, R.M. and Pencharz, P.B. 2001. The effect of graded intake of glycyl-L-tyrosine on phenylalanine and tyrosine metabolism in parenterally fed neonates with an estimation of tyrosine requirement. *Pediatr. Res.* 49: 111–119.

Sampson, H.A. and Anderson, J.A. 2000. Summary and recommendations: Classification of gastrointestinal manifestations due to immunologic reactions to foods in infants and young children. *J. Pediatr. Gastroent. Nutr.* 30(Suppl.): S87–S94.

Satake, M., Enjoh, M., Nakamura, Y. et al. 2002. Transepithelial transport of the bioactive tripeptide, Val-Pro-Pro, in human intestinal Caco-2 cell monolayers. *Biosci. Biotechnol. Biochem.* 66: 378–384.

Sicherer, S.H. and Sampson, H.A. 2008. Food allergy: Recent advances in pathophysiology and treatment. *Annu. Rev. Med.* 60: 261–277.

Stehle, P., Albers, S., Pollack, L., and Furst, P. 1988. *In vivo* utilization of cystine-containing synthetic short-chain peptides after intravenous bolus injection in the rat. *J. Nutr.* 118: 1470–1474.

Sütas, Y., Hurme, M., and Isolauri, E. 1996. Down-regulation of anti-CD3 antibody-induced IL-4 production by bovine caseins hydrolysed with Lactobacillus GG-derived enzymes. *Scand. J. Immunol.* 43, 687–689.

Suzuki, Y., Motoi, H., and Sato, K. 1999. Quantitative analysis of pyroglutamic acid in peptides. *J. Agric. Food Chem.* 47: 3248–3251.

Wu, G. 2009. Amino acids: Metabolism, functions, and nutrition. *Amino Acids.* 37: 1–17.

Wang, M.F., Yamamoto, S., Chung, H.M. et al. 1995. Antihypercholesterolemic effect of undigested fraction of soybean protein in young female volunteers. *J. Nutr. Sci. Vitaminol.* 41: 187–195.

Young, G.A. and Parsons, F.M. 1973. Impairment of phenylalanine hydroxylation in chronic renal insufficiency. *Clin. Sci.* 45: 89–97.

8

Hypolipidemic and Hypocholesterolemic Food Proteins and Peptides

Chibuike C. Udenigwe and Rotimi E. Aluko

CONTENTS

8.1 Introduction

Lipids and cholesterol serve several structural and metabolic functions pertinent to the survival of living organisms. However, elevated concentrations of some lipids in the plasma have been associated with increased risk for heart diseases in humans (Durrington, 2003; Fletcher et al., 2005). These increases are elicited by changes in the lipid metabolism pathways induced by environmental and genetic factors leading to hyperlipidemia and hypercholesterolemia, which are not diseases but health conditions characterized by elevated serum lipids and cholesterol, respectively, that could potentially lead to cardiovascular disease. Hypercholesterolemia arises when the total plasma and low-density lipoprotein (LDL)-cholesterol levels of an individual exceed 240 and 160 mg/dl, respectively, compared to their expected normal levels of below 200 and 100 mg/dl, respectively (National Institute of Health, 2001; Chen, Jiao, and Ma, 2008). Moreover, elevated blood triacylglycerol (TAG) level (hypertriglyceridemia) is a form of hyperlipidemia that has been recognized as an independent risk factor for atherosclerosis and coronary heart disease in humans (Tirosh et al., 2007). The major causes of elevated blood cholesterol in humans include environmental factors such as high levels of exogenous cholesterol, and saturated and *trans*-fatty acids in the diet. In addition, familial hypercholesterolemia (FH) arises as a result of mutation on the LDL receptor (LDL-R) gene responsible for the synthesis of LDL-R protein that mediates cellular uptake of LDL from the body fluid (Durrington, 2003). FH may also result from an abnormal apolipoprotein B (Apo B) gene that codes Apo B protein, a major structural component of LDL that recognizes and binds LDL-R (Durrington, 2003). These genetic abnormalities result in aberrant cholesterol metabolism that could potentially lead to elevated blood levels in subjects. Thus, lowering of elevated cholesterol and TAG levels in the blood is important in management of hypercholesterolemia and hyperlipidemia.

Although several drugs have been developed to manage severe cases of hypercholesterolemia in humans, dietary and lifestyle modification approaches have been recommended for maintaining low blood level of proatherogenic cholesterol (Fletcher et al., 2005). In addition, various research investigations have focused on the use of less toxic functional foods and nutraceuticals to treat and manage mild cases of hypercholesterolemia (Chen et al., 2008). These functional foods include whole and purified food proteins, their enzymatic hydrolysates, and constituent peptides, which have been investigated as potential cholesterol- and lipid-lowering dietary agents in animals and humans. Depending on their sources, structural properties and the presence of other bioactive compounds, these food proteins exhibit hypocholesterolemic and hypolipidemic properties through various mechanisms in cell cultures and *in vivo*.

8.2 Therapeutic Approaches to the Development of Lipid-Lowering Food Proteins and Peptides

A number of physiological molecules have been explored as possible targets for the discovery of cholesterol-lowering agents. Among the most widely studied is 3-hydroxy-3-methylglutaryl-CoA (HMG-CoA) reductase, an enzyme that catalyzes the rate-limiting step in the mevalonate pathway for hepatic cholesterol synthesis. The statin class of drugs has been discovered as effective agents for the treatment of severe cases of hypercholesterolemia due to their role as HMG-CoA reductase inhibitors that could decrease total endogenous cholesterol synthesis. However, these enzyme inhibitors may not be able to modulate hypercholesterolemia induced by consumption of high cholesterol diets. In addition, compounds that enhance gene expression of LDL-R can also be applied as cholesterol-lowering agents. Because LDL-R plays a role in removing LDL cholesterol from the blood, therapeutic agents that upregulate its expression could typically exert a hypocholesterolemic property.

Acyl CoA:cholesterol acyltransferase (ACAT) and cholesteryl ester transport protein (CETP) are also potential targets for the development of hypocholesterolemic agents. ACAT catalyzes the esterification of cholesterol in the intestine prior to its incorporation into chylomicrons for circulation. Inhibition of ACAT activity potentially lowers intestinal cholesterol absorption, reduces blood cholesterol levels, and could delay the onset or progression of atherosclerosis (Kusunoki et al., 2001). On the other hand, CEPT mediates the transfer of cholesteryl ester from HDL to LDL or VLDL in exchange for TAG; thus, CEPT inhibition decreases blood LDL-cholesterol and increases blood HDL-cholesterol in human subjects (Barter et al., 2003). In addition, bile acid binding agents can also contribute in modulating blood cholesterol level in hypercholesterolemia. Bile acids are metabolites of cholesterol produced in the liver and reabsorbed in the intestine. Thus, inhibitors of bile acid absorption or bile acid sequestrants bind and prevent reabsorption of bile acids in the intestine. The resulting water-insoluble complex formed is subsequently removed through the feces, which leads to upregulation of bile acid synthesis from cholesterol with concomitant decreases in hepatic and blood cholesterol levels. These strategies have been used toward the search for potent hypolipidemic and hypocholesterolemic food proteins and peptides.

8.3 Hypocholesterolemic and Hypolipidemic Food Proteins

It is well established that consumption of certain food proteins modulates cholesterol and lipid concentrations in diet, chemical-induced or genetic animal models of hypercholesterolemia, and in humans. Generally, it is widely accepted that proteins of plant origin possess better effects on plasma total cholesterol than animal proteins (Carrol and Hamilton, 1975). Thus, most studies that evaluate the hypocholesterolemic and hypolipidemic activities of dietary proteins use casein-fed rats as a control. It was recently reported that gastrointestinal-stimulated casein digest downregulated *cyp7a* expression in Hep G2 cells (Nass et al., 2008). Cyp7α catalyzes the rate-limiting step

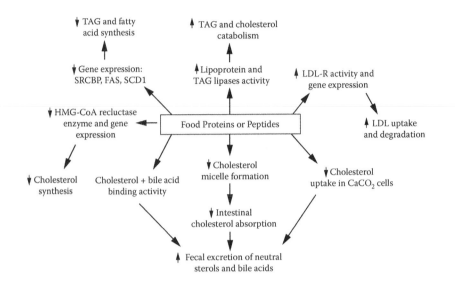

FIGURE 8.1
Proposed mechanisms of hypolipidemic and hypocholesterolemic properties of food proteins and peptides in cell culture, animal models, and humans. Abbreviations: TAG, triacylglycerol; LDL, low-density lipoprotein; LDL-R, LDL-receptor; SREBP, sterol regulatory element binding proteins; FAS, fatty acid synthase; SCD1, steroly-CoA desaturase-1.

of bile acid synthesis; its downregulation decreases bile acid synthesis from cholesterol and potentially increases plasma cholesterol. However, Sautier et al. (1983) reported that the lipid-lowering property of selected food proteins is independent of origin but dependent on their amino acid profiles. Indeed, a number of factors could determine the lipid-lowering property of food proteins in addition to their origin and amino acid profile. The proposed mechanisms of hypolipidemic and hypocholesterolemic activity of food proteins in cell culture, animals and humans are shown in Figure 8.1. Generally, hypocholesterolemic proteins and peptides work by means of increased cholesterol catabolism, reduced cholesterol synthesis, increased expression of LDL reception genes, or increased excretion of cholesterol in the feces.

8.3.1 Soybean Proteins

In the last three decades, a wide range of common food proteins has been investigated for their potential lipid-lowering properties. Dietary soy protein been of particular interest to many researchers due to its consistent hypocholesterolemic activity in several studies with animals and humans that received high or normal levels of dietary cholesterol (Sirtori et al., 1977; Aoyoma et al., 2000; Mackey, Ekangaki, and Eden, 2000; Chen et al., 2005; Sirtori et al., 2009). In a randomized double-blind placebo-controlled study using hyperlipidemic dialysis patients, daily intakes of 30 g soy protein as a beverage for 84 days showed cardioprotective properties characterized by decreases in plasma cholesterol (–18%), TAG (–43%), non-HDL-cholesterol (–23%), and LDL-cholesterol (–26%) with significant increase in HDL-cholesterol (Chen et al., 2005). It was earlier thought that the major isoflavone constituents of soy protein isolates, especially genistein and diadzein, contributed to its lipid-lowering activities. However, an animal study using isoflavone-free soy protein failed to support this hypothesis (Fukui et al., 2002). Moreover, clinical studies in human beings have also shown that the level of isoflavones in soy protein isolate did not influence its effects on elevated plasma lipid and lipoprotein profiles (Sirtori et al., 1997, 1998; Mackey et al., 2000). In addition, consumption of soy protein isolate lowered blood cholesterol in hypercholesterolemic postmenopausal women by increasing mononuclear cell LDL-R messenger RNA (mRNA) independent of the isoflavone contents of the protein (Baum et al., 1998).

These results as well as other studies in cultured hepatocytes (Lovati et al., 1998, 2000), mice (Moriyama et al., 2004), and human subjects (Kohno et al., 2006) indicate that the hypocholesterolemic and hypolipidemic property of soy protein isolate is mainly associated with its protein components. Thus, normal and genetically obese mice that were fed 20% of purified β-conglycinin fraction of soy protein daily for two weeks showed lower serum TAG concentrations than rats that received similar amounts of soy 11S glycinin or casein (Moriyama et al., 2004). This observation was attributed to altered gene expressions by the β-conglycinin treatment that resulted in increased lipid degradation and fatty acid synthesis in rats.

8.3.2 Other Food Proteins

In addition to soy protein, a number of other dietary proteins have been studied for potential hypocholesterolemic and hypolipidemic properties. These food proteins include whey protein (Sautier et al., 1983), egg white protein (Matsuoka et al., 2008), fava bean protein (Macarulla et al., 2001), black gram protein (Kurup and Kurup, 1982), buckwheat protein (Kayashita et al., 1997; Tomotake et al., 2001), cowpea protein (Frota et al., 2008), white (Sirtori et al., 2004) and blue lupin proteins (Bettzieche et al., 2008a), fish protein (Zhang and Beynen, 1993), and *Rhus verniciflua* Stokes fruit protein (Oh et al., 2006). Table 8.1 shows the effects of these food proteins on blood and tissue levels of cholesterol and TAG in different animal models that received normal and increased levels of dietary cholesterol, and their proposed mechanisms of action. It could be observed that these common and underutilized food proteins also possess hypolipidemic and hypocholesterolemic activities similar to or better than the activity reported for dietary soy proteins.

TABLE 8.1

Dietary Proteins with Hypolipidemic and Hypocholesterolemic Properties in Animals and Humans[a]

Protein	Animal Model/Cell Culture/Human Subject	Dose	Duration (Days)	Outcome/Mechanism	Reference
Soy protein	Energy-restricted Sprague–Dawley rats fed high fat diet to induce obesity	35%	28	Reduced plasma TAG by 25% compared to casein-fed group but no significant effects on plasma cholesterol	Aoyoma et al., 2000
	Hyperlipidemic and normolipidemic hemodialysis patients	30 g/day as beverage	84	Significantly decreased plasma cholesterol (−18%), TAG (−43%), non-HDL cholesterol (−23%), apo B (−15%) and LDL-cholesterol (−26%) with increase in HDL cholesterol (17%) in hyperlipidemic subjects; no effects of treatments on these parameters in normolipidemic subjects	Chen et al., 2005
Cowpea protein	Diet-induced hypercholesterolemic male Golden Syrian hamster	19.5%	28	Reduced plasma total and non-HDL cholesterol by 20% and 22%, respectively, without any significant effects on fecal excretion of cholesterol and bile acids, compared to casein-fed control group	Frota et al., 2008
Egg white protein	Male Sprague–Dawley rats fed cholesterol-rich diet	20%	21	Decreased serum (−17%) and hepatic cholesterol (−36%) without any effects on TAG and HDL cholesterol, compared to casein-fed group; reduced cholesterol level in liver and intestine; decreases cholesterol absorption by inhibiting intestinal micelle formation	Matsuoka et al., 2008

Fava bean protein	Diet-induced hypercholesterolemic male Wistar rats	18.1%	14	Serum (LDL-VLDL) and total cholesterol decrease by 36% and 29%, respectively, with a 42% reduction in atherogenic index;[b] significant decrease in diet-induced elevated hepatic total, free cholesterol, and TAG; increased fecal excretion of total fat and cholesterol compared to casein-fed group	Macarulla et al., 2001
Blackgram protein	Male Sprague–Dawley rats fed cholesterol-rich diet	16%	30	Significantly reduced cholesterol and TAG in serum, aorta, and liver (with increased activity of lipoprotein and TAG lipases), and increased fecal excretion of bile acid and neutral sterols compared to casein-fed group; activity was reduced by removal of the carbohydrate moiety of the protein	Kurup and Kurup, 1982
Buckwheat protein	Diet-induced hypercholesterolemic male Sprague–Dawley rats	20%	21	Reduced plasma and liver cholesterol by 35% and 48%, respectively; induced a twofold fecal excretion of total neutral sterols compared to casein-fed rats; activity was due to its low digestibility; activity was better than the effects of equal amount of soy protein after a 56-day study especially in the excretion of fecal neutral steroids and bile acids	Kayashita et al., 1997; Tomotake et al., 2001
Whey protein	Male Wistar rats	23%	49	Decreased serum (−26%) and liver cholesterol (−16%) compared to rats that received casein	Sautier et al., 1983
White lupin (*Lupinus albus* L.) protein	Male Sprague–Dawley rats fed cholesterol-rich diet	50 mg/day	14	Reduced plasma total and (LDL+VLDL)-cholesterol and TAG by 21%, 30%, and 15%, respectively; purified protein component (conglutin γ) upregulated LDL-R activity in HepG2 (hepatic) cell culture with 53% and 21% increases in LDL uptake and degradation, respectively compared to whole protein isolate	Sirtori et al., 2004

(continued)

TABLE 8.1 (continued)

Dietary Proteins with Hypolipidemic and Hypocholesterolemic Properties in Animals and Humans[a]

Protein	Animal Model/Cell Culture/Human Subject	Dose	Duration (Days)	Outcome/Mechanism	Reference
Blue lupin (*L. angustifolius* L.) protein	Male Sprague–Dawley rats fed cholesterol-rich diet	5%	17	Vitabor cultivar decreased plasma total and VLDL TAG by 24% and 40%, respectively, and LDL cholesterol by 37% due to decreased expressions of genes (SREBP-1, FAS, SCD1, HMG-CoA R genes) related to endogenous TAG and cholesterol synthesis; α-/β-conglutin from Boregine cultivar also decreased plasma total and VLDL TAG by 20% and 29%, respectively; activity not observed in other cultivars compared to casein-fed control rats	Bettzieche et al., 2008a
Fish protein	Female Wistar rats	2.4–7.2%	21	Cod and plaice meals showed dose-dependent significant decreases in plasma and hepatic cholesterol levels compared to casein; whiting meal decreased only liver cholesterol; all effects lower than those observed for soy protein; no proposed mechanism of action	Zhang and Beynen, 1993
Rhus verniciflua Stokes (RVS) fruit glycoprotein	Trixton WR-1339-induced hyperlipidemic male ICR mice	100 mg/kg BW/day	14	Decreased plasma total cholesterol, TAG and LDL in normal mice; dose-dependent decreases (30%, 49%, and 26%, respectively, at 100 mg/kg BW) observed in hyperlipidemic rats; modulated decreased plasma HDL level and inhibited the activity of HMG-CoA R in hepatocytes compared to control	Oh et al., 2006

[a] For a list of other dietary proteins and their cholesterol-lowering activities in animals and humans, see Sirtori et al. (2009).

[b] Atherogenic index in this study was defined as: (VLDL+LDL)-cholesterol:HDL-cholesterol

Abbreviations: TAG, triacylglycerol; HDL, high-density lipoprotein; LDL, low-density lipoprotein; LDL-R, LDL-receptor; VLDL, very low-density lipoprotein; SREBP-1, sterol regulatory element binding protein-1; FAS, fatty acid synthase; SCD1, steroly-CoA desaturase-1; HMG-CoA R, HMG-CoA reductase; BW, body weight.

8.3.2.1 Cowpea, Blackgram, Rhus verniciflua, and Buckwheat Proteins

In addition to soy, other legume seed proteins have exhibited lipid-lowering properties in rats fed high-fat diets. Recently, in a diet-induced hypercholesterolemia hamster, Frota et al. (2008) reported that consumption of about 20% cowpea protein isolate for 28 days resulted in a decrease in both plasma total and non-HDL-cholesterol by 20% and 22%, respectively. However, the observed effects were more pronounced in rats that received the cowpea seed-based diet and were probably due to its other bioactive components. Moreover, daily intake of similar amounts of fava bean protein isolate for two weeks resulted in a more pronounced reduction in serum total cholesterol as well as LDL+VLDL-cholesterol in hypercholesterolemic rats; this was partly due to the ability of fava bean protein isolate to increase fecal cholesterol removal from the intestine of the rats (Macarulla et al., 2001). This mechanism was earlier proposed for the cholesterol-lowering activity reported for blackgram protein, which reduced cholesterol in the serum, liver, and aorta of diet-induced hypocholesterolemic rats (Kurup and Kurup, 1982). Moreover, the authors also reported that the activity of blackgram protein was associated with increased activity of lipoprotein and TAG lipases, which contributed to enhanced cholesterol and TAG catabolism in these tissues.

In a more recent study, a 36 kDa glycoprotein isolated from *Rhus verniciflua* Stokes fruit showed a dose-dependent reduction in plasma cholesterol, TAG and LDL concentrations, and modulated the decreased plasma HDL level in Trixton WR-1339-induced hyperlipidemic rats (Oh et al., 2006). This activity was attributed to the inhibition of hepatic HMG-CoA reductase activity by the glycoprotein isolate even though the liver concentration of cholesterol was not measured. However, this observed activity might not be due to the protein component alone as the glycoprotein contains a significant amount of carbohydrate moiety, which might reduce the digestibility of the protein. It has been shown that digestibility plays an important role in determining the ability of food proteins to modify lipid profiles in hypercholesterolemic rats. Kayashita et al. (1997) reported that buckwheat protein induced a decrease in total plasma and liver cholesterol by 35% and 48%, respectively, due to increased fecal excretion of neutral sterols, and this was partially attributed to the low digestibility of the protein isolate compared to casein. It could be that the resulting large peptides from the partial *in vivo* digestion of the protein participated in the removal of the neutral sterols from the intestine, inasmuch as buckwheat protein contains high amounts of hydrophobic amino acids compared to casein and soy protein; it was also proposed that the enhanced transit time of the poorly digestible protein may have increased the fecal secretion of neutral sterols in rats (Kayashita et al., 1997).

8.3.2.2 Lupin Proteins

A recent review noted that both short- and long-term consumption of grain legumes exhibited positive attributes in the lipid profiles of hypercholester- olemic human subjects (Sirtori et al., 2009) and this could also be due to their constituent protein fractions. Among the well-known grain legume pro- teins, a number of studies have shown that lupin proteins possess excellent hypocholesterolemic and hypolipidemic properties in animals (Sirtori et al., 2004; Spielmann et al., 2007; Bettzieche et al., 2008a,b). In a rat model that received a cholesterol-rich diet, daily administration of 50 mg isoflavone- poor white lupin (*Lupinus albus* L.) by gavage resulted in the reduction of elevated plasma (–21%) and LDL+VLDL-cholesterol (–30%) and TAG (–15%) during a two-week study (Sirtori et al., 2004). This cholesterol-lowering activ- ity was attributed to one of its purified glycoprotein components, conglutin γ (Cγ), which potently induced upregulation of LDL-R activity in cultured hepatocyte leading to increased uptake and degradation of LDL. Other stud- ies have shown that white lupin proteins also possess *in vivo* influence on the expression of genes relevant to lipid metabolism.

Thus, feeding 5% white lupin protein to hypercholesterolemic rats resulted in the downreglation of liver mRNA levels of genes for lipid syn- thesis including sterol regulatory element binding protein-1c (SREBP-1c), ACAT and fatty acid synthase (FAS), and upregulation of genes for TAG catabolism including lipoprotein and hepatic lipases leading to a decrease in hepatic TAG concentration (Bettzieche et al., 2008b). SREBP-1c is a transcription factor that regulates the expression of lipogenic enzymes, thus its downregulation will lead to decreased lipid synthesis. Similarly, low levels of blue lupin proteins (*L. angustifolius* L.) induced a decrease in the expression of these genes and another lipogenic enzyme, steroly- CoA desaturase-1 (SCD1), with a corresponding significant reduction in plasma total cholesterol and LDL-cholesterol when compared to the effects of casein (Bettzieche et al., 2008b). An understanding of the molecular mechanism of hypocholesterolemic and hypolipidemic activities of lupin proteins will potentially increase its application as a health-promoting dietary protein. However, it is recommended that human clinical studies be conducted in order to validate these activities considering the differ- ences in the regulation of lipid metabolism in rats and humans.

8.3.3 Animal Proteins: Milk, Egg White, and Fish

Despite the general opinion about negative effects on lipid profile, some animal proteins such as milk and egg white proteins have also shown some potential lipid-lowering properties. In a pioneering study with a rat model that received 0.4–3.4 mg cholesterol daily, Sautier et al. (1983) reported that consumption of 23% whey proteins for 49 days lowered both total serum and hepatic cholesterol when compared to rats that were fed casein. Moreover, even though egg consumption has been associated with hypercholesterolemia due to the high cholesterol contents of egg yolk, some of the egg protein components could normalize physiological levels of cholesterol. A number of previous works have reported that egg white proteins possess hypocholesterolemic activity in animals and human beings (Asato et al., 1996; Nagaoka et al., 2002; Matsuoka et al., 2008). It is thought that the ability of egg white proteins to reduce cholesterol levels in rats fed high-cholesterol diets is related to their activity in interfering with the formation of micelles in the intestine, which subsequently decreases the rate of cholesterol absorption (Nagaoka et al., 2002; Matsuoka et al., 2008).

In these studies, this observation was supported by the increased excretion of neutral sterols and bile acids with the concomitant decrease in cholesterol levels in the serum, liver, and intestinal mucosa of rats. Due to its hydrophobicity, absorption of cholesterol in the intestine requires the formation of micelles with bile acids prior to transport across the intestinal epithelium. Thus, inhibition of micelle formation decreases the total absorbed exogenous cholesterol. In addition, a minor insoluble component of egg white proteins, ovomucin, was reported to decrease absorption of cholesterol in a Caco-2 human intestinal cell culture consistent with the observation of decreased cholesterol absorption in the jejenum of rats that received dietary cholesterol (Nagaoka et al., 2002). In contrast, fish protein had a negative influence on lipid profiles of rats by potentially inducing hypercholesterolemia.

In an animal study, isolated fish proteins from Alaska pollock induced an elevated expression of SREBP-2 with a concomitant increase in the activation of HMG-CoA reductase leading to elevated liver cholesterol in hypercholesterolemic rats; however, the protein possessed hypotriglycerodemic activity compared to rats that received a casein diet (Shukla et al., 2006). These results indicate that the potential lipid-lowering property of food proteins might indeed be dependent on their origin. This was evident in a previous study that showed a dose of fish proteins (cod and plaice meal) dependently reduced plasma and hepatic cholesterol levels compared to casein, but these beneficial properties were smaller than the effects of an equal amount of soy protein (Zhang and Beynen, 1993). Overall, the lipid-lowering activities of dietary proteins are promising toward the management or treatment of mild cases of hyperlipidemia and hypercholesterolemia. However, because dietary proteins are acted upon by intestinal proteases during digestion, the resulting peptides are excellent candidates that could be responsible for the observed activity; thus, several studies have also investigated the potential cholesterol- and lipid-lowering properties of enzymatically prepared hydrolyzed food proteins and their constituent peptides.

8.4 Hypocholesterolemic and Hypolipidemic Food Protein-Derived Peptides

Food protein-derived bioactive peptides have been extensively investigated for potential application in the management and treatment of some human diseases. These peptide sequences are inactive within the intact protein structure but could exert several pharmacological properties upon release by the activity of proteases. Thus, the structure and activity of the peptides released depend largely on the specificity of the proteolytic enzyme used in hydrolysis. Of all the bioactive peptides derived from food proteins, inhibitors of angiotensin I-converting enzyme have received the most attention for their ability to lower blood pressure in hypertensive rats and in humans. Other reported bioactive peptides from food sources include peptides with antioxidant, anticancer, and immunomodulatory activites. Moreover, protease-assisted controlled hydrolysis of food proteins could also release peptide sequences that possess hypocholesterolemic and hypolipidemic activities.

Some food protein sources reported to contain bioactive peptide sequences with lipid and cholesterol-lowering properties include soy protein (Doi et al., 1986; Nagaoka et al., 1999; Aoyama et al., 2000; Chen et al., 2006; Cho, Juillerat, and Lee, 2007; Zhong et al., 2007a), milk β-lactoglobulin (Kirana et al., 2005), buckwheat protein (Kayashita et al., 1997), blackgram protein (Kurup and Kurup, 1982), egg white protein (Manso et al., 2008), fish protein (Wergedahl et al., 2004), pork liver protein (Shimizu et al., 2006), and *Brassica carinata* protein (Pedroche et al., 2007). Table 8.2 shows the effects of these food protein hydrolysates on blood and tissue cholesterol and lipid levels in animals and human subjects. Thus, it could be assumed that the lipid-lowering activities observed for food proteins may be partly due to their constituent peptides released by the activity of intestinal proteases during *in vivo* digestion. However, a number of studies have indicated that enzymatic hydrolysis could also result in reduced lipid-lowering activity of food proteins (Kurup and Kurup, 1982; Kayashita et al., 1997; Ueda, 2000). In some studies, physiological digestive proteases were used to hydrolyze these proteins whereas others used food-grade microbial enzymes. The proteases and reaction conditions (pH, temperature, duration, and enzyme-substrate ratio, E/S) employed in the production of hypocholesterolemic and hypolipidemic food protein hydrolysates and peptides are shown in Table 8.2.

8.4.1 Soybean Protein-Derived Peptides

Numerous studies have established the hypocholesterolemic and hypolipidemic properties of soy protein hydrolysates in animals (Yashiro, Oda, and Sugano, 1985; Doi et al., 1986; Aoyoma et al., 2000; Chen et al., 2006; Tamaru et al., 2007; Yang et al., 2007) and human beings (Hori et al., 2001). The soy 7S globulin

TABLE 8.2

Enzymatic Food Protein Hydrolysates and Peptides with Potential Hypolipidemic and Hypocholesterolemic Properties

Protein Source	Protease	Hydrolysis Conditions	Model	Dose/ Duration	Outcome/Mechanism	Reference
Cell Culture						
Soy 7S globulin-derived peptides	N/A	N/A	Human hepatoma (Hep G2) cells	10^{-4} M	Peptide fragment 127–150 (MW, 2.27 kDa) stimulated increase in LDL uptake and degradation in hepatocytes	Lovati et al., 2000
Croksoy®70 (isoflavone-poor) digest	Porcine pepsin and bovine pancreas trypsin	Pepsin: 37°C, pH 2.0, 1 h, E/S 1:100 Trypsin: 37°C, pH 7.0, 1 h, E/S 1:100	Human hepatoma (Hep G2) cells	0.05–1 mg/ml	Dose-dependent upregulation of LDL-R activity compared to undigested protein leading to increased uptake and degradation of LDL; activity was enhanced in the high MW (>3 kDa) fraction of the digest and absent in lower MW fractions	Lovati et al., 2000
Soy protein	Neutral proteases from *Bacillus amyloliquefaciens* FSE-68	45°C, pH 7.0, E/S 0.1:100, DH = 5–15%	Human hepatocytes (Hep T9A4 cells)	0.2 mg/ml	Stimulated LDL-R transcription from 100% to 126.8–238%, which might lead to increased *in vivo* cholesterol catabolism; both intact protein and isoflavones extract of the digest did not show any effects on LDL-R transcription; thus, the observed activity was solely due to the soy peptides	Cho et al., 2007

(continued)

TABLE 8.2 (continued)

Enzymatic Food Protein Hydrolysates and Peptides with Potential Hypolipidemic and Hypocholesterolemic Properties

Protein Source	Protease	Hydrolysis Conditions	Model	Dose/ Duration	Outcome/Mechanism	Reference
Soy protein-derived peptides	Neutral proteases from *Bacillus amyloliquefaciens* FSE-68	45°C, pH 7.0, E/S 0.1:100, DH = 15%	Human hepatocytes (Hep T9A4 cells)	100 _M	Identification of active peptides from Cho et al. (2007); peptides were fractionated by RP-HPLC; fractions stimulated LDL-R transcription, identified and sequenced; of all nine sequenced peptides, FVVNATSN (7S conglycinin β-subunit origin) displayed the best activity by increasing hepatic cell LDL-R transcription by 248.8%	Cho et al., 2008
Bovine milk β-lactoglobulin (heat treated)	Porcine trypsin	37°C, pH 8.0, 3 h, E/S 0.4:100	Male Wistar rats and human enterocytes (Caco-2 cells)	1.8 mg/mL, 1 h	Cardioprotective effects by reducing *in vivo* cholesterol uptake into serum, liver, and intestine compared to casein tryptic hydrolysate; decreased cholesterol uptake (~40%) in cell culture; peptide IIAEK identified as active agent that decreased total serum cholesterol and increased HDL-cholesterol	Nagaoka et al., 2001

Animal

Soy protein	Protease from *Bacillus subtilis*	N/A	Energy-restricted obese Sprague–Dawley rats fed high fat diet to induce obesity	35% for 28 days	Significantly reduced plasma total cholesterol (−29%) and TAG (−33%) compared to casein-fed group, which led to reduction in body fat accumulation; also lowered HDL-cholesterol	Aoyoma et al., 2000
Soy protein (heat-treated)	Alcalase	60°C, pH 8.5, E/S 1:80, DH = 18%	Diet-induced hypercholesterolemic male Kunming mice	0.1, 0.5, 2.5 g/kg BW/day for 30 days	Time- and dose-dependent decrease in serum total and LDL+VLDL-cholesterol; no change in serum TAG; peptide WGAPSL identified as active component that inhibited cholesterol micellar solubility *in vitro*	Zhong et al., 2007a
Buckwheat protein	Porcine trypsin	45°C, pH 8.0, 5 h, E/S 1:100	Diet-induced hypercholesterolemic male Sprague–Dawley rats	20% for 14 days	Decrease in plasma cholesterol compared to casein; however, intact buckwheat protein exhibited better activity; trypsin digest lost activity due to increased digestibility; thus, low protein digestibility is partly responsible for cholesterol-lowering activity	Kayashita et al., 1997

(continued)

TABLE 8.2 (continued)

Enzymatic Food Protein Hydrolysates and Peptides with Potential Hypolipidemic and Hypocholesterolemic Properties

Protein Source	Protease	Hydrolysis Conditions	Model	Dose/ Duration	Outcome/Mechanism	Reference
Globin	Acid protease	N/A	Mice (Slc:ICR strain) and rats (Scl:Wistar strain)	7.7 to 100 mg for 8 h	Dose-dependent suppression of total serum and chylomicron TAG (up to −85%) induced by oral intake of olive oil; peptide VVPY (0.37% of globin digest) displayed 7,000 times better TAG-lowering activity than globin hydrolysate; maybe due to increased hepatic TAG lipase activity	Kagawa et al., 1996
Soy protein	Porcine pepsin	37°C, pH 1.6, 2 h, E/S 1:100	Growing male Wister rats fed cholesterol-free diet	20% for 30 days	Significantly lowered serum total cholesterol and TAG by 27% and 46%, respectively, compared to casein; intact soy protein showed reduced effect on serum cholesterol and no effect on TAG	Doi et al., 1986
Blackgram protein	Papain	65-70°C, pH 6.5, 48 h, E/S 1:100	Male Sprague–Dawley rats fed cholesterol-rich diet	16% for 30 days	Decreased total cholesterol in serum, aorta and liver of rats compared to the casein-fed group; hydrolysis reduced the cholesterol- and TAG-lowering ability of the intact blackgram protein	Kurup and Kurup, 1982

Protein	Enzyme	Conditions	Model	Dose/Duration	Effects	Reference
Egg white protein	Pepsin	37°C, pH 2.0, 3 h, DH_{max}[a]	Male spontaneously hypertensive rats	5–10% for 20 weeks	Long-term feeding significantly reduced plasma TAG and total cholesterol without any effects on HDL cholesterol; related to their antioxidant activity; intact protein showed no effects	Manso et al., 2008
Fish protein	Protamex™ (E.C. 3.4.21.62, 3.4.24.28)	55°C, pH 6.5, 1 h, E/S 11 AU/kg crude protein	Male lean Wistar rats and Male obese *fa/fa* Zucker rats	20% for 11–12 days; 20% for 22–23 days	Increased plasma HDL/total cholesterol ratio compared to casein; reduced plasma total (−33%) and HDL cholesterol (−21%) and hepatic lipids (−36%); decreased mRNA levels of lipogenic enzymes:_5 and _6 desaturases and ACAT	Wergedahl et al., 2004
Pork liver protein	Proteinase (E.C. 3.4.21.63)	45°C, pH 7.0, 4 h, E/S 1:28	OLETF (genetically obese) rats	20% for 14 weeks	Significantly decreased plasma free fatty acids, TAG, and cholesterol with increased fecal excretion of total fat; associated with inhibition of hepatic lipogenic enzymes: G6PDH and FAS compared to casein, even without changes in hepatic TAG and cholesterol	Shimizu et al., 2006

[a] Enzyme added until maximum degree of hydrolysis (DH).

Abbreviations: G6PDH, glusoce-6-phosphate dehydrogenase; FAS, fatty acid synthase; ACAT, acyl-CoA:cholesterol acyltransferase; OLETF, Otsuka Long-Evans Tokushima Fatty; DH, degree of hydrolysis.

(β-conglycinin) has been implicated as a source of the hypocholesterolemic property of soy protein. The α+α′ subunit of this protein strongly upregulated the expression of LDL-R in cultured hepatocytes leading to an increase in LDL uptake and degradation (Table 8.3); however, the intact β-conglycinin and its β subunit were thought to be less active (Lovati et al., 1998).

Based on the differences between the amino acid sequences of the active α+α′ subunit and the less active β subunit of 7S β-conglycinin, the peptide region responsible for the activity has been identified from the α′ subunit and sequenced (Lovati et al., 2000). This 24-amino acid peptide (MW 2.27 kDa) that corresponds to position 127-150 of the α′ subunit displayed potential in modulating cholesterol homeostasis by increasing LDL-R-mediated LDL uptake in Hep G2 cells (Lovati et al., 2000). Moreover, Cho, Juillerat, and Lee (2008) recently identified an octapeptide (FVVNATSN) from the enzymatic digest of soy protein belonging to the purportedly less active 7S β-conglycinin-β chain as the most active stimulator of LDL-R transcription in Hep T9A4 human hepatic cells (Figure 8.2). In their study, HPLC fractionation of soy protein hydrolysates based on hydrophobicity yielded peptide fractions that showed different activities in the upregulation of LDL-R transcription in Hep T9A4 cells (Table 8.3). One of the fractions (RP14) strongly increased the LDL-R transcription level by 268%; the observed activity was independent of hydrophobicity of the peptide fractions but may depend on their amino acid sequences.

Thus, proteolytic digestion of soy protein was important for releasing more active small peptides with improved potential cardioprotective properties. This has also been demonstrated in a study by Mochizuki et al. (2009) that produced bioactive peptides from purified isoflavone-free soy 7S β-conglycinin using bacterial proteases (thermoase, bioprase, and sumizyme FP). The resulting 7S-peptides (MW < 10 kDa) showed potential hypotriglyceridemic properties by altering gene expressions related to TAG synthesis and also decreased Apo B-100 accumulation in Hep G2 cells partly due to an increase in LDL-R mRNA expression (Mochizuki et al., 2009). Apo B-100 is a functional component of VLDL and its degradation reduces VLDL synthesis. These observations supported a previous study that showed soy β-conglycinin possesses beneficial effects on plasma TAG in human subjects (Kohno et al., 2006).

In addition to alterations of gene expressions, soy protein hydrolysates and constituent peptides also exhibited hypocholesterolemic activity by binding bile acids and neutral sterols in the intestine leading to increased fecal removal (Pak et al., 2005; Cho et al., 2007; Yang et al., 2007). The ability of the soy protein hydrolysates to bind bile acids may depend in part on their insoluble high MW (HMW) peptide fraction rich in hydrophobic amino acids (Higaki et al., 2006), as earlier observed for a HMW fraction from a tryptic digest of buckwheat protein (Kayashita et al., 1997). This shows that even though large bioactive peptides may not be able to cross the intestinal epithelium into circulation to exert their beneficial effects in the liver, they

TABLE 8.3

Peptide Sequences with Potential Cholesterol- and Triglyceride-Lowering Activities and Their Food Protein Sources

Peptide Sequence	Food Protein Source	Position	Activity	Reference
LRVPAGTTFYVV-NPDNDENLRMIA	Soy 7S globulin α′ subunit	127–150	Increase LDL-R mediated LDL uptake	Lovati et al., 2000
FVVNATSN	Soy 7S β-conglycinin β subunit	–	Stimulates LDL-R transcription	Cho et al., 2008
FKTNDRPSIGN	Soy 11S glycinin G2 or G3 basic subunit	–	Stimulates LDL-R transcription	Cho et al., 2008
SSPDIYNPQAGS	Soy 11S glycinin G1 or G2 basic subunit	–	Stimulates LDL-R transcription	Cho et al., 2008
DTPMIGT	Soy 11S glycinin G1 basic subunit	–	Stimulates LDL-R transcription	Cho et al., 2008
LPYP	Soy glycinin	–	HMG-CoA R inhibitor, cholesterol-lowering activity *in vivo*	Kwon et al., 2002
LPYPR	Soy glycinin	–	HMG-CoA R inhibitor, cholesterol-lowering activity *in vivo*	Kwon et al., 2002
SPYPR	Synthetic peptide	N/A	HMG-CoA R inhibitor, cholesterol-lowering activity *in vivo*	Kwon et al., 2002
IAVPGEVA	Soy 11S globulin	–	Bile acid-binding activity	Pak et al., 2005
IIAEK	Milk β-lactoglobulin	71–75	Inhibits cholesterol uptake	Nagaoka et al., 2001
GLDIQK	Milk β-lactoglobulin	9–14	Inhibits cholesterol uptake	Nagaoka et al., 2001
ALPMH	Milk β-lactoglobulin	142–146	Inhibits cholesterol uptake	Nagaoka et al., 2001
VYVEELKPTPEGD-LEILLQK	Milk β-lactoglobulin	41–60	Inhibits cholesterol uptake	Nagaoka et al., 2001
VVPY	Bovine blood globin	–	TAG-lowering activity *in vivo*	Kagawa et al., 1996
WGAPSL	Soy protein	–	Inhibits cholesterol micellar solubility	Zhong et al., 2007b

Abbreviations: LDL-R, LDL-receptor; HMG-CoA R, HMG-CoA reductase

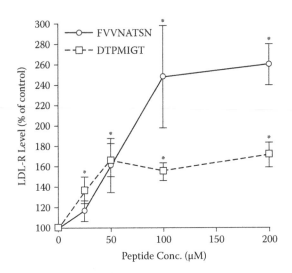

FIGURE 8.2
Effect of synthetic soybean peptides on LDL-R transcription. Changes in LDL-R transcription are expressed as percentage of untreated controls (mean (SD) (n) 3). *$P < 0.05$ versus control. (Reproduced from Cho, S.-J. et al., *J. Agric. Food Chem.* 56, 4372–4376, 2008, with permission of the American Chemical Society.)

might be useful in cholesterol homeostasis by enhancing fecal removal of bile acids and exogenous cholesterol from the intestine depending on their hydrophobic properties.

In a recent study, Zhong et al. (2007b) identified the hydrophobic peptide WGAPSL from Alcalase-prepared soy protein hydrolysate as the major hypocholesterolemic component based on *in vitro* inhibition of cholesterol micellar solubility (Table 8.3). Moreover, the soy protein hydrolysate that yielded the WGAPSL exhibited *in vivo* hypocholesterolemic activity by decreasing the elevated serum total and VLDL+LDL-cholesterol in hyper-cholesterolemic mice (Zhong et al., 2007a). Other attempts have been made to identify the amino acid sequences of soy protein responsible for its lipid-lowering activity especially in binding bile acids. Thus, a hexapep-tide (VAWWMY, f129-134) has been identified in soy glycinin A1aB1b subunit as the bile acid-binding region (Choi, Adachi, and Utsumi, 2002). Moreover, the bile acid-binding property of soy glycinin subunit has been improved by gene modification. Choi, Adachi, and Utsumi (2004) reported that introducing the nucleotide sequence that codes for VAWWMY into soy glycinin resulted in a modified protein that possesses an improved bile acid-binding property. This demonstrates the feasibility of improving the functional and bioactive properties of these food proteins using gene technology.

8.4.2 Egg White Protein-Derived Peptides

Oxidative stress plays a major role as a mechanism for the initiation and pathology of atherosclerosis, therefore antioxidant food protein-derived peptides may also participate in modifying blood lipid profiles in hypercholesterolemia and hyperlipidemia. In a recent study, long-term feeding of 10% pepsin-hydrolyzed egg white protein to spontaneously hypertensive rats (SHR) for 20 weeks resulted in a significant decrease in both plasma TAG and total cholesterol without altering plasma HDL-cholesterol concentration (Manso et al., 2008). Although pepsin activity could potentially release these bioactive peptides from egg white protein *in vivo*, SHR that received an equal amount of the intact protein in this study did not show any beneficial changes in plasma lipids. This observation could be related to increased intestinal absorption of bioactive peptides from the protein hydrolysates compared to peptides that resulted due to gastrointestinal digestion of the protein. Because other studies reported hypocholesterolemic activity for egg white proteins in rats fed cholesterol-rich diets (Matsuoka et al., 2008; Nagaoka et al., 2002), the lack of hypocholesterolemic activity for egg white protein in SHR may be related to the low basal plasma cholesterol level (approximately 60 mg/dl) inasmuch as SHR did not receive additional exogenous cholesterol. Moreover, in an attempt to identify the active protein components, Matsuoka et al. (2008) suggested that the inhibitory effects of egg white protein on micellar cholesterol absorption may be due to the water-soluble fractions comprising of pepsin hydrolysates of egg ovalbumin and ovotransferrin contrary to the hydrophobicity requirements for cholesterol-lowering soy peptides. However, the detailed lipid-lowering mechanisms of egg white protein remain to be completely elucidated.

8.4.3 Milk Protein-Derived Peptides

In addition to egg proteins, milk protein has also yielded bioactive peptides with lipid-lowering properties. *In vitro* hydrolysis of milk β-lactoglobulin using 0.4% trypsin (w/w protein) for 3 h yielded peptides that potently exerted cardioprotective effects in rats by decreasing cholesterol uptake in the liver, serum, and intestine (Nagaoka et al., 2001). From this hydrolysate, four peptides were identified as the most active components belonging to peptide positions 9-14, 41-60, 71-75, and 142-146 of the β-lactoglobulin primary sequence (Table 8.3). The activities of the peptides in inhibiting cholesterol absorption in the intestinal cell culture tended to depend on peptide hydrophobicity after separation using RP-HPLC. One of the peptides (IIAEK, f71-75) also displayed potent *in vivo* hypocholesterolemic activity comparable to the activity of β-sitosterol; this peptide decreased total serum and LDL+VLDL-cholesterol, and increased serum HDL-cholesterol in hypercholesterolemic rats (Nagaoka et al., 2001).

8.5 Effects of Amino Acid Compositions

Hydrophobicity plays a major role in the hypocholesterolemic activity of peptides especially in binding bile acids (Kwon et al., 2002; Higaki et al., 2006). Kwon et al. (2002) observed that replacement of the N-terminal leucine residue of a peptide (LPYP) with serine resulted in a peptide (SPYP) with decreased cholesterol-lowering activity in hypercholesterolic mice (Table 8.3). Moreover, a previous study reported that another peptide (LPYPR) isolated from trypsin-treated soy glycinin also displayed cholesterol-lowering activity *in vivo* (Yoshikawa, Yamamoto, and Takemaka, 1999), but the additional C-terminal arginine residue did not alter its bioactivity (Kwon et al., 2002). The hydrophobic amino acids of the hypocholesterolemic peptides and proteins are thought to interact with bile acids by hydrophobic interactions (Higaki et al., 2006) leading to the formation of insoluble complexes for fecal removal. Kwon et al. (2002) reported that hydrophobicity of a synthetic peptide (SPYPR) originally identified in tryptic soy glycinin hydrolysate may not be entirely responsible for their potential hypocholesterolemic activity because it failed to bind bile acid but exhibited *in vivo* cholesterol-lowering effects in rats (Table 8.3). Thus, other factors such as amino acid sequences of peptides may also contribute to cholesterol-lowering activity.

Kritchevsky et al. (1982) considered that a high arginine–lysine ratio in food proteins might contribute to increased cholesterol-lowering ability. This hypothesis was supported by the hypocholesterolemic activity reported for fish protein hydrolysate, which had a high arginine–lysine ratio when compared to casein (Wergedahl et al., 2004). However, an experiment with buckwheat protein and supplemental arginine in diet-induced hypercholesterolemic rats failed to show any relationship between an increased arginine–lysine ratio and decrease in plasma cholesterol (Kayashita et al., 1997). In general, based on several studies, amino acid compositions of dietary proteins identified to induce hypercholesterolemia include methionine (Muramatsu and Sugiyama, 1990; Zhang and Beynen, 1993; Kirana et al., 2005), glutamate and tyrosine (Sautier et al., 1983) whereas glycine (Muramatsu and Sugiyama, 1990; Zhang and Beynen, 1993), cysteine (Matsuoka et al., 2008; Sautier et al., 1983), arginine (Kritchevsky et al., 1982), and alanine (Sautier et al., 1983) tend to be hypocholesterolemic.

8.6 Commercial Cholesterol-Lowering Proteins and Peptides

Two soybean protein-derived products, LunaSoy™ and Lunasin XP® have recently been commercialized as suitable ingredients for the formulation of cholesterol-lowering foods (Soy Labs, 2010). The two products are made from lunasin, a bioactive soy protein component that has been shown to be responsible for the cholesterol-lowering effects associated with consumption of soybean foods. Lunasin acts by reducing the level of HMG-CoA reductase, which is similar to the action of statins, the popular cholesterol-lowering drugs. The cellular mechanism of action of lunasin involves reduction in the rate of gene expression for HMG-CoA reductase, therefore, less enzyme protein is made by the liver, which leads to reduced production of cholesterol. In addition to modulating production of HMG-CoA reductase, lunasin increases the transcription levels of LDL receptor mRNA, which enhances clearance of plasma LDL cholesterol. LunaSoy™ is presented as a protein complex, which on a per gram basis, delivers twice the bioactivity of lunasin as soybean protein isolates and is suitable for the formulation of functional foods and beverages whereas Lunasin XP is a peptide extract formulated for use as a dietary supplement (Soy Labs, 2010).

In another development, a new soy peptide, CSPHP (C-fraction soy protein hydrolysate with bound phospholipids) has been recently granted Generally Recognized As Safe (GRAS) status, allowing it to be sold as an ingredient for the formulation of cholesterol-powering foods (functional foods and beverages) or dietary supplements (Kyowa Hakko USA, 2009). In human clinical trials, daily consumption of 3 g of CSPHP for three consecutive months led to reductions in total cholesterol by about 38 mg/dl and LDL-cholesterol level by 46 mg/dl in hypercholesterolemia patients (Soy Science, 2010). The patients also benefited from an increase in HDL-cholesterol as a result of the treatment with CSPHP. Also important is the finding that CSPHP did not reduce cholesterol levels in people with normal cholesterol levels. The mechanism of action is believed to involve suppression in absorption of dietary cholesterol from the intestinal tract, which enhances lowering of plasma cholesterol levels (Hori et al., 2001; Nagaoka et al., 1999). Therefore, CSPHP is able to suppress plasma cholesterol levels associated with consumption of diets rich in cholesterol. CSPHP differs from lunasin peptides in mode of action: the former acts in the intestine to limit absorption of dietary cholesterol whereas the latter acts at the molecular level by limiting cholesterol synthesis in the liver. However, the two products have not been used simultaneously in a similar experiment and therefore, cannot be compared in terms of efficacy. It should be noted that no side effects have been reported for these commercial cholesterol-lowering protein products.

8.7 Conclusion

The increased incidence of cardiovascular diseases, especially in Western countries, has contributed to the continuing search for diet- or natural supplement-based therapeutics. Food proteins contribute to the cholesterol-lowering ability of diets by reducing bioavailability of dietary cholesterol or through the release of bioactive peptide sequences during digestion. Indigestible proteins can bind to cholesterol within the intestinal lumen, prevent absorption into the blood circulatory system, and ultimately contribute to enhanced fecal cholesterol content. In contrast, the bioactive peptides are absorbed into the blood and are transported to the liver where they reduce the activity of HMG-CoA or upregulate gene expression of LDL receptors. The availability of commercial forms of these hypocholesterolemic proteins and peptides suggests an increase in the choice of tools that can be used as dietary interventions to reduce cholesterol-induced cardiovascular diseases.

References

Aoyama, T., Fukui, K., Takamatsu, K., Hashimoto, Y., and Yamamoto, T. 2000. Soy protein isolate and its hydrolysate reduce body fat of dietary obese rats and genetically obese mice (yellow KK). *Nutrition (N.Y.)* 16: 349–354.

Asato, L., Wang, M.F., Chan, Y.C., Yeh, S.H., Chung, H.M., Chung, S.Y., Chida, S., Uezato, T., Suzuki, I., Yamagata, N., Kokubu, T., and Yamamoto, S. 1996. Effects of egg white on serum cholesterol concentration in young women. *J. Nutr. Sci. Vitaminol.* 42: 87–96.

Barter, P.J., Brewer, H.B., Jr, Chapman, M.J., Hennekens, C.H., Rader, D.J., and Tall, A.R. 2003. Cholesteryl ester transfer protein: A novel target for raising HDL and inhibiting atherosclerosis. *Arterioscler. Thromb. Vasc. Biol.* 23: 160–167.

Baum, J.A., Teng, H., Erdman, J.W., Weigel, R.M., Klein, B.P., Persky, V.W., Freels, S., Surya, P., Bakhit, R.M., Ramos, E., Shay, N.F., and Potter, S.M. 1998. Long-term intake of soy protein improves blood lipid profiles and increases mononuclear cell low-density-lipoprotein receptor messenger RNA in hypercholesterolemic, postmenopausal women. *Am. J. Clin. Nutr.* 68: 545–51.

Bettzieche, A., Brandsch, C., Schimdt, M., Weisse, K., Eder, K., and Stangl, G.I. 2008a. Differing effect of protein isolates from different cultivars of blue lupin on plasma lipoproteins of hypercholesterolemic rats. *Biosci. Biotechnol. Biochem.* 72: 3114–3121.

Bettzieche, A., Brandsch, C., Weisse, K., Hirche, F., Eder, K., and Stangl, G.I. 2008b. Lupin protein influences the expression of hepatic gene involved in fatty acid synthesis and triacylglycerol hydrolysis of adult rats. *Br. J. Nutr.* 99: 952–962.

Carrol, K.K. and Hamilton, R.M.G. 1975. Effects of dietary protein and carbohydrate on plasma cholesterol levels in relation to atherosclerosis. *J. Food Sci.* 40: 18–23.

Chen, S.T., Ferng, S.H., Yang, C.S., Peng, S.J., Lee, H.R., and Chen, J.R. 2005. Variable effects of soy protein on plasma lipids in hyperlipidemic and normolipidemic hemodialysis patients. *Am. J. Kidney Dis.* 46: 1099–1106.

Chen, S.T., Yang, H.Y., Huang, H.Y., Peng, S.J., and Chen, J.R. 2006. Effects of various soya protein hydrolysates on lipid profile, blood pressure and renal function in five-sixth nephrectomised rats. *Br. J. Nutr.* 96: 435–441.

Chen, Z.Y., Jiao, R., and Ma, K.Y. 2008. Cholesterol-lowering nutraceuticals and functional foods. *J. Agric. Food. Chem.* 56: 8761–8773.

Cho, S., Juillerat, M.A., and Lee, C. 2007. Cholesterol lowering mechanism of soybean protein hydrolysate. *J. Agric. Food Chem.* 55: 10599–10604.

Cho, S.J., Juillerat, M.A., and Lee, C.H. 2008. Identification of LDL-receptor transcription stimulating peptides from soybean hydrolysate in human hepatocytes. *J. Agric. Food Chem.* 56: 4372–4376.

Choi, S.K., Adachi, M., and Utsumi, S. 2002. Identification of the bile acid-binding region in the soy glycinin A1aB1b subunit. *Biosci. Biotechnol. Biochem.* 66: 2395–2401.

Choi, S.K., Adachi, M., and Utsumi, S. 2004. Improved bile acid-binding ability of soybean glycinin A1a polypeptide by the introduction of a bile acid-binding peptide (VAWWMY). *Biosci. Biotechnol. Biochem.* 68: 1980–1983.

Doi, H., Iwami, K., Ibuki, F., and Kanamori, M. 1986. Effect of feeding peptic digest of soy protein isolate on rat serum cholesterol. *J. Nutr. Sci. Vitaminol.* 32: 373–379.

Durrington, P. 2003. Dyslipidemia. *Lancet* 362: 717–731.

Fletcher, B., Berra, K., Ades, P. et al. 2005. Managing abnormal blood lipids: A collaborative approach. *Circulation* 112: 3184–3209.

Frota, K.M.G., Mendonça, S., Saldiva, P.H.N., Cruz, R.J., and Arêas, J.A.G. 2008. Cholesterol-lowering properties of whole cowpea seed and its protein isolate in hamsters. *J. Food Sci.* 73: H235–H240.

Fukui, K., Tachibana, N., Wanezaki, S., Tsuzaki, S., Takamatsu, K., Yamamoto, T., Hashimoto, Y., and Shimoda, T. 2002. Isoflavone-free soy protein prepared by column chromatography reduces plasma cholesterol in rats. *J. Agric. Food Chem.* 50: 5717–5721.

Higaki, N., Sato, K., Suda, H., Suzuka, T., Komori, T., Saeki, T., Nakamura, Y., Ohtsuki, K., Iwami, K., and Kanamoto, R. 2006. Evidence for the existence of a soybean resistant protein that captures bile acid and stimulates its fecal excretion. *Biosci. Biotechnol. Biochem.* 70: 2844–2852.

Hori, G., Wang, M.-F., Chan, Y.-C., Komatsu, T., Wong, Y., Chen, T.-H., Yamamoto, K., Nagaoka, S., and Yamamoto, S. 2001. Soy protein hydrolyzate with bound phospholipids reduces serum cholesterol levels in hypercholesterolemic adult male volunteers. *Biosci. Biotechnol. Biochem.* 65: 72–78.

Kagawa, K., Matsutaka, H., Fukuhama, C., Watanabe, Y., and Fujino, H. 1996. Globin digest, acidic protease hydrolysate, inhibits dietary hypertriglyceridemia and Val-Val-Tyr-Pro, one of its constituents, possesses most superior effect. *Life Sci.* 58: 1745–1755.

Kayashita, J., Shimaoka, I., Nakajoh, M., Yamazaki, M., and Kato, N. 1997. Consumption of buckwheat protein lowers plasma cholesterol and raises fecal neutral sterols in cholesterol-fed rats because of its low digestibility. *J. Nutr.* 127: 1395–1400.

Kirana, C., Rogers, P.F., Bennett, L.E., Abeywardena, M.Y., and Patten, G.S. 2005. Naturally derived micelles for rapid *in vitro* screening of potential cholesterol-lowering bioactives. *J. Agric. Food Chem.* 53: 4623–4627.

Kohno, M., Hirotsuka, M., Kito, M., and Matsuzawa, Y. 2006. Decreases in serum triacylglycerol and visceral fat mediated by dietary soybean beta-conglycinin. *J. Atheroscler. Thromb.* 13: 247–255.

Kritchevsky, D., Tepper, S.A., Czarnecki, S.K., and Klurfeld, D.M. 1982. Atherogenicity of animal and vegetable protein: Influence of the lysine to arginine ratio. *Atherosclerosis* 41: 429–431.

Kurup, G.M. and Kurup, P.A. 1982. Hypolipidemic action of blackgram protein—Effect of protein hydrolysate. *Indian J. Biochem. Biophys.* 19: 208–212.

Kusunoki, J., Hansoty, D.K., Aragane, K., Fallon, J.T., Badimon, J.J., and Fisher, E.A. 2001. Acyl-CoA:cholesterol acyltransferase inhibition reduces atherosclerosis in apolipoprotein E–deficient mice. *Circulation* 103: 2604–2609.

Kwon, D.Y., Oh, S.W., Lee, J.S., Yang, H.J., Lee, S.H., Lee, J.H., Lee, Y.B., and Sohn, H.S. 2002. Amino acid substitution of hypocholesterolemic peptide originated from glycinin hydrolyzate. *Food Sci. Biotechnol.* 11: 55–61.

Kyowa Hakko USA. 2009. SoyScience, the frontier of soy for health. www.kyowa-usa. com/brands/soyscience.html. Accessed February 25.

Lovati, M.R., Manzoni, C., Gianazza, E., Arnoldi, A., Kurowska, E., Carroll, K.K., and Sirtori, C.R. 2000. Soy protein peptides regulate cholesterol homeostasis in Hep G2 cells. *J. Nutr.* 130: 2543–2549.

Lovati, M.R., Manzoni, C., Gianazza, E., Sirtori, C.R. 1998. Soybean protein products as regulators of liver low-density lipoprotein receptors. I. Identification of active b-conglycinin subunits. *J. Agric. Food. Chem.* 46: 2474–2480.

Macarulla, T.M., Medina, C., De Diego, M.A., Chaávarri, M., Zulet, M.A., Martinez, J.A., Nöel-Suberville, C., Higueret, P., and Portillo, M.P. 2001. Effects of the whole seed and a protein isolate of fava bean (*Vicia fava*) on the cholesterol metabolism of hypercholesterolemic rats. *Br. J. Nutr.* 85: 607–614.

Mackey, R., Ekangaki, A., and Eden, J.A. 2000. The effects of soy protein in women and men with elevated plasma lipids. *BioFactors* 12: 251–257.

Manso, M.A., Miguel, M., Even, J., Hernández, R., Aleixandre, A., and López-Fandiño, R. 2008. Effects of the long-term intake of an egg white hydrolysate on the oxidative status and blood lipid profile of spontaneously hypertensive rats. *Food Chem.* 109: 361–367.

Matsuoka, R., Kimura, M., Muto, A., Masuda, Y., Sato, M., and Imaizumi, K. 2008. Mechanism for the cholesterol-lowering action of egg white proteins in rats. *Biosci. Biotechnol. Biochem.* 72: 1506–1512.

Mochizuki, Y., Maebuchi, M., Kohno, M., Hirotsuka, M., Wadahama, H., Moriyama, T., Kawada, T., and Urade, R. 2009. Changes in lipid metabolism by soy β-conglycinin-derived peptides in HepG2 cells. *J. Agric. Food Chem.* 57: 1473–1480.

Moriyama, T., Kishimoto, K., Nagai, K., Urade, R., Ogawa, T., Utsumi, S., Maruyama, N., and Maebuchi, M. 2004. Soybean beta-conglycinin diet suppresses serum triglyceride levels in normal and genetically obese mice by induction of beta-oxidation, downregulation of fatty acid synthase, and inhibition of triglyceride absorption. *Biosci. Biotechnol. Biochem.* 68: 352–359.

Muramatsu, K. and Sugiyama, K. 1990. Relationship between amino acid composition of dietary protein and plasma cholesterol level in rats. *Monographs on Atherosclerosis* 16: 97–109.

Nagaoka, S., Futamura, Y., Miwa, K., Awano, T., Yamauchi, K., Kanamaru, Y., Tadashi, K., and Kuwata, T. 2001. Identification of novel hypocholesterolemic peptides derived from bovine milk beta-lactoglobulin. *Biochem. Biophys. Res. Commun.* 281: 11–17.

Nagaoka, S., Masaoka, M., Zhang, Q., Hasegawa, M., and Watanabe, K. 2002. Egg ovomucin attenuates hypercholesterolemia in rats and inhibits cholesterol absorption in Caco-2 cells. *Lipids* 37: 267–272.

Nagaoka, S., Miwa, K., Eto, M., Kuzuya, Y., Hori, G., and Yamamoto, K. 1999. Soy protein peptic hydrolysate with bound phospholipids decreases micellar solubility and cholesterol absorption in rats and Caco-2 cells. *J. Nutr.* 129: 1725–1730.

Nass, N., Schoeps, R., Ulbrich-Hofmann, R., Simm, A., Hohndorf, L., Schmelzer, C., Raith, K., Neubert, R.H., and Eder, K. 2008. Screening for nutritive peptides that modify cholesterol 7alpha-hydroxylase expression. *J. Agric. Food Chem.* 56: 4987–4994.

National Institutes of Health (NIH). 2001. Third report of the National Cholesterol Education Program (NCEP); Expert panel on detection, evaluation, and treatment of high blood cholesterol in adults (Adult Treatment Panel III), May, NIH Publication No. 01-3670.

Oh, P.S., Lee, S.J., and Lim, K.-T. 2006. Hypolipidemic and antioxidant effects of the plant glycoprotein (36 kDa) from *Rhus verniciflua* Stokes fruit in Trixton WR-1339-induced hyperlipidemic mice. *Biosci. Biotechnol. Biochem.* 70: 447–456.

Pak, V.V., Koo, M.S., Kasymova, T.D., and Kwon, D.Y. 2005. Isolation and identification of peptides from soy 11S-globulin with hypocholesterolemic activity. *Chem. Nat. Compd.* 41: 710–714.

Pedroche, J., Yust, M.M., Lqari, H., Megias, C., Giron-Calle, J., Alaiz, M., Vioque, J., and Millan, F. 2007. Obtaining of *Brassica carinata* protein hydrolysates enriched in bioactive peptides using immobilized digestive proteases. *Food Res. Int.* 40: 931–938.

Sautier, C., Dieng, K., Flament, C., Doucet, C., Suquet, J.P., and Lemonnier, D. 1983. Effects of whey protein, casein, soya-bean and sunflower proteins on the serum, tissue and faecal steroids in rats. *Br. J. Nutr.* 49: 313–319.

Shimizu, M., Tanabe, S., Morimatsu, F., Nagao, K., Yanagita, T., Kato, N., and Nishimura, T. 2006. Consumption of pork-liver protein hydrolysate reduces body fat in Otsuka Long-Evans Tokushima Fatty rats by suppressing hepatic lipogenesis. *Biosci. Biotechnol. Biochem.* 70: 112–118.

Shukla, A., Bettzieche, A., Hirche, F., Brandsch, C., Stangl, G.I., Eder, K. 2006. Dietary fish protein alters blood lipid concentrations and hepatic genes involved in cholesterol homeostasis in the rat model. *Br. J. Nutr.* 96: 674–682.

Sirtori, C.R., Agradi, E., Conti, F., Mantero, O., and Gatti, E. 1977. Soybean protein diet in the treatment of type-II hyperlipoproteinaemia. *Lancet* 1: 275–277.

Sirtori, C.R., Galli, C., Anderson, J.W., and Arnoldi, A. 2009. Nutritional and nutraceutical approaches to dyslipidemia and atherosclerosis prevention: Focus on dietary proteins. *Atherosclerosis* 203: 8–17.

Sirtori, C.R., Gianazza, E., Manzoni, C., Lovati, M.R., and Murphy, P.A. 1997. Role of isoflavones in the cholesterol reduction by soy proteins in the clinic. *Am. J. Clin. Nutr.* 65: 166–167.

Sirtori, C.R., Lovati, M.R., Manzoni, C., Castiglioni, S., Duranti, M., Magni, C., Morandi, S., D'Agostina, A., and Arnoldi, A. 2004. Proteins of white lupin seed, a naturally isoflavone-poor legume, reduce cholesterolemia in rats and increase LDL receptor activity in HepG2 cells. *J. Nutr.* 134: 18–23.

Sirtori, C.R., Lovati, M.R., Manzoni, C., Gianazza, E., Bondioli, A., Staels, B., and Auwerx, J. 1998. Reduction of serum cholesterol by soy proteins. *Nutr. Metab. Cardiovasc. Dis.* 8: 334–340.

Soy Labs, LLC. 2010. Lunasin™. www.lunasin.com. Accessed November 29.

Soy Science. 2010. Soy Science™, a complex of soy protein hydrolysate and phospholipids. http://www.kyowa-usa.com/brands/soyscience.html. Accessed November 29.

Spielmann, J., Shukla, A., Brandsch, C., Hirche, F., Stangl, G.I., and Eder, K. 2007. Dietary lupin protein lowers triglyceride concentrations in liver and plasma in rats by reducing hepatic gene expression of sterol regulatory element-binding protein-1c. *Ann. Nutr. Metab.* 51: 387–392.

Tamaru, S., Kurayama, T., Sakono, M., Fukuda, N., Nakamori, T., Furuta, H., Tanaka, K., and Sugano, M. 2007. Effects of dietary soybean peptides on hepatic production of ketone bodies and secretion of triglyceride by perfused rat liver. *Biosci. Biotechnol. Biochem.* 71: 2451–2457.

Tirosh, A., Rudich, A., Shochat, T., Tekes-Manova, D., Isreali, E., Henkin, Y., Kochba, I., and Shai, I. 2007. Changes in triglyceride levels and risk of coronary heart disease in young men. *Ann. Intern. Med.* 147: 377–385.

Tomotake, H., Shimaoka, I., Kayashita, J., Yokoyama, F., Nakajoh, M., and Kato, N. 2001. Stronger suppression of plasma of plasma cholesterol and enhancement of the fecal excretion of steroids by buckwheat protein product than by a soy protein isolate in rats fed on a cholesterol-free diet. *Biosci. Biotechnol. Biochem.* 65: 1412–1414.

Ueda, H. 2000. Cholesterol-lowering effect of soybean protein isolate in chicks is partly lost by microbial protease digestion. *Anim. Sci. J. (Tokyo)* 71: 57–62.

Wergedahl, H., Liaset, B., Gudbrandsen, O.A., Lied, E., Espe, M., Muna, Z., Mork, S., and Berge, R.K. 2004. Fish protein hydrolysate reduces plasma total cholesterol, increases the proportion of HDL cholesterol, and lowers acyl-CoA:Cholesterol acyltransferase activity in liver of Zucker rats. *J. Nutr.* 134: 1320–1327.

Yang, S., Liu, S., Yang, H., Lin, Y., and Chen, J. 2007. Soybean protein hydrolysate improves plasma and liver lipid profiles in rats fed high-cholesterol diet. *J. Am. Coll. Nutr.* 26: 416–423.

Yashiro, A., Oda, S., and Sugano, M. 1985. Hypocholesterolemic effect of soybean protein in rats and mice after peptic digestion. *J. Nutr.* 115: 1325–1336.

Yoshikawa, M., Yamamoto, T., and Takemaka, Y. 1999. Study on a low molecular weight peptide derived from soybean protein having hypocholesterolemic activity. *Soy Protein Res.* 2: 125–128.

Zhang, X. and Beynen, A.C. 1993. Influence of dietary fish proteins on plasma and liver cholesterol concentrations in rats. *Br. J. Nutr.* 69: 767–777.

Zhong, F., Liu, J., Ma, J., and Shoemaker, C.F. 2007a. Preparation of hypocholesterol peptides from soy protein and their hypocholesterolemic effect in mice. *Food Res. Int.* 40: 661–667.

Zhong, F., Zhang, X., Ma, J., and Shoemaker, C.F. 2007b. Fractionation and identification of a novel hypocholesterolemic peptide derived from soy protein Alcalase hydrolysates. *Food Res. Int.* 40: 756–762.

9

Peptides and Proteins Increase Mineral Absorption and Improve Bone Condition

Hiroshi Kawakami

CONTENTS

9.1 Introduction

In our aging society, it is becoming a social problem that the number of elderly people bedridden due to lifestyle-related diseases and who need nursing care is rapidly growing (Ray et al., 1997; Looker et al., 1995; Hui, Slemenda, and Johnston, 1988). In a survey on the causes of requirements for nursing care, bone disease from fracture and fall were reported to be placed third, after cerebral stroke and senility. There were over 10 million Japanese patients with osteoporosis in 2000, estimated to reach 16 million if potential patients who had just begun to lose bone mass were included (Sone and Fukunaga, 2004). Moreover, these patients with osteoporosis are now estimated to total 200 million worldwide (Lin and Lane, 2004). As remarked above, it is considered an issue common to aging societies, including not only Japan but also Western countries, to decrease the number of osteoporotic patients by improving eating habits. At present, women account for 80% of patients with osteoporosis, but men are at risk equally to women when they are over 80 years of age (Peate, 2004; Ebeling 1998). Osteoporosis is attributed to aging, but its occurrence can be delayed by changing dietary habits (Wei et al., 2003). In addition, it is also becoming an issue that young women lose bone mineral density because of unreasonable and imbalanced dieting (Flynn, 2003). In our society, where aging is rapidly progressing, it is presumed that bone-related diseases will increase, suggesting that, from the viewpoint of dietary lifestyle, it is more important to pay attention to bone health from people's earlier years by maintaining a healthy and active lifestyle (Silverman et al., 2001; Prince, 1997).

Bone tissue, which resembles a lump of calcium, has an internal network of vessels and is actively remodeled like other organs (Bryant, Endo, and Gardiner, 2002; Pittenger et al., 1999). Present inside the bones are osteoblasts, osteoclasts, and osteocytes, which mutually interact to control bone remodeling (Boyle, Simonet, and Lacey, 2003; Erlebacher et al., 1995; Aarden, Burger, and Nijweide, 1994). Namely, bone formation by osteoblasts, which produce the bone matrix protein collagen to adsorb bone minerals such as calcium, and bone resorption by osteoclasts, which dissolve collagen and calcium, are constantly in balance. Imbalance between the bone formation and resorption results in osteoporosis (Matsumoto, 2004).

A large number of scientists have long been working to fathom the mysteries of milk, focusing on cow's milk, which is essential in keeping people healthy, and mother's milk, which is the only nutrient able to support a baby's development. Animal (mammalian) babies grow quickly on mother's milk alone during a period of several months after their birth. Although animals feed on various foods, milk is the only entity in the world that must be ingested. This means that milk contains components that are most critical and ideal for the life of humans. In particular, daily intake of calcium is critical to suppress development of osteoporosis (Dawson-Hughes, 1999; McKane et al., 1996), although the current daily calcium intake in Japan is not sufficient to reach the level of the recommended dietary allowances for the Japanese. It is well known that cow's milk is useful to keep bones healthy, and the reason is because it contains plenty of calcium, a nutrient essential for bone formation (Kalkwarf, Burger, and Nijweide, 2003; Teegarden et al., 1999; Sandler et al., 1985). However, the excellence of cow's milk for bone health might not be limited to abundance of calcium, and recent research has proposed several components that may enhance the absorption of calcium in the intestine, and directly increase the strength of bones. In neonates, in whom the only source of nutrition is mother's milk, bones are actively remodeled, suggesting the possibility that mother's milk may contain some functional components other than calcium that affect bone metabolism. To maintain healthy bones, it is ideal not only to supplement with calcium as a material for bone formation but also simultaneously to assist the formation of bones and to suppress loss of bone mass. This chapter introduces several substances in milk that regulate bone remodeling to enhance bone mineral density.

9.2 Peptides

9.2.1 Casein Phosphopeptides (CPPs)

The casein-derived phosphorylated peptides (CPPs) have been known to possess the ability of calcium solubilization *in vitro* and calcium absorption in the gastrointestinal tract. Phosphorous is bound to the caseins via monoester linkages to seryl residues. The extent of phosphorylation is dependent on casein type; that is, bovine αs2-casein can have up to 13 phosphate groups, whereas κ-casein has only one phosphate. Table 9.1 represents the sequences of the different CPPs found in bovine milk. Many of these peptides contain a common motif: a sequence of three phosphoseryls followed by two glutamic acid residues. The highly polar acidic domains represent the binding sites for calcium. All of the available serine is not phosphorylated in the caseins; this is presumably related to the specificity of the kinase activities in the mammary gland. In addition, dephosphorylated peptides do not bind minerals. The role of phosphorylated residues in calcium binding is further illustrated by the observation that chemical phosphorylation of αs1- and β-casein increased the binding capacity and stability of these proteins in the presence of calcium ions. The ability of CPPs to bind calcium is intrinsic to their potential role as functional food ingredients for bone health.

TABLE 9.1

Phosphopeptides Derived from Caseins in Bovine Milk

Casein		Amino Acid Sequence
αs1	(f43-58)	-Asp-Ile-Gly-**SerP**-Glu-**SerP**-Thr-Glu-Asp-Gln-Ala-Met-Glu-Asp-Ile-Lys-
	(f45-115)	-Gly-**SerP**-Glu-**SerP**-Thr-Glu-····-Glu-**SerP**-Ile-**SerP**-**SerP**-Glu-Glu-····-Val-Pro-Asn-**SerP**-
	(f59-79)	-Gln-Met-Glu-Ala-Glu-**SerP**-Ile-**SerP**-**SerP**-Glu-Glu-Ile-Val-Pro-Asp-**SerP**-Val-Glu-Gln-Lys-
	(f112-119)	-Val-Pro-Asn-**SerP**-Ala-Glu-Glu-Arg-
αs2	(f5-18)	-Glu-His-Val-**SerP**-**SerP**-**SerP**-Glu-Glu-Ser-Ile-Ile-**SerP**-Gln-Glu-
	(f7-62)	-Val-**SerP**-**SerP**-**SerP**-Glu-Glu-Ser-Ile-Ile-**SerP**-Gln-Glu-····-Gly-**SerP**-**SerP**-**SerP**-Glu-Glu-**SerP**-Ala-
	(f15-144)	-Ile-**SerP**-Gln-Glu-····-Gly-**SerP**-**SerP**-**SerP**-Glu-Glu-**SerP**-Ala-····-Leu-**SerP**-Thr-**SerP**-Glu-Glu-····-Met-Glu-**SerP**-Thr-
	(f29-34)	-Asn-Pro-**SerP**-Lys-Glu-Asn-
	(f55-64)	-Gly-**SerP**-**SerP**-**SerP**-Glu-Glu-**SerP**-Ala-Glu-Val-
	(f127-147)	-Gln-Leu-**SerP**-Thr-**SerP**-Glu-Glu-Asn-Ser-Lys-Thr-Val-Asp-Met-Glu-**SerP**-Thr-Glu-Val-Phe-
β	(f1-25)	-Arg-Glu-Leu-Glu-Glu-Leu-Asn-Val-Pro-Gly-Glu-Ile-Val-Glu-**SerP**-Leu-**SerP**-**SerP**-**SerP**-Glu-Glu-Ser-Ile-Thr-
	(f7-24)	-Asn-Val-Pro-Gly-Glu-Ile-Val-Glu-**SerP**-Leu-**SerP**-**SerP**-**SerP**-Glu-Glu-Ser-Ile-
	(f12-23)	-Ile-Val-Glu-**SerP**-Leu-**SerP**-**SerP**-**SerP**-Glu-Glu-Ser-Ile-Thr-
	(f14-36)	-Glu-**SerP**-Leu-**SerP**-**SerP**-**SerP**-Glu- Glu-Ser-Ile-Thr-Arg-Ile-Asn-Lys-Lys-Ile-Glu-Lys-Phe-Gln-**SerP**-Glu-
κ	(f147-151)	-Glu-Ala-**SerP**-Pro-Glu-
	(f147-153)	-Glu-Ala-**SerP**-Pro-Glu-Val-Ile-

SerP: phosphrylated serine.

9.2.2 Effect of CPPs on Calcium Bioavailability

The excellent bioavailability of calcium from milk has been attributed to CPPs mediated enhanced solubility of calcium ion in the intestine. CPPs are supposed to be formed *in vivo* following digestion of casein by gastrointestinal proteases, relatively resistant to further proteolytic degradation, and able to accumulate in the distal ileum. Again, CPPs form soluble complexes with calcium phosphate at alkaline pH *in vitro*, and CPP-calcium complexes can lead to enhanced calcium absorption across the intestinal mucosa by limiting the precipitation of calcium ion in the distal ileum. However, the *in vivo* studies in animals and humans on the effect of CPPs on calcium metabolism have provided inconsistent results (Narva et al., 2003; Scholz-Ahrens and Schrezenmeir, 2000; Hansen et al., 1997; Heaney Saito, and Orimo, 1994; Scholz-Ahrens, de Vrese, and Barth, 1991; Sato, Noguchi, and Naito, 1986). Sato et al. (1986) reported that CPPs accelerated the absorption of instilled $^{45}CaCl_2$ from the intestine and stimulated femur calcification in rats.

Scholz-Ahrens et al. (1991) showed in long-term experiments over 12 weeks with mini pigs, a model that is closer to human than rat, small changes occurred in calcium absorption and bone metabolism by feeding casein, compared to whey protein. Effects on calcium absorption, calcium parameters, and bone metabolism could be observed at specific experimental conditions, that is, at a lower content of dietary calcium or during vitamin D deficiency. In the presence of low dietary calcium, soluble calcium in the intestine and calcium retention were lower, and concentration of calcium and parathyroid hormone in plasma were higher in mini pigs fed casein compared to whey protein. In the presence of a diet with high calcium content but free of vitamin D, bone mineral density was higher in animals fed casein compared to whey protein. Narva et al. (2003) examined the effect of additional CPPs in milk and fermented milk on acute calcium metabolism by measuring parathyroid hormone and calcium in serum. There were no statistically significant differences in parathyroid hormone and calcium between the groups receiving control milk and CPPs-fortified milk. Although CPPs increased calcium absorption from a rice-based infant gruel in human adults by approximately 30%, no change was seen when CPPs were ingested in either high or low phytate wholegrain cereal meals. These results serve to emphasize the complex interactions between meal constituents and calcium bioavailability. The positive effect of CPPs needs further investigation to clarify their role in calcium bioavailability.

9.2.3 Production and Application of CPPs

CPP production from casein was generally performed by enzymatic digestion with trypsin, chymotrypsin, pancreatin, papain, pepsin, thermolysin, and pronase. In industrial production, proteolytic enzymes from a range of bacterial and fungal sources have recently been used to generate CPPs from sodium caseinate. Following proteolytic digestion, insoluble or unhydrolyzed casein material is removed by centrifugation of the hydrolysate, which has been adjusted to pH 4.6. CPPs are then aggregated using calcium ions, and separated from nonphosphorylated peptides by ultrafiltration. Aggregated phosphopeptides recovered by ultrafiltration can be diafiltered to remove excess aggregating salt. Nakano et al. (2004) prepared micellar calcium phosphate as a complex with CPPs from rennet casein using ultrafiltration. The micellar calcium phosphate-CPPs complex was easily dissolved in water and contained 63.1% peptides, 9.2% calcium, 3.5% inorganic phosphorus, and 1.8% organic phosphorus on a dry weight basis. Again, CPPs were also obtained from enzymatic hydrolysates by ion exchange chromatography. CPPs mixtures are commercially available as spray-dried peptide powders.

The major applications of CPPs center on their ability to solubilize calcium. Certain sectors of the population, such as preterm infants, young females, postmenopausals, and the elderly, are at risk from ingestion of low calcium levels. Several multinational companies currently market CPPs-containing products aimed at enhancing the bioavailability of calcium-fortified foods. CPPs have been proposed for use in dietary products such as beverages, chewing gum, bread, cake, flour, and in pharmaceutical preparations such as tablets, toothpaste, and dental filling material. The pharmaceutical products are intended for use in the treatment of dental disease and for rarefying bone disease (Meisel, 1998).

9.3 Proteins

9.3.1 Osteoprotective Proteins in Milk

The protein fraction of milk contains many valuable components and biologically active substances. The main milk proteins are whey protein and casein, and both proteins have a significant role in calcium absorption in the intestine and bone metabolism. In addition to major whey proteins, such as α-lactalbumin and β-lactoglobulin, several bioactive proteins with alkaline isoelectric points are present in trace amounts in whey. The basic proteins can be obtained by processing skimmed milk or whey by cation exchange chromatography. Skimmed milk was first fractionated into casein and whey, and the whey fraction was further fractionated into components of proteins, lactose, and so on, to investigate their bone metabolism-regulating actions. The study results indicated that whey protein stimulated osteoblasts so as to remodel bones and simultaneously suppressed osteoclasts so as not to dissolve bones. Further studies focusing on active components among whey proteins discovered that a protein aggregation in whey with an alkaline isoelectric point was milk basic proteins (MBP), which stimulated bone formation and simultaneously suppressed bone resorption. The study results eventually revealed that MBP stimulated proliferation of osteoblasts involved in bone formation and collagen production and also suppressed the activity of osteoclasts involved in bone resorption (Kawakami, 2005).

MBP was further fractionated by various types of chromatography to probe for active components using the osteoblast-proliferation activity as an indicator, and the following substances were identified: fragment 1-2 with MW 23 kDa of high molecular weight kininogen, a blood coagulating factor (Yamamura et al., 2006, 2000), and high mobility group (HMG)-like protein with MW 10 kDa (Yamamura et al., 1999). On the other hand, functional components were researched using the bone resorption-suppressing activity in an isolated osteoclast culture system as an indicator, and a cystatin C with MW 12 kDa was identified (Matsuoka et al., 2002). Cystatin is produced by osteoblasts and inhibits cysteine protease, which is produced by osteoclasts and destroys bones, so that bone resorption is regulated (Johansson et al., 2000; Lerner et al., 1997). Angiogenin is also the substance mainly responsible for the inhibitory effect of MBP on osteoclast-mediated bone resorption, and it exerts activity by acting directly on osteoclasts (Morita et al., 2008). Angiogenin is a member of the ribonuclease superfamily and a normal constituent of circulating blood.

In addition, it has been reported in an *in vitro* cell culture study that lactoferrin, one of the basic proteins in milk, stimulated the growth and differentiation of osteoblast (Cornish et al., 2004) and suppressed the differentiation and resorbing activity of osteoclast (Lorget et al., 2002). Cornish et al. (2004) also demonstrated that local injection of lactoferrin into the calvarium resulted in increased bone growth of mice. Furthermore, several growth factors in milk are well known to possess alkaline isoelectric points (Francis et al., 1995). It is supposed that composite reactions of the various basic protein

TABLE 9.2

Osteoprotective Proteins with Alkaline Isoelectric Point in Bovine Milk

Protein	Function	Reference
Lactoferrin (MW: 80 kDa)	Inhibition of osteoblast apoptosis	Grey et al. (2006)
	Promotion of osteoblast growth	Cornish et al. (2004)
	Suppression of osteoclast differentiation	Lorget et al. (2002)
Kininogen fragment 1-2 (23 kDa)	Promotion of osteoblast growth	Yamamura et al. (2006)
Angiogenin (15 kDa)	Inhibition of osteoclast-mediated bone resorption	Morita et al. (2008)
Cystatin C (12 kDa)	Inhibition of osteoclast-mediated bone resorption	Matsuoka et al. (2002)
HMG-like protein (10 kDa)	Promotion of osteoblast growth	Yamamura et al. (1999)

MW: molecular weight

components (Table 9.2) comprehensively exert improvement in bone metabolism and increase in bone mineral density. Recently, Bharadwaj et al. (2009) reported that milk angiogenin-enriched lactoferrin induces positive effects on bone turnover markers in postmenopausal women. This angiogenin-enriched lactoferrin supplementation demonstrated a statistically significant increase in osteoblastic bone formation and reduction in bone resorption, restoring the balance of bone turnover within a short period.

9.3.2 Effects on Improvement of Bone Condition

9.3.2.1 *Proliferation and Differentiation of Osteoblasts*

Mouse osteoblast cell lines, MC3T3-E1 and MG63, were used to study the effects of MBP on DNA synthesis and collagen production in a bone matrix (Kawakami, 2005). MBP stimulated proliferation of osteoblasts in a concentration-dependent manner and production of PICP (procollagen I carboxy-terminal propeptide), which is an indicator of collagen synthesis in bone matrix. Yamamura et al. (2006) demonstrated that the enzymatically digested fragments of bovine high molecular weight kininogen, considered naturally occurring active proteins, promoted the proliferation of osteoblasts. One of the fragments was similar to the MBP fragment. In addition, lactoferrin produced a dose-related increase in thymidine incorporation in primary or cell line cultures of human or rat osteoblast-like cells at physiological concentrations (1-100 µg/ml), and increased osteoblast differentiation (Cornish et al., 2004). Local injection of lactoferrin above the hemicalvaria of adult mice resulted in substantial increases in the dynamic histomorphometric indices of bone formation and bone area. Recently, Huang et al. (2008) reported that iron-bound recombinant lactoferrin, derived from rice, promoted the growth of primary osteoblasts from rats.

9.3.2.2 *Actions to Suppress Bone Resorption by Osteoclasts*

The effects of MBP on the bone-resorbing action of osteoclasts were studied (Kawakami, 2005). Osteoclasts isolated from rabbit femur were cultured on dentine slices in media containing or not containing MBP for 48 h, and the number of pits formed by osteoclasts indicating destruction of bone tissue was counted. The number of pits formed by osteoclasts decreased depending on MBP concentration, suggesting that MBP was effective in suppressing the bone-resorbing action of osteoclasts. MBP appeared to contain cystatin C, a cysteine protease inhibitor, and angiogenin as the factors inhibiting bone resorption (Matsuoka et al., 2002). Cystatin is produced by osteoblasts and inhibits cysteine protease, which is produced by osteoclasts and destroys bones, so that bone resorption is regulated (Johansson et al., 2000; Lerner et al., 1997). Ohashi et al. (2003) reported inhibitory function of lactoferrin against cysteine proteases, cathepsin L. The intramolecular peptide,

Tyr_{679}-Glu-Lys-Tyr-Leu-Gly-Pro-Gln-Tyr-Val-Ala-Gly-Ile-Thr-Asn-Leu-Lys_{695}, of lactoferrin is an active domain, with a 90% homology of the sequence for the common active site of the cystatin family, and the synthesized peptide inhibited the activity of cysteine proteases. Moreover, Lorget et al. (2002) reported that lactoferrin reduced *in vitro* osteoclast differentiation and resorbing activity. Lactoferrin is a key modulator of inflammatory response. Increasing evidence indicates that bone and immune systems are genetically and functionally linked. Lactoferrin inhibited *in vitro* bone-resorbing activity in a rabbit mixed bone cell culture, consisting of authentic osteoclasts in an environment of osteoblasts and stromal cells.

Recently, Morita et al. (2008) demonstrated that angiogenin is also the substance mainly responsible for the inhibitory action of MBP on osteoclast-mediated bone resorption, and that it exerts its activity by acting directly on the osteoclasts. The inhibitory activity was confirmed in mice both *in vitro* and *in vivo*. Treatment of osteoclasts with angiogenin resulted in an impairment of the formation of F-actin ring and a reduction in the mRNA levels of tartrate-resistant acid phosphatase and cathepsin K, both known to be essential for bone resorption activity of osteoclasts.

9.3.2.3 Stimulation of Bone Formation in Rats During the Growth Period

The bone mineral density and strength of femur were significantly increased in rats (female SD rats, 5 weeks old) orally administered MBP for 4 weeks during their growth period. The activity of serum alkaline phosphatase, a bone formation marker, was significantly increased in the rat fed with MBP, suggesting that MBP accelerated bone formation. Another study on tibial morphology found that administration of MBP increased cancellous bones and grew epiphysial plates. These results suggested MBP accelerated bone formation during the growth period of animals.

9.3.2.4 Bone Matrix-Increasing and Bone-Strengthening Actions in a Rat Model of Bone Resorption

Ovariecomized female rats (SD, 10 weeks old) were fed a low calcium diet in order to reduce bone mass and study the action of MBP on enhancement of bone mass and strength (Kato et al., 2000). This animal model is presumed to resemble young women who impose undesirable dieting, which leads to calcium deficiency and hormone imbalance. A 3-week feeding with a diet containing MBP significantly increased femoral strength in the ovariecomized rats. In addition, hydroxyproline and hydroxylysine, characteristic of matrix collagen and helpful for making bones robust and pliable, were significantly higher after administration of MBP. These results suggested that MBP increased bone matrices such as collagen to enhance bone strength.

9.3.2.5 Bone Mass Reduction-Suppressing Action in Animal Models of Osteoporosis

A rat model of postmenopausal osteoporosis (SD, 55 weeks old) was used to study whether MBP suppresses reduction in bone mass (Toba et al., 2000). The results showed that intake of MBP significantly suppressed reduction in femoral bone mineral density. Additionally, observation of nondecalcificated tibial specimens confirmed that intake of MBP significantly suppressed reduction in cancellous bone. Furthermore, urinary excretion of deoxypyridinoline, a degradation product of collagen in bone matrix, was significantly lowered under administration of MBP, suggesting that bone resorption was suppressed. Thus, MBP was suggested to be effective in suppressing bone mass reduction due to postmenopausal acceleration of bone resorption. Morita et al. (2008) investigated that angiogenin, a member of the ribonuclease superfamily, suppressed reduction of bone mineral density in a mouse model. Eight-week-old female ddY mice were sham-operated or ovariectomized, and fed with a modified AIN-76 diet containing 0.3% calcium. Angiogenin was intravenously administered six times during the test period for two weeks. Finally, the femur was removed for the measurement of bone mineral density. The results showed that intake of angiogenin significantly suppressed reduction in femoral bone mineral density.

9.3.2.6 Alveolar Bone Formation in Rat Experimental Periodontitis

Alveolar bone resorption is a major problem in advanced-stage periodontitis. Tissue regeneration therapy, using growth factors such as bone morphogenetic protein and fibroblast growth factor, have been examined in periodontitis patients. Seto et al. (2007) investigated the bone-formative effects of MBP in rat experimental periodontitis by assessing morphological data obtained from microcomputerized tomography and histological sections. Although alveolar bone resorption was severely induced around the molar by the 20-day ligature procedure, feeding with a diet containing 1% MBP recovered ligature-induced alveolar bone resorption after 45 days. Histological examination clarified that the osteoid thickness of alveolar bone was dose-dependently increased by the feeding for 90 days.

9.3.2.7 Bone Formation-Stimulating and Bone Resorption-Suppressing Actions in Humans

An MBP intake study was conducted in healthy male adult volunteers who were given MBP-containing beverages (Toba et al., 2001). Urine and blood were sampled before and 16 days after initiation of beverage intake, to determine markers of bone metabolism. As depicted in Figure 9.1, the study results showed that intake of MBP significantly increased serum concentration of osteocalcin, a marker indicating bone formation, and significantly decreased

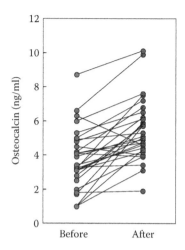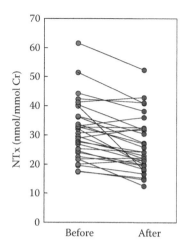

FIGURE 9.1

Individual changes in serum osteocalcin concentrations (left) and urinary NTx excretion (right) before and after 16 days of ingesting an experimental beverage containing MBP. Increase in serum osteocalcin concentration was found in 28 of the 30 subjects, and decrease in urinary NTx excretion was in 24 of the 30 subjects. (From Toba, Y., Matsuoka, Y., Morita, Y. et al. 2001. *Biosci. Biotechnol. Biochem.* 65: 1353–1357. With permission from Japan Society for Bioscience, Biotechnology and Agrochemistry.)

urinary excretion of cross-linked *N*-teleopeptides of type-I collagen (NTx), a marker indicating bone resorption.

In addition, there was a significant correlation between serum concentration of osteocalcin and urinary excretion of NTx after initiation of MBP intake, indicating that bone formation was well balanced with bone resorption (Figure 9.2).

These results suggested that MBP regulated bone metabolism, maintaining a balance in bone remodeling. Recently, Bharadwaj et al. (2009) demonstrated that milk angiogenin-enriched lactoferrin induced positive action on bone turnover markers in postmenopausal women. Thirty-eight healthy postmenopausal women, 45–60 years old, were randomly divided into placebo or ribonuclease-enriched lactoferrin supplement groups. Bone health status was monitored by assessing bone resorption markers, NTx and deoxypyridinoline, and bone formation markers, bone-specific alkaline phosphatase and osteocalcin, for 180 days. Angiogenin-enriched lactoferrin supplementation demonstrated a decrease in deoxypyridinoline levels by 14% (19% increase for placebo) and NTx was maintained at 24% of baseline (41% for placebo) and bone-specific alkaline phosphatase and osteocalcin levels showed a 45% and 16% elevation (25% and 5% for placebo), respectively. Thus, milk angiogenin-enriched lactoferrin supplementation in human clinical trials has shown promising and favorable action on biomarkers of bone turnover in postmenopausal women.

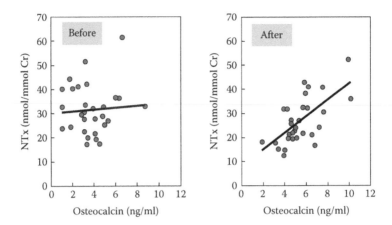

FIGURE 9.2
Relationship between urinary NTx excretion and serum osteocalcin concentration before (left) and after 16 days (right) of ingesting an experimental beverage containing MBP. The correlation coefficient before ingestion was 0.0641 ($P = 0.7366$), and after 16 days was 0.6457 ($P < 0.0001$). Differences are considered significant if $p < 0.05$. (From Toba, Y., Matsuoka, Y., Morita, Y. et al. 2001. *Biosci. Biotechnol. Biochem.* 65: 1353–1357. With permission from Japan Society for Bioscience, Biotechnology and Agrochemistry.)

9.3.2.8 Bone Mineral Density-Increasing and Bone Metabolism-Improving Actions in Humans

A six-month MBP intake study was conducted in healthy female adult volunteers who were given MBP-containing beverages (Yamamura et al., 2002). Participants were divided into MBP and placebo groups under double-blind study conditions. Measurements of bone metabolic markers revealed that urinary excretion of NTx, a bone resorption marker, was significantly lower three and six months after initiation of MBP intake. In addition, the increasing rate of bone mineral density at six months was significantly higher in the MBP group, as compared to the placebo group. On the other hand, analysis of records on food intake during the study period demonstrated no correlation between dietary intake of calcium, magnesium, and vitamins D and K, and bone mineral density (Aoe et al., 2001). In addition, a one-year MBP intake study of healthy female aged volunteers (65–86 years old) who were given MBP-containing beverages, demonstrated that MBP administration significantly suppressed the increase in urinary excretion of NTx and deoxypyridinoline (Park et al., 2007). The study results in human volunteers demonstrated that MBP increased bone mineral density of calcaneus in elderly women, as shown in Figure 9.3.

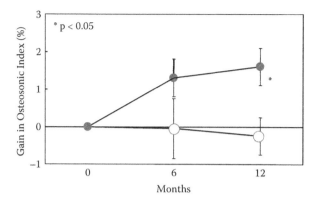

FIGURE 9.3
Gain in osteosonic index of calcaneus in healthy aged women (65–86 years old) by ingesting an experimental beverage containing MBP. The gain in osteosonic index was significantly higher in the MBP group after 12 months ($p < 0.05$). MBP group ($n = 44$; filled circle); control group ($n = 35$; unfilled circle).

9.3.3 Safety Evaluation and Application of Milk Basic Proteins Fraction

MBP has been evaluated for its use as an ingredient in food and concluded to be safe for its intended use (Kruger et al., 2007). Safety evaluation studies were performed in accordance with *Ordinance on Standard of Conduct of Non-clinical Studies of Drug Safety,* the Ministry of Health, Labor and Welfare Ordinance No. 21, Japan, March 26, 1997, and *Guidelines for Designation of Food Additives and for Revision of Standards for Use of Food Additives,* Notification No. 29 of the Environmental Health Bureau, the Ministry of Health, Labor and Welfare, Japan, March 22, 1996. In a single-dose oral toxicity study, a 4-week oral repeated dose toxicity study, a 13-week oral repeated dose toxicity study, and a teratogenicity study of MBP, there was no change attributable to MBP in males or females at any dose group in clinical observation, body weights, food consumption, urinalysis, hematology, blood chemistry, ophthalmology, autopsy, organ weights, and histopathology. In addition, mutagenic activity of MBP was not observed in a reverse mutation-screening test using *Salmonella typhimurium.* On the other hand, cow's milk is a well-studied common allergenic food. Although the basic proteins in milk are not identified as milk allergens, food products containing MBP will be labeled as containing milk as a caution for milk-allergic consumers as MBP has not been demonstrated to be free of milk allergens. The potential allergenicity of the identified proteins in MBP has been evaluated (Goodman et al., 2007). Based on molecular characteristics and expected exposure, the protein components in MBP are unlikely to present any increased risk of allergy for milk allergic subjects or of cross-reactivity for other allergic subjects. Finally, MBP was determined as a substance Generally Recognized As Safe (GRAS) by the Food and Drug Administration (FDA) in September 2006.

MBP, highly water soluble and without a specific flavor, can be blended into various types of food products (Kawakami, 2007). Although the active ingredients of MBP are proteins, the activities are practically not affected by heat or coexisting components, and are resistant to manufacturing processes for food production (homogenization, heating sterilization, and spray drying, etc.). Therefore, MBP could be applied to a wide variety of food products, including chilled foods, long shelf-life foods, and foods distributed at ambient temperature regardless of the type of food, fluid or solid.

At present, MBP is supplemented in various milk products, and the enriched products, including processed cheese, skim milk, follow-up milk, milk beverage, yogurt, drinkable yogurt, lactic acid bacteria beverage, ice cream, and milky beverage, and so on, are marketed both in Japan and abroad (Kawakami, 2007). The product, Mainichi Hone Kea MBP®, which is a Food for Specified Health Use (FOSHU), was allowed by the Ministry of Health, Labor and Welfare in February, 2002, to label an advertisement with the following statement, "The product contains milk basic proteins fraction, effective in increasing bone mineral density, and hence is a beverage suitable for persons who are concerned about their bone health." This is the only product that was approved for the label "increases bone mineral density" among foods specified for bone health use. Mainichi Hone Kea MBP is served as a handy drinking supplement and widely distributed so that people can use the product even though they do not like cow's milk or milk products because of preference or lactose intolerance.

9.4 Conclusion

Entering the twenty-first century, it is becoming more and more important to prevent various types of lifestyle-related diseases by consuming ordinary healthy meals. In addition, it is becoming a social problem in our aging society that the number of elderly people bedridden due to lifestyle-related diseases and who need nursing care is rapidly growing. A survey on the causes of the requirements for nursing care has reported that bone diseases from fracture and fall are third after cerebral stroke and senility. In our society, where aging is rapidly progressing, it is presumed that bone-related diseases will continue to increase, suggesting that, from the viewpoint of dietary lifestyle, it is more important to pay attention to bone health during people's earlier years by maintaining an active healthy living style (Silverman et al., 2001).

The social phenomenon of the growing number of elderly patients with bone disease or fractures attributable to osteoporosis is closely associated with an increase in the number of bedridden elderly (Kanis et al., 1994; Melton et al., 1992), and is a cause for the rise in the total cost of national health care

(Simon and Mack, 2003; Kanis and Johnell, 1999). Until now, nutritional measures such as recommending an increase in calcium intake by administering calcium agents have so far been taken to prevent such problems (Heaney, 2000). However, supplementation of partial daily foods with casein phosphopeptides or milk basic proteins has made it possible to enhance calcium absorption and maintain bone remodeling. This will allow exploitation of foods containing milk-derived ingredients to prevent the future occurrence of senior bone disorders represented by osteoporosis. The popularity of these food products may be expected to help curtail medical costs and also to exercise a major social influence over healthcare policy for the elderly. Additionally, reduction in bone mineral density often occurs not only in the elderly but also in young women, in particular, those on diets (Flynn, 2003). To maintain healthy bones in such young women and not to have an increase of patients with osteoporosis in the future, it is considered quite important for young people to eat foods containing milk-derived ingredients.

References

Aarden, E.M., Burger, E.H. and Nijweide, P.J. 1994. Function of osteocytes in bone. *J. Cell. Biochem.* 55: 287–299.

Aoe, S., Yamamura, J., Kawakami, H. et al. 2001. Controlled trial of the effects of milk basic protein supplementation on bone metabolism in healthy adult women. *Biosci. Biotechnol. Biochem.* 65: 913–918.

Bharadwaj, S., Naidu, A.G.T., Betageri, G.V., Prasadarao, N.V., and Naidu, A.S. 2009. Milk ribonuclease-enriched lactoferrin induces positive effects on bone turnover markers in postmenopausal women. *Osteoporos. Intern.* (January)

Boyle, W.J., Simonet, W.S., and Lacey, D.L. 2003. Osteoclast differentiation and activation. *Nature* 423: 337–342.

Bryant, S.V., Endo, T., and Gardiner, D.M. 2002. Vertebrate limb regeneration and the origin of limb stem cells. *Int. J. Dev. Biol.* 46: 887–896.

Cornish J., Callon, K.E., Naot, D. et al. 2004. Lactoferrin is a potent regulator of bone cell activity and increases bone formation *in vivo*. *Endocrinology* 145: 4366–4374.

Dawson-Hughes, B. 1999. Calcium and vitamin D nutrition. In *Osteoporosis in Men*, Volume 1 (E. Orwoll, Ed.), San Diego: Academic Press, pp. 197–209.

Ebeling, P.R. 1998. Osteoporosis in men: New insights into aetiology, pathogenesis, prevention and management. *Drugs Aging* 13: 421–434.

Erlebacher, A., Filvaroff, E.H., Gitelman, S.E., and Derynck, R. 1995. Toward a molecular understanding of skeletal development. *Cell* 80: 371–378.

Flynn, A. 2003. The role of dietary calcium in bone health. *Proc. Nutr. Soc.* 62: 851–858.

Francis, G.L., Regester, G.O., Webb, H.A., and Ballard, F.J. 1995. Extraction from cheese whey by cation exchange chromatography of factors that stimulate the growth of mammalian cells. *J. Dairy Sci.* 23: 1209–1218.

Goodman, R.E., Taylor, S.L., Yamamura, J. et al. 2007. Assessment of the potential allergenicity of a milk basic protein fraction. *Food Chem. Toxicol.* 45: 1787–1794.

Grey, A., Zhu, Q., Watson, M., Callon, K., and Cornish, J. 2006. Lactoferrin potently inhibits osteoblast apoptosis, via an LRP1-independent pathway. *Mol. Cell. Endocrinol.* 251: 96–102.

Hansen, M., Sandstrom, B., Jensen, M., and Sorensen, S.S. 1997. Casein phosphopeptides improve zinc and calcium absorption from rice-based but not from whole-grain infant cereal. *J. Pediatr. Gastroenterol. Nutr.* 24: 56–62.

Heaney, R.P. 2000. Calcium, dairy products, and osteoporosis. *J. Am. Coll. Nutr.* 19: 83S–99S.

Heaney, R.P., Saito, Y., and Orimo, H. 1994. Effect of caseinphosphopeptide on absorbability of co-ingested calcium in normal postmenopausal women. *J. Bone Miner. Met.* 12: 77–81.

Huang, N., Bethell, D., Card, C. et al. 2008. Bioactive recombinant human lactoferrin, derived from rice, stimulates mammalian cell growth. *In Vitro Cell. Dev. Biol. Animal* 44: 464–471.

Hui, S.L., Slemenda, C.W., and Johnston, C.C., Jr. 1988. Age and bone mass as predictors of fracture in a prospective study. *J. Clin. Invest.* 81: 1804–1809.

Johansson, L., Grubb, A., Abrahamson, M. et al. 2000. A peptidyl derivative structurally based on the inhibitory center of cystatin C inhibits bone resorption *in vitro*. *Bone* 26: 451–459.

Kalkwarf, H.J., Khoury, J.C., and Lamphear, B.P. 2003. Milk intake during childhood and adolescence, adult bone density, and osteoporotic fractures in US women. *Am. J. Clin. Nutr.* 77: 257–265.

Kanis, J.A. and Johnell, O. 1999. The burden of osteoporosis. *J. Endocrinol. Invest.* 22: 583–588.

Kanis, J.A., Melton, L.J., III, Christiansen, C., Johnston, C.C., and Khaltaev, N. 1994. The diagnosis of osteoporosis. *J. Bone Miner. Res.* 8: 1137–1141.

Kato, K., Toba, Y., Matsuyama, H. et al. 2000. Milk basic protein enhances the bone strength in ovariectomized rats. *J. Food Biochem.* 24: 467–476.

Kawakami, H. 2005. Biological significance of milk basic protein for bone health. *Food Sci. Technol. Res.* 11: 1–8.

Kawakami, H. 2007. Case study: Milk basic protein – Biological significance for bone health and product applications. *Bull. Intern. Dairy Fed.* 413: 40–47.

Kruger, C.L., Marano, K.M., Morita, Y. et al. 2007. Safety evaluation of a milk basic protein fraction. *Food Chem. Toxicol.* 45: 1301–1307.

Lerner, U.H., Johansson, L., Ranjso, M. et al. 1997. Cystatin C, an inhibitor of bone resorption produced by osteoblasts. *Acta Physiol. Scand.* 161: 81–92.

Lin, J. T. and Lane, J. M. 2004. Osteoporosis: A review. *Clin. Orthop. Relat. Res.* 425: 126–134.

Looker, A.C., Johnston, C.C., Jr., Wahner, H.W. et al. 1995. Prevalence of low femoral bone density in older US women from NHANES III. *J. Bone Miner. Res.* 10: 796–802.

Lorget, F., Clough, J., Oliveira, M., Daury, M.C., Sabokbar, A., and Offord, E. 2002. Lactoferrin reduces *in vitro* osteoclast differentiation and resorbing activity. *Biochem. Biophys. Res. Commun.* 296: 261–266.

Matsumoto, T. 2004. Recent advances in the regulation of bone remodeling. *Nippon Rinsho* 62: 759–763.

Matsuoka, Y., Serizawa, A., Yamamura, J. et al. 2002. Cystatin C in milk basic protein and its inhibitory effect on bone resorption *in vitro*. *Biosci. Biotechnol. Biochem.* 66: 2531–2536.

McKane, W. R., Khosla, S., Egan, K.S., Robins, S.P., Burritt, M.F., and Riggs, B.L. 1996. Role of calcium intake in modulating age-related increases in parathyroid function and bone resorption. *J. Clin. Endocrinol. Metab.* 81: 1699–1703.

Meisel, H. 1998. Overview on milk protein-derived peptides. *Int. Dairy J.* 8: 363–373.

Melton, L.J., III, Chrischilles, E.A., Cooper, C., Lane, A.W., and Riggs, B.L. 1992. Perspective: How many women have osteoporosis? *J. Bone Miner. Res.* 7: 1005–1010.

Morita, Y., Matsuyama, H., Serizawa, A., Takeya, T., and Kawakami, H. 2008. Identification of angiogenin as the osteoclastic bone resorption-inhibitory factor in bovine milk. *Bone* 42: 380–387.

Nakano, T., Sugimoto, Y., Ibrahim, H.R., Toba, Y., Kawakami, H. and Aoki, T. 2004. Preparation of micellar calcium phosphate-casein phosphopeptide complex from rennet casein using ultrafiltration. *Milk Sci.* 53: 63–69.

Narva, M., Karkkainen, M., Poussa, T., Lamberg-Allardt, C., and Korpela, R. 2003. Caseinphosphopeptides in milk and fermented milk do not affect calcium metabolism acutely in postmenopausal women. *J. Am. Coll. Nutr.* 22: 88–93.

Ohashi, A., Murata, E., Yamamoto, K. et al. 2003. New functions of lactoferrin and beta–casein in mammalian milk as cysteine protease inhibitors. *Biochem. Biophys. Res. Comm.* 306: 98–103.

Park, K., Aoyagi, Y., Kawakami, H. et al. 2007. Effect of milk basic protein ingestion on bone health in the elderly, a one-year randomized controlled trial from the Nakanojo study. *J. Bone Miner. Res.* 22: 443.

Peate, I. 2004. A review of osteoporosis in men: Implications for practice. *Br. J. Nurs.* 13: 300–306.

Pittenger, M.F., Mackay, A.M., Beck, S.C. et al. 1999. Multilineage potential of adult human mesenchymal stem cells. *Science* 284: 143–147.

Prince, R.L. 1997. Diets and the prevention of osteoporotic fractures. *New Engl. J. Med.* 337: 701–702.

Ray, N.F., Chan, J.K., Thamer, M., and Melton, L.J., III. 1997. Medical expenditure for the treatment of osteoporotic fractures in the United States in 1995: Report from the National Osteoporosis Foundation. *J. Bone Miner. Res.* 12: 24–35.

Sandler, R.B., Slemenda, C.W., LaPorte, R.E. et al. 1985. Postmenopausal bone density and milk consumption in childhood and adolescence. *Am. J. Clin. Nutr.* 42: 270–274.

Sato, R., Noguchi, T., and Naito, H. 1986. Casein phosphopeptide (CPP) enhances calcium absorption from the ligated segment of rat small intestine. *J. Nutr. Sci. Vitaminol.* 32: 67–76.

Scholz-Ahrens, K.E. and Schrezenmeir, J. 2000. Effects of bioactive substances in milk on mineral and trace element metabolism with special reference to casein phosphopeptides. *Br. J. Nutr.* 84: S147–S153.

Scholz-Ahrens, K.E., de Vrese, M., and Barth, C.A. 1991. Influence of casein-derived phosphopeptides on the bioavailability of calcium in vitamin D-deficient miniature pigs. In *Gene Regulation, Structure-Function Analysis and Clinical Application* (A.W. Norman, R. Bouillon, and M. Thomasset, Eds.), Berlin: W. de Gruyter, pp. 724–725.

Seto, H., Toba, Y., Takada, Y., Kawakami, H., and Nagata, T. 2007. Milk basic protein increases alveolar bone formation in rat experimental periodontitis. *J. Periodont. Res.* 42: 85–89.

Silverman, S.L., Minshall, M.E., Shen, W., Harper, K.D., and Xie, S. 2001. The relationship of health-related quality of life to prevalent and incident vertebral fractures in postmenopausal women with osteoporosis. *Arthritis Rheum.* 44: 2611–2619.

Simon, J.A. and Mack, C.J. 2003. Prevention and management of osteoporosis. *Clin. Cornerstone* suppl. 2: S5–S12.

Sone, T. and Fukunaga, M. 2004. Prevalence of osteoporosis in Japan and the international comparison. *Nippon Rinsho* 62: 759–763.

Teegarden, D., Lyle, R.M., Ploulx, W.R., Johnston, C.C., and Weaver, C.M. 1999. Previous milk consumption is associated with greater bone density in young women. *Am. J. Clin. Nutr.* 69: 1014–1017.

Toba, Y., Matsuoka, Y., Morita, Y. et al. 2001. Milk basic protein promotes bone formation and suppresses bone resorption in healthy adult men. *Biosci. Biotechnol. Biochem.* 65: 1353–1357.

Toba, Y., Takada, Y., Yamamura, J. et al. 2000. Milk basic protein: A novel protective function of milk against osteoporosis. *Bone* 27: 403–408.

Wei, G.S., Jackson, J.L., Hatzigeorgiou, C., and Tofferi, J.K. 2003. Osteoporosis management in the new millennium. *Prim. Care* 30: 711–741.

Yamamura, J., Morita, Y., Takada, Y., and Kawakami, H. 2006. The fragments of bovine high molecular weight kininogen promote osteoblast proliferation *in vitro*. *J. Biochem.* 140: 825–830.

Yamamura, J., Motouri, M., Kawakami, H. et al. 2002. Milk basic protein increases radial bone mineral density in healthy adult women. *Biosci. Biotechnol. Biochem.* 66: 702–704.

Yamamura, J., Takada, Y., Goto, M. et al. 1999. High mobility group-like protein in bovine milk stimulates the proliferation of osteoblastic MC3T3-E1 cells. *Biochem. Biophys. Res. Commun.* 261: 113–117.

Yamamura, J., Takada, Y., Goto, M. et al. 2000. Bovine milk kininogen fragment 1.2 promotes the proliferation of osteoblastic MC3T3-E1 cells. *Biochem. Biophys. Res. Commun.* 269: 628–632.

10

Proteins and Peptides for Complicated Disease Types

Kenji Sato

CONTENTS

10.1 Introduction

Developments in diagnosis and therapeutic techniques have reduced mortality of a majority of acute life-threatening diseases with the exception of some infectious diseases. In contrast to the acute diseases, it is still difficult to treat and control chronic degenerative disease, which is characterized by progressive deterioration of structure and function of affected tissues or organs. Cancer, atherosclerosis, diabetes, chronic hepatitis, cirrhosis, Parkinson's disease, Alzheimer's disease, inflammatory bowel disease, prostatitis, osteoarthritis, osteoporosis, and rheumatoid arthritis, among others, can be classified as degenerative diseases. These are caused by normal aging, genetics, lifestyle such as eating habits and exercise, and so on.

Due to the increased longevity of industrialized populations, there are increasing numbers of people suffering from these degenerative diseases worldwide. Frequently, excess inflammation is an underlying phenomenon in these diseases. Inflammation is part of the complex biological response to infection, tissue damage, irritants, and the like. Inflammation is a protective response to initiate the healing process. However, chronic inflammation can also cause diseases such as inflammatory bowel disease, hepatitis, and rheumatoid arthritis. In those diseases, expression of proinflammatory cytokines, tumor necrosis factor (TNF)-α, interleukin (IL)-6, IL-8, IL-17, interferon (IFN)-γ, and so on, are frequently elevated and play a significant role in the progression of these diseases. Currently inflammatory diseases have therapeutic agents such as corticosteroids and aminosalicylates that show limited efficacy and potential long-term toxicity.

Some commercial food-derived peptide-based products have been successfully used to promote health and reduce risk of some of these complicated diseases. As mentioned and discussed in this book, peptides and proteins have been used to control mild hypertension (Chapter 6), hyperlipidemia (Chapter 8), and so on, which are risk factors for atherosclerosis and diabetes. Some of these products have been approved by governments to present health claims on labels. In addition, *in vitro* and animal experiments have implied that the progression of cancer could be suppressed by food-derived peptides; see the chapters by Panda et al. (Chapter 12) and Hsieh et al. (Chapter 13). However, there are few data on the control of other degenerative diseases on joint, skin, bone, liver, digestive bowel, neuron, and so on by food-derived peptides and proteins. Recently, some animal experiments and preclinical trials have suggested the potential of food-derived proteins and peptides for improvement of these complicated diseases. In this chapter, recent knowledge of moderation of some degenerative and inflammatory diseases by proteins and peptides is introduced and discussed for future prospect.

10.2 Degenerative Diseases of Connective Tissues

Progressive deterioration of function of joint, bone, skin, and so on, occur by aging and other pathological conditions, which degrade the ability of movement, consequently worsening the quality of life. To control these degenerative diseases, gelatin and its partial hydrolysates have been used in folk medicine. However, little is known about the underlying mechanism for its beneficial effect. In this section, possible protective actions of peptides on connective tissue and its underlying mechanism are introduced.

The main protein constituents of connective tissues are collagen and elastin. Collagen forms consist of different but closely related gene products (van der Rest and Garrone, 1991; Heino, 2007). Type I and II collagens are the main constituents of skin, tendon, bone (type I), and cartilage (type II). Other collagens are contained as minor constituents. In addition to protein components, hyaluronan, a nonsulfated glycosaminoglycan, and proteoglycans that have a core protein with one or more covalently attached glycosaminoglycan, are also distributed in connective tissues, which contribute to tissue hydrodynamics and movement (Gandhi and Mancera, 2008). Progressive deterioration of these components induces connective tissue disorders including osteoarthritis and osteoporosis. These diseases are prevalent in advanced countries and have a great impact on one's lifestyle (Melton, 2001; Quintana et al., 2008).

To improve joint and skin conditions, gelatin, a denatured form of collagen, or its enzymatic hydrolysate, collagen peptide, have been used in Asian and Western countries. There are some case study reports suggesting the beneficial activity of gelatin and collagen peptides on joint and skin conditions (Moskowitz, 2002; Matsumoto et al., 2006). In Japan, annual sales of collagen peptide as a supplement reached approximately US \$300 million in 2007. Until recently, these beneficial effects have not been confirmed by animal experiments and well-designed human trials. In addition, the underlying mechanism for these episodes could not be explained by conventional nutritional concepts. Therefore, people in academia tended to be suspicious of such episodes.

However, recent studies using animal models have demonstrated that ingestion of gelatin and collagen peptide shows protective actions on connective tissue damage. Nomura et al. (2005) demonstrated that ingestion of shark gelatin increases bone mineral density of rat having ovariectomy, whereas milk protein has no positive effect. A similar result was reported for a collagen peptide (Wu et al., 2004). Ingestion of collagen peptide also enhances the bone fracture healing process (Tsuruoka et al., 2007; Hata et al., 2008) and ultraviolet B-induced skin damage (Tanaka, Koyama, and Nomura, 2009) in rat models. More recently, Nakatani et al. (2009) demonstrated that ingestion of collagen peptide suppresses high phosphorous diet-induced rat osteoarthritis. Furthermore, two double-blind placebo-controlled human trials demonstrated that moisture content and subjective symptoms on skin of women are improved by collagen peptide ingestion in comparison to placebo control (Ohara et al., 2009; Koyama, 2009).

These studies suggested that the effect of collagen peptide on human skin condition depends on the subject's age. Higher age groups (>30 years old) significantly responded to collagen peptide ingestion, whereas younger groups (<30 years old) did not significantly respond. These facts indicate that ingestion of collagen peptide has some protective effect on the skin of humans with certain backgrounds. Further studies on optimization of dose, interval for administration of collagen peptide, and suitable group for application are necessary.

Collagen predominantly consists of nonessential amino acids: Gly, Ala, Pro, Gln, Asn, and so on. The beneficial action by collagen peptide could not be explained by supplementation of amino acids for collagen synthesis. In the 1960s, Prockop, Keiser, and Sjoerdsma (1962) observed an increase of hydroxyproline (Hyp) containing peptide in human urine after gelatin ingestion. As Hyp is specifically distributed in collagen and only negligible amounts of Hyp-containing peptide were detected in the human urine before gelatin ingestion, it can be concluded that food-derived collagen peptide consisting of Hyp can be absorbed into the circulation system and excreted into urine. Recently, our group found the occurrence of food-derived collagen peptide in human peripheral blood after ingestion of collagen peptide (Iwai et al., 2005; Ohara et al., 2007; Ichikawa et al., 2010). By ingestion of fish scale collagen peptide (0.385 g/kg of body weight), the content of peptide as Hyp reached approximately 140 µM, which is higher than that of free Hyp and it can be detected 7 h after ingestion (Figure 10.1).

So far, occurrences of some angiotensin converting enzyme inhibitory peptides, Val-Tyr (Matsui et al., 2002) and Ile-Pro-Pro (Foltz et al., 2007), in human blood after ingestion of an enzymatic digest of sardine muscle protein and yogurt have been reported. However, their content in human plasma is approximately less than 1 nM. Compared to those angiotensin converting enzyme inhibitory peptides, unexpectedly high concentrations of food-derived collagen peptide in the human circulation system have been demonstrated, which may change the conventional nutritional concept of peptide and amino acid absorption; food peptides are rapidly and completely degraded into amino acids during digestion and absorption processes.

Some food-derived collagen peptides in human plasma have been isolated and identified as listed in Table 10.1. In all cases, Pro-Hyp is a major component. After finding these food-derived collagen peptides in the human circulatory system, much effort has been focused on examining the biological activity of these food-derived collagen peptides to explain how collagen peptide exerts beneficial activities on connective tissues. Effects of Pro-Hyp on growth of fibroblasts, the cells responsible for collagen synthesis, have been examined. As is well known, fibroblasts rapidly grow on a plastic plate, whereas it almost stops growing in the skin and connective tissues under physiological conditions. In a culture system, fibroblasts also stop growing on collagen gel, which can mimic the fibroblasts in connective tissues (Kono

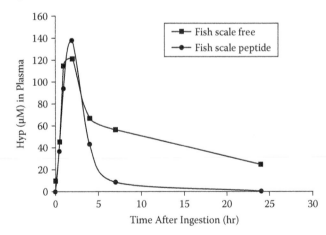

FIGURE 10.1
Contents of free and peptide forms of hydroxyproline in the human plasma before and after ingestion of fish scale gelatin hydrolysate. (Data from Ohara, H., Matsumoto, H., Ito, K., Iwai, K., Sato, K. 2007. *J. Agric. Food Chem.* 55: 1532–1535.)

TABLE 10.1

Structure of Food-Derived Collagen Peptides in Human Peripheral Blood after Ingestion of Gelatin Hydrolysates from Different Sources (%)

	Fish Scale	Fish Skin	Porcine Skin
Ala-Hyp	15	15	N.D.
Ala-Hyp-Gly	16	N.D.	N.D.
Ser-Hyp-Gly	12	N.D.	N.D.
Pro-Hyp	39	42	95
Pro-Hyp-Gly	5	3	N.D
Ile-Hyp	2	7	1
Leu-Hyp	10	27	3
Phe-Hyp	3	7	1

Source: Data compiled from Ohara, H., Matsumoto, H., Ito, K., Iwai, K., Sato, K. 2007. *J. Agric. Food Chem.* 55: 1532–1535.

N.D.: not detected.

et al., 1990). As shown in Figure 10.2, Pro-Hyp enhances the growth of mouse primary cultured fibroblasts on the collagen gel in a dose-dependent manner (Shigemura et al., 2009).

On the other hand, Pro-Hyp does not enhance fibroblast growth on collagen gel in the absence of other growth factors. Our group also examined the effect of Pro-Hyp on the number of fibroblasts migrating from mouse skin. Pro-Hyp increases the number of fibroblasts migrating from the skin to the plate (Shigemura et al., 2009). On the basis of these results, it was concluded that Pro-Hyp can abolish the suppression of fibroblast growth by attaching

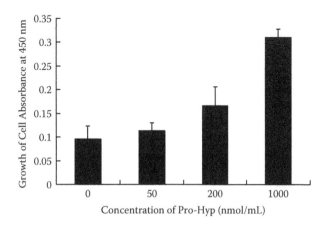

FIGURE 10.2

Effect of Pro-Hyp on the growth of primary cultured mouse skin fibroblast on collagen gel. Fibroblast grows by supplementation of Pro-Hyp in dose-dependent manner. (Data from Shigemura, Y., Iwai, K., Morimatsu, F., Iwamoto, T., Mori, T., Oda, C., Taira, T., Park, E.Y., Nakamura, Y., and Sato, K. 2009. *J. Agric. Food Chem.* 57: 444–449.)

collagen fibrils rather than acting as a growth factor, which suggested that food-derived Pro-Hyp in human peripheral blood promotes wound healing by enhancing fibroblast proliferation in the damaged tissues in the presence of growth factors. This hypothesis could at least partially explain the improvement of damaged skin.

After reporting our findings, two groups also revealed the effect of Pro-Hyp on the fibroblast and chondrocyte. Nakatani et al. (2009) demonstrated that Pro-Hyp increases glycosaminoglycan synthesis by chondrocyte and also suppresses calcification of chondrocyte in a cell culture system and also confirmed that Pro-Hyp suppresses progress of high-phosphate diet-induced osteoarthritis in a rat model. More recently, Ohara et al. (2010) revealed that Pro-Hyp increases hyaluronic acid synthesis by fibroblasts in a culture system. These facts clearly demonstrate that Pro-Hyp can affect growth of fibroblasts and synthesis of extracellular components *in vitro*, which might be linked to suggested protective effects on connective tissues.

Little is known about the molecular mechanism behind Pro-Hyp-mediated enhancement of fibroblast growth and synthesis of connective tissue components. Pro-Hyp might act by binding with the cell surface receptor or by passing into the target cell via the transporter. Alternatively, Pro-Hyp might affect the interaction between the extracellular matrix and cell.

To the best of our knowledge, it has not been reported that ingestion of collagen peptide or gelatin mediates pathological fibrosis and excess glycosaminoglycan synthesis. Underlying mechanisms for control of excessive synthesis of connective tissue components in the presence of food-derived collagen peptides is unknown. To solve these problems, further efforts are necessary.

Until now, a couple of placebo-controlled double-blind human trials have revealed that ingestion of collagen peptide has a significant impact on skin condition for a specific group of subjects (Ohara et al., 2009; Koyama et al., 2009). The *in vitro* and animal studies have also suggested that ingestion of collagen peptide or gelatin might have protective effects on osteoarthritis and osteoporosis, age-associated degenerative diseases of the joint and bone, respectively. To confirm these suggested effects, well-designed human trials are necessary.

10.3 Degenerative Disease of Muscle–Sarcopenia

In general, it takes a longer time to cure a bone fracture in an older person compared to a younger one, thus causing the older person suffering from the bone fracture pain and requiring him or her to be bedridden for long periods. Stroke and other diseases damaging the nervous system also confine patients to bed. Extensive immobility by such diseases induces muscle atrophy referred to as sarcopenia, which worsens the quality of life (Abate et al., 2007). Prevention and recovery from sarcopenia is one of the important issues in modern science.

To study the underlying mechanism of sarcopenia, rat muscles conditioned by spaceflight and tail-suspension unloading models have been used. Sarcopenia is characterized by decreased response to myogenetic growth factors such as insulin-like growth factor-1 and insulin, and also increased proteolysis (Nakao et al., 2009). DNA microarray analysis revealed upregulation of genes associated with the ubiquitin-proteasome pathway, such as ubiqutin ligase (cble-b) up to eight times in the unloaded rat muscle (Nikawa et al., 2004).

Nikawa and associates have demonstrated that insulin-like growth factor-1 signaling, which is responsible for muscle growth, is suppressed by ubiqutination and following degradation of insulin receptor substrate-1 (IRS-1). Ubiquitin ligase (Cbl-b) binds to target protein (IRS-1) for the following degradation by proteasome. In addition, suppressing Cbl-b expression in mice led them to become resistant to unloading-induced atrophy and the loss of muscle function (Nakao et al., 2009). These facts indicate that upregulation of Cble-b plays a critical role in the development of sarcopenia in the rat unloading model. Therefore, inhibition of ubiqutination of IRS-1 can be a good target for prevention of sarcopenia. Cbl-b preferentially binds to phosphorylated tyrosine residues on substrate protein. IRS-1 requires some tyrosine residues to be phosphorylated for IGF-1-stimulated responses (Greene et al., 2004). Therefore, there is a possibility that oligopeptides corresponding to tyrosine phosphorylation domains of IRS-1 could inhibit binding of Cbl-b to IRS-1.

By using an *in vitro* ubiqutination system, Asp-Gly-phospho-Tyr-Met-Pro (DGpYMP) and Leu-Asn-phospho-Tyr-Ile-Asp (LNpYID), Tyr phosphorylation domains of IRS-1, were identified as inhibitory peptides against ubiqutination of IRS-1 (Nakao et al., 2009). These authors also demonstrated that intramuscular injection of these peptides suppresses muscle loss in the rat model (Nakao et al., 2009). Nikawa and coworkers also have tested food protein hydrolysates for inhibitory activity against IRS-1 ubiquitination and found soy glycinin hydrolysate showing the inhibitory activity against ubiqutination of IRS-1 by intramuscular administration (Nikawa, 2009). These facts indicate that oligopeptides have the potential for improvement of quality of life of patients suffering from muscle loss. To exert an antisarcopenia effect based on ubiqutin ligase (Cbl-1) inhibition by oral administration, the peptide must be absorbed into the circulation system and then transported into the muscle cell for activity. For development of peptide-based formulae, which can suppress sarcopenia, further studies on the bioavailability and transportation of food-derived Cbl-b inhibitory peptide and clinical trials are necessary.

10.4 Inflammatory Bowel Disease

Inflammatory bowel disease (IBD) includes Crohn's disease and ulcerative colitis. Major symptoms of IBD are abdominal pain, vomiting, diarrhea, rectal bleeding, and weight loss (Baumgart and Carding, 2007). Both diseases are characterized by elevated inflammation and deregulation of the mucosal immune system (Sartor, 2006; Baumgart and Carding, 2007). Crohn's disease preferentially occurs in the ileum but can occur anywhere along the gastrointestinal tract. On the other hand, ulcerative colitis, in contrast, is restricted to the colon and rectum.

Although the causes of IBD are still unknown, it has been indicated that overproduction of proinflammatory cytokines plays a significant role in the development of IBD (Sartor, 2006). For treatment of IBD, anti-inflammatory agents such as corticosteroids and immunosuppressive agents are used. However, they have limited therapeutic efficacy and frequently show adverse side effects. Alternative safe and effective therapeutic agents for the treatment of IBD are needed for treatment. To improve the nutritional state of the patient, enteral nutrition including tube feeding is the first-line therapy for both active and quiescent Crohn's disease. Proteins and peptides are used for the enteral nutrition as described in Chapter 7, "Therapeutic Peptides as Amino Acid Source." Beyond the conventional nutritional aspect, specified amino acids, peptides, and proteins have been suggested to have therapeutic potential for IBD.

Amino acids can be classified as essential amino acids and nonessential amino acids. Essential amino acids cannot be synthesized to meet the demand of the human body under physiological conditions. On the other hand, enough of the nonessential amino acids can be synthesized under physiological conditions. However, some nonessential amino acids such as Gln cannot be provided in sufficient amounts by synthesis under pathological conditions (Lacey and Wilmore, 1990). Animal studies have demonstrated that Gln supplementation decreases the LPS-induced inflammatory response in infant rat intestine and attenuate intestinal cytokine-induced neutrophil chemoattractant mRNA and plasma TNF-α (Li et al., 2004). In addition, Kim et al. (2009) reported that Cys supplementation also attenuated weight loss and intestinal permeability and improved colon histology in a DSS-induced pig model. Reduced expression of proinflammatory cytokines (TNF-α, IL-6, IL-1β) and local chemokine expression and neutraphil influx were also observed. More recently, it has been demonstrated that supplementation of Cys and Thr reduces dextran sulfate sodium (DSS)-induced gene expression of inflammation markers (IL-1β, calprptectin, inducible nitric oxide synthase (iNOS)) and moderates diarrhea and fecal blood loss (Sprong, Schonewille, and van der Meer, 2010).

Gln is a good energy source and also indispensable for nuclear acid synthesis. Thus, rapidly growing cells such as enterocyte, leucocyte and so on require Gln. Gln and Cys are also used for synthesis of glutathione (γ-L-glutamyl-L-cysteinylglycine), which plays a significant role in the antioxidant system. Katayama and Mine (2007) demonstrated that Cys enhanced glutathione biosynthesis enzyme activity and increased cellular GSH levels and inhibited H_2O_2-induced IL-8 secretion from Caco-2 cells. In addition, Cys, Ser, Thr, and Pro are the main constituents of intestinal mucin, a family of high molecular weight, heavily glycosylated proteins, serving mucosal barriers. Supplementation of these amino acids increases fecal mucin excretion, suggesting stimulation of intestinal mucin synthesis (Sprong et al., 2010). These facts suggest that supplementation of specific amino acids including nonessential ones may enhance growth of enterocyte and also enhance synthesis of peptides and proteins, which have intestinal protective activity against oxidation and inflammation. However, some human trials revealed that a Gln-enriched polymeric diet offers no advantage over a standard low-Gln diet in the treatment of active Crohn's disease (Akobeng et al., 2000).

Supplementation of some proteins has also demonstrated the improvement of gastrointestinal condition in animal models. Sprong et al. (2010) demonstrated that cheese whey protein reduces DSS-induced inflammation and symptoms in comparison to a casein-based diet. These authors suggested that high content of Thr and Cys in cheese might contribute to the intestinal protective effect of the cheese whey. On the other hand, Ozawa et al. (2009) demonstrated that commercially available pasteurized cow's milk retains transforming growth factor (TGF)-β activity and oral administration of pasteurized cow's milk increases intestinal TGF-β level and also TGF-β-induced intracellular signaling (phosphoryzation of Smad2). Oral administration of 500 µL cow's milk containing 3 µg/L TGF-β for two weeks ameliorates DSS and LPS-induced tissue damage and mortality in mice. In addition, ingestion of 10 mL/kg body weight cow's milk containing TGF-β 3 µg/mL, increased plasma TGF-β levels in humans. This study suggests that TGF-β in commercially available milk could provide protection against intestinal inflammation. As cow's milk is rich in immunoglobulins, antimicrobial peptides, and growth factors, it has therapeutic potential for IBD. However, it might also exert unexpected effects on patients. Thus, standardization of milk products and evaluation of their therapeutic nature on IBD are necessary.

Daddaoua et al. (2005) demonstrated that casein macropeptide (another name; glycomacropeptide) exerts anti-inflammatory action in trinitrobenzenesulfonic acid-induced rat colitis and decreases expression of iNOS and IL-1β. More recently, Lee et al. (2009) reported that hen egg lysozyme supplementation reduces the gut permeability caused by DSS and improves intestinal epithelial barrier function in a porcine model. They also demonstrated hen egg lysozyme reduces the expression of proinflammatory cytokines TNF-α, INF-γ, IL-8, and IL-17 and increases the expression of the anti-inflammatory mediators IL-4 and TGF-β. They proposed a possibility that hen egg lysozyme directly modulates cytokines involved in inflammation. There is another possibility that the lysozyme or its degradation product might attenuate the DSS-induced IBD through modulation of intestinal microbial flora.

These data indicate that specific food protein has therapeutic potential against IBD possibly by providing intestinal protective amino acids or by directly modulating proinflammatory cytokines. The proposed mechanism for modulating inflammation by food protein could be summarized as follows. The food protein and peptide can be used for specific amino acid sources in the synthesis of endogenous intestinal protective peptides and proteins. The intact protein such as TGF-β in food might directly modulate inflammation. Alternatively, the smaller peptide might modulate intestinal microflora or suppress oxidation and inflammation. However, detailed molecular mechanisms for the anti-inflammatory activity of protein remains unclear. With the present knowledge, the efficacy of supplementation of amino acids and proteins on IBD has not been confirmed by well-designed human trials.

10.5 Hepatitis

Hepatitis implies inflammation of the liver characterized by the presence of inflammatory cells and hepatocyte necrosis. Viral infection, excess alcohol consumption, toxins, fatty liver, and so on cause hepatitis. Chronic hepatitis frequently develops fibrosis in the liver and consequently cirrhosis. Cirrhosis is believed to irreversibly progress and cause life-threatening damage to the liver. The therapy for hepatitis is primarily based on the treatment of the underlying diseases by alimentary, exercise, and drug therapies. In addition to these conventional therapies, alternative therapies, which are suitable for chronic use and show high compliance, have been required. There is great interest in food functionality to control hepatitis and the following fibrosis.

As well as IBD, it has been demonstrated that supplementation of some amino acids such as Gln, Asn, Gly, Ser, His, Tyr, or Lys at 10% of diet attenuate D-galactosamine-induced hepatic injury in rat model (Wang et al., 1999). It has been proposed that large amounts of D-galactosamine activate mast cells to release histamine, which increases gut permeability. Then bacterial endotoxin increases in blood, which activates a Kupffer cell, a macrophage-like cell in the liver, to produce inflammatory cytokines such as TNF-α, IL-8, and the like. In addition, D-galactosamine decreases uridine triphosphate in the hepatocyte, which suppresses mRNA and protein synthesis. Under such conditions, the hepatocyte increases sensitivity to inflammatory cytokines and starts to die via apoptosis and subsequently necrosis (Stachlewitz et al., 1999). The Kupffer cell-TNF-α pathway is based on the facts that D-galactosamine-induced liver injury can be attenuated by destroying the Kupffer cell by gadolinium chloride or neutralizing TNF-α by specific antisera. Stachlewitz et al. (1999) also demonstrated that Gly supplement suppresses the increase in D-galactosamin–induced endotoxin and consequently liver inflammation, and subsequent hepatocyte death.

Recently, Komano et al. (2009) demonstrated that the suppressive effect of Gln and Gly on D-galactosamine-induced hepatocyte death is stronger than gadolinium chloride pretreatment, suggesting the presence of another hepatoprotective pathway of such amino acids rather than suppression of TNF-α production by Kupffer cell. Gln and Gly are sources for synthesis of nucleotides and glutathione. Then it can be anticipated that supplementation of them might increase uridine triphosphate and glutathione levels in hepatocyte, which may partially contribute to the hepatoprotective effect of these amino acids. In addition, Komano, Egashira, and Sanada (2008) demonstrated that supplementation of Gln and Ser suppresses the D-galactosamine-induced elevated expression of IL-18, a proinflammatory cytokine that also plays a critical role in inflammation leading to liver damage. However, the protective action of these amino acids against D-galactosamine-induced liver damage and efficacy of supplementation of these amino acids to human hepatitis remain to be examined. In those studies, these free amino acids have been supplemented to 5 or 10% of the diet. It may be difficult to ingest such high doses of free amino acids by patients for daily consumption.

Recently, some food-derived peptides have been demonstrated to have hepatoprotective activity in animal models. He et al. (2008) demonstrated that peptides derived from Chinese medical mushroom (*Ganoderma lucidum*) moderate D-galactosamine-induced liver injury. More recently, Guo et al. (2009) demonstrated that corn gluten peptide also moderates LPS-induced liver injury. In both studies, supplementation of these peptides suppressed formations of malondialdehyde, NO, and also suppressed the LPS-induced decreases of superoxide dismutase and glutathione peroxidase activities and glutathione content in liver. On the basis of these facts, these authors suggested that the hepatoprotective effect of these peptides might be attributed to antioxidative activity, suppression of NO synthesis, and enhancement of synthesis of antioxidant component (glutathione) and antioxidant enzymes.

Manabe et al. (1996) have demonstrated that wheat gluten, which is rich in Gln residue, also suppresses galactosamine-induced hepatic injury. On the basis of these findings, wheat gluten hydrolysate has been prepared as a stable Gln source on an industrial scale and is commercially available. As shown in Figure 10.3, supplementation of wheat gluten hydrolysate at 4–15 g/day to out-patients suffering from hepatitis with different backgrounds improved hepatic condition at an unexpectedly higher efficacy (Horiguchi et al., 2004).

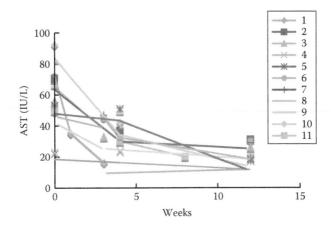

FIGURE 10.3

Effect of ingestion of wheat gluten hydrolysate on the serum asparatate amino transferase (AST) activity of patients suffering from hepatitis with different backgrounds. (Data from Horiguchi, N., Horiguchi, H., and Suzuki, Y. 2004. *Jpn. Pharmacol. Therapeutics* 32: 415–420.)

The authors also examined hepatoprotective action of free Gln at 6 and 8 g/day and found no positive effect. These facts suggest that some specific peptide in wheat gluten hydrolysate may suppress hepatitis beyond a traditional nutritional aspect, source of amino acids. Recently our group has demonstrated that wheat gluten hydrolysate suppressed carbon tetrachloride-induced rat liver fibrosis and galactosamine-induced acute hepatitis (unpublished data). To identify the active peptide, peptides in the wheat gluten hydrolysate were fractionated by large-scale (5 L) ampholyte-free preparative isoelectric focusing based on the amphoteric nature of the sample peptide (Hashimoto et al., 2005). The acidic fraction showed significant liver protective effect against D-galactosamine-induced rat liver injury, in which free pyroGlu, pyroGlu-Gln, pyroGlu-Gln-Gln, pyroGlu-Tyr, pyroGlu-Leu, and pyroGlu-Phe were identified. By using synthesized peptides, pyroGlu-Leu was identified as the hepatoprotective peptide. The pyroGlu-Leu can suppress endotoxin-induced production of inflammatory mediators including inducible NO synthetase and TNF-α from macrophage (unpublished data). The molecular mechanism for how to suppress inflammatory cytokines by pyroGlu-Leu remains unclear. It is worthwhile to stress that pyroGlu-Leu can be produced from Gln-Lys during preparation of wheat gluten hydrolysate (Hettiarachchy et al., 2012).

The wheat and corn gluten peptides can be prepared from by-products of the starch industry. If the efficacy is confirmed by a double-blind placebo control study that these peptides have a therapeutic potency for hepatitis, a process method for large-scale preparation of hepatoprotective fractions from these starting materials should be developed.

10.6 Future Prospect for Other Complicated Diseases

As described in this chapter, some peptides are stable and soluble sources for relatively instable and insoluble amino acids; for example, gluten and milk whey hydrolysates are good sources for Gln and Cys, respectively. These amino acids are used for some endogenous active peptides and proteins such as glutathione, mucin, and the like, which could explain part of the beneficial activity of some food peptides. However, some peptides may have specific activity beyond their amino acid source. As shown in this chapter, specific food-derived collagen peptides are absorbed into the circulatory system and enhance cell proliferation and synthesis of extracellular components. In addition, wheat gluten peptides show hepatoprotective action, suppression of carbon tetrachloride-induced fibrosis, and reduced galactosamine-induced acute liver injury.

These activities could not simply be explained by the source of amino acids, as peptides show higher efficacy in comparison to Gln and intact gluten at the same dose. *In vivo* activity-guided fractionation revealed that a pyroGlu-Leu is one of the hepatoprotective peptides and it can be absorbed into the circulation system. This peptide has no antioxidant activity based on the ORAC assay. On the other hand, it suppresses LPS-induced production of iNOS, IL-8, and TNF-α from macrophage, which indicates that some food-derived peptides can directly modulate inflammation. The molecular mechanism of food-derived peptides for proliferation and anti-inflammatory activities still remains to be examined. Some peptides may be incorporated into target cells and affect intracellular signaling by inhibition of phosphorylation of intracellular proteins as suggested by Nakao et al. (2009). Some peptides may inhibit extracellular peptidases which are involved in processing of cytokines: for example, angiotensin converting enzyme, tumor necrosis factor converting enzyme, and plasmin for producing active forms of angiotensin II, TNF-α, and TGF-β, respectively. Some peptides may interfere or enhance cytokine-receptor or cell adhesion molecule on cell surface-extracellular matrix interactions. Much effort should be concentrated to identify food-derived peptides in the target cells and elucidate the molecular mechanism.

Excess inflammation underlies not only IBD and hepatitis but also many chronic diseases such as rheumatoid arthritis, Parkinson's disease, and Alzheimer's disease, among others. Food-derived peptides have potential for controlling and modulating other complicated diseases.

References

Abate, M., Di Iorio, A., Di Renzo, D., Paganelli, R., Saggini, R. and Bate, G. 2007. Frailty in the elderly: The physical dimension. *Eura Mediciphys* 43: 407–4015.

Akobeng, A.K., Miller, V., Stanton, J., Elbadri, A.M., and Thomas, A.G. 2000. Double-blind randomized controlled trial of glutamine-enriched polymeric diet in the treatment of active Crohn's disease. *J. Pediatr. Gastroent. Nutri.* 30: 78–84.

Baumgart, D.C. and Carding, S.R. 2007. Inflammatory bowel disease: Cause and immunobiology. *Lancet* 369: 1627–1640.

Daddaoua, A., Puerta, V., Zarzuelo, A., Suárez, M.D., Sánchez de Medina, F., and Martinez-Augustin, O. 2005. Bovine glycomacropeptide is anti-inflamamatory in rats with hapten-induced colitis. *J. Nutri.* 135: 1164–1170.

Foltz, M., Meynen, E.E., Bianco, V., van Platerink, C., Koning, T.M.M.G., and Kloek, J. 2007. Angiotensin converting enzyme inhibitory peptides from a lactotripeptide-enriched milk beverage are absorbed intact into the circulation. *J. Nutr.* 137: 953–958.

Gandhi, N.S. and Mancera, R.L. 2008. The structure of glycosaminoglycans and their interactions with proteins. *Chem. Biol. Drug Des.* 72: 455–482.

Greene, M.W., Morrice, N., Garofalo, R.S., and Roth, R.A. 2004. Modulation of human insulin receptor substrate-1 tyrosine phosphorylation by protein kinase C. *Biochem. J.* 378: 105–116.

Guo, H., Sun, J., He, H., Yu, G.-C., and Du, J. 2009. Antihepatotoxic effect of corn peptides against *Bacillus calmette-guerin*/lipopolysaccharide-induced liver injury in mice. *Food Chem. Toxicol.* 47: 2431–2435.

Hashimoto, K., Sato, K., Nakamura, Y., and Ohtsuki, K. 2005. Development of a large-scale (50 L) apparatus for ampholyte-free isoelectric focusing (autofocusing) of peptides in enzymatic hydrolysates of food proteins. *J. Agric. Food Chem.* 53: 3801–3806.

Hata, S., Hayakawa, T., Okada, H., Hayashi, K., Akimoto, Y., and Yamamoto, H. 2008. Effect of oral administration of high advanced-collagen tripeptide (HACP) on bone healing process in rat. *J. Hard Tissue Biol.* 17: 17–22.

He, H., He, J.-P., Sui, Y.-J., Zhou, S.-Q., and Wang, J. 2008. The hepatoprotective effects of *Ganoderma lucidum* peptides against carbon tetrachloride-induced liver injury in mice. *J. Food Biochem.* 32: 628–641.

Heino, J. 2007. The collagen family members as cell adhesion proteins. *Bioessays* 29: 1001–1010.

Hettiarachchy, N.S., Kannan, A., Marshall, M., and Sato, K. (2012). Food protein-peptides: Structure function relationship. In *Food Proteins and Peptides: Chemistry, Functionality, Interactions, and Commercialization* (N.S. Hettiarachchy, K. Sato, and M. Marshall, Eds.) Boca Raton, FL: CRC Press.

Horiguchi, N., Horiguchi, H., and Suzuki, Y. 2004. Out-hospital patients with hyperlipidemia and hepatitis with various backgrounds improved by wheat protein hydrolysat (glutamine peptide) administration. *Jpn. Pharmacol. Therapeutics* 32: 415–420.

Ichikawa, S., Morifuji, M., Ohara, H., Matsumoto, H., Takeuchi, Y., and Sato, K. 2010. Hydroxyproline-containing dipeptides and tripeptides quantified at high concentration in human blood after oral administration of gelatin hydrolysate. *Int. J. Food Sci. Nutri.* 61: 52–60.

Iwai, K., Hasegawa, T., Taguchi, Y., Morimatsu, F., Sato, K., Nakamura, Y., Higashi, A., Kido, Y., Nakabo, Y., and Ohtsuki, K. 2005. Identification of food-derived collagen peptides in human blood after oral ingestion of gelatin hydrolysates. *J. Agric. Food Chem.* 53: 6531–6536.

Katayama, S. and Mine, Y. 2007. Antioxidative activity of amino acids on tissue oxidative stress in human intestinal epithelial cell model. *Agric. Food Chem.* 55: 8458–8464.

Kim, C.J., Kovacs-Nolan, J., Yang, C., Archbold, T., Fan, M.Z., and Mine, Y. 2009. L-cysteine supplementation attenuates local inflammation and restores gut homeostasis in an porcine model of colitis. *Biochim. Biophys. Acta* 1790: 1161–1169.

Komano, T., Egashira, Y., and Sanada, H. 2008. L-Gln and L-Ser suppress the D-galactosamine-induced IL-18 expression of hepatitis. *Biochem. Biophys. Res. Commun.* 372: 688–690.

Komano, T., Funakoshi, R., Egashira, Y., and Sanada, H. 2009. Mechanism of the suppression against D-galactosamine-induced hepatic injury by dietary amino acids in rats. *Amino Acids* 37: 239–247.

Kono, T., Tamii, T., Furukawa, M., Mizuno, N., Kitajima, J., Ishi, M., and Hamada, K., 1990. Cell cycle analysis of human dermal fibroblasts cultured on or in hydrated type I collagen lattices. *Arch. Dermatol. Res.* 282: 258–262.

Koyama, Y. 2009. Effect of collagen peptide ingestion on the skin. *Shokuhinto Kaihatsu* 44: 10–12.

Lacey, J.M. and Wilmore, D.W. 1990. Glutamine a conditionally essential amino acid? *Nutri. Rev.* 48: 297–309.

Lee, M., Kovacs-Nolan, J., Yang, C., Archbold, T., Fan, M.Z., and Mine, Y. 2009. Hen egg lysozyme attenuates inflammation and modulates local gene expression in a porcine model od dextran sodium sulfate (DSS)-induced colitis. *J. Agric. Food Chem.* 57: 2233–2240.

Li, N., Liboni, K., Fang, M.Z., Samuelson, D., Lewis, P., Patel, R., and Neu, J. 2004. Glutamine decreases lipopolysaccharide-induced intestinal inflammation in infant rats. *Am. J. Physiol. Gastrointest. Liver Physiol.* 286: 914–921.

Manabe, A., Cheng, C.C., Egashira, Y., Ohta, T., and Sanada, H. 1996. Dietary wheat gluten alleviates the elevation of serum transaminase activities in D-galactosamine-injected rats. *J. Nutr. Sci. Vitaminol.* 42: 121–132.

Matsui, T., Tamaya, K., Seki, E., Osajima, K., Matsumo, K., and Kawasaki, T. 2002. Absorption of Val-Tyr with *in vitro* angiotensin I-converting enzyme inhibitory activity into the circulating blood system of mild hypertensive subjects. *Biol. Pharm. Bull.* 25: 1228–1230.

Matsumoto, H., Ohara, H., Ito, K., Nakamura, Y., and Takahashi, S. 2006. Clinical effects of fish type I collagen hydrolysate on skin properties. *ITE Lett.* 7: 386–390.

Melton, L.J. 2001. The prevalence of osteoporosis: Gender and racial comparison. *Calcif. Tissue Int.* 69: 179–81.

Moskowitz, R.W. 2002. Role of collagen hydrolysate in bone and joint disease. *Semin. Arthritis Rheum.* 30: 87–99. 2004.

Nakao, R., Hirasaka, K., Goto, J., Ishidoh, K., Yamada, C., Ohno, A., Okumura, Y. et al. 2009. Ubiquitin ligase Cbl-b is a negative regulator for IGF-1 signaling during muscle atrophy caused by unloading. *Mol. Cell Biol.* 17: 4798–4811.

Nakatani, S., Mano, H., Sampei, C., Shimizu, J., and Wada, M. 2009. Chondroprotective effect of the bioactive peptide prolylhydroxyproline in mouse articular cartilage *in vitro* and *in vivo*. *Osteoarthritis Cartilage* 17: 1620–1627.

Nikawa, T. 2009. Development of ubiquitin ligase inhibitor as a drug against unloading-mediated muscle atrophy. *Seikagaku* 81: 614–618.

Nikawa, T., Ishidoh, K., Hirasaka, K., Ishihara, I., Ikemoto, M., Kano, K., Kominami, E. et al. 2004. Skeletal muscle gene expression in space-flown rats. *FASEB J.* 18: 522–524.

Nomura, Y., Oohashi, K., Watanabe, M., and Kasugai, S. 2005. Increase in bone mineral density through oral administration of shark gelatin to ovariectomized rats. *Nutrition* 21: 1120–1126.

Nonaka, I., Yasutomo, K., Baldwin, K.M., Kominami, E., Higashibata, A., Nagano, K., Tanaka, K. et al. 2009. Ubiquitin ligase Cbl-b is a negative regulator for insulin-like growth factor 1 signaling during muscle atrophy caused by unloading. *Molec. Cell. Biol.* 29: 4798–4811.

Ohara, H., Ichikawa, S., Matsumito, H., Akiyama, M., Fujimoto, N., Kobayashi, T., and Tajima, S. 2010. Collagen-derived dipeptide, proline-hydroxyproline. Stimulates cell proliferation and hyaluronic acid synthesis in cultured human dermal fibroblast. *J. Dermatol.* 37: 330–338.

Ohara, H., Ito, K., Iida, H., and Matsumoto, H. 2009. Improvement in the moisture content of the stratum corneum following 4 weeks of collagen hydrolysate ingestion. *Nippon Shokuhin Koaku Kaishi* 56: 137–145.

Ohara, H., Matsumoto, H., Ito, K., Iwai, K., and Sato, K. 2007. Comparison of quantity and structures of hydroxyproline-containing peptides in human blood after oral ingestion of gelatin hydrolysates from different sources. *J. Agric. Food Chem.* 55: 1532–1535.

Ozawa, T., Miyata, M., Nishimura, M., Ando, T., Ouyang, Y., Ohba, T., Shimokawa, N. et al. 2009. Transforming growth factor-β activity in commercially available pasteurized cow's milk provides protection against inflammation in mice. *J. Nutri.* 139: 69–75.

Prockop, D.J., Keiser, H.R., and Sjoerdsma, A. 1962. Gastrointestinal absorption and renal excretion of hydroxyproline peptides. *Lancet* 2: 527–528.

Quintana, J.M., Arostegui, I., Escobar, A., Azkarate, J., Goenaga, J.I. and Lafuente, I. 2008. Prevalence of knee and hip osteoarthritis and the appropriateness of joint replacement in an older population. *Arch. Intern. Med.* 168: 1576–1584.

Sartor, R.B. 2006. Mechanisms of disease: Pathogenesis of Crohn's disease and ulcerative colitis. *Gastroent. Hepatol.* 3: 390–407.

Shigemura, Y., Iwai, K., Morimatsu, F., Iwamoto, T., Mori, T., Oda, C., Taira, T., Park, E.Y., Nakamura, Y., and Sato, K. 2009. Effect of prolyl-hydroxyproline (Pro-Hyp), a food-derived collagen peptide in human blood, on growth of fibroblasts from mouse skin. *J. Agric. Food Chem.* 57: 444–449.

Sprong, R.C., Schonewille, A.J., and van der Meer, R. 2010. Dietary cheese whey protein protects rats against mild dextran sulfate sodium-induced colitis: Role of mucin and microbiota. *J. Dairy Sci.* 93: 1364–1371.

Stachlewitz, R.F., Seabra, V., Bradford, B., Bradham, C.A., Rusyn, I., Germolec, D., and Thurman, R.G. 1999. Glycine and uridine prevent D-galactosamine hepatotoxicity in the rat: Role of Kupffer cells. *Hepatology* 29: 737–745.

Tanaka, M., Koyama, Y., and Nomura, Y. 2009. Effects of collagen peptide ingestion on UV-B-induced skin damage. *Biosci. Biotechnol. Biochem.* 73: 930–932.

Tsuruoka, N., Yamamoto, R., Sakai, Y., Yoshitake, Y., and Yonekura, H. 2007. Promotion by collagen tripeptide of type I collagen gene expression in human osteoblastic cell and fracture healing of rat femur. *Biosci. Biotechnol. Biochem.* 71: 2680–2687.

Van der Rest, M. and Garrone, R. 1991. Collagen family of proteins. *FASEB J.* 5: 2814–2823.

Wang, B., Ishihara, M., Egashira, Y., Ohta, T., and Sanada, H. 1999. Effects of various kinds of dietary amino acids on the hepatotoxic action of D-galactosamine in rats. *Biosci. Biotechnol. Biochem.* 63: 319–322.

Wu, J., Fujioka, M., Sugimoto, K., Mu, G., and Ishimi, Y. 2004. Assessment of effectiveness of oral administration of collagen peptide on bone metabolism in growing and mature rats. *J. Bone Miner. Metab.* 22: 547–553.

11

Proteins and Peptides Improve Mental Health

Kenji Sato

CONTENTS

11.1 Introduction

The World Health Organization has defined health as being "a state of complete physical, mental, and social well-being and not merely the absence of disease or infirmity." Mental health may include cognitive and emotional well-being and an absence of a mental disorder. Now cognitive decline of aged people and extensive stress, anxiety, and depression have become a serious social problem in many countries and communities. It has been demonstrated that serotonin 5-HT$_{1A}$ receptor, dopamine D$_1$ receptor, and γ-amino butyric acid type A (GABAa) receptor play important roles in anxiolytic activity (Short et al., 2006; Hirata et al., 2007; Leonardo and Hen, 2008; Jacob, Moss, and Jurd, 2008). Therefore, medicine targeting dopaminergic, GABAergic, and serotonergic pathways have been developed and used for treatment of anxiety disorders, depression, and neurological diseases. However, potential side effects increase interest in the use of dietary supplements and functional foods to manage normal stress and anxiety.

Some plant extracts such as valerian (*Valeriana officinalis*), skullcap (genus of *Scutellaria*), hops (*Humuluus lupulus*), lemon balm (*Melissa officinalis*), St. John's wort (*Hypericum perforatum*), and so on have been traditionally used in folk medicine to control depression and reduce anxiety (Weeks, 2009). The efficacy of some plant products has been checked by human trials and some positive effects have been reported (Weeks, 2009). In addition to these plant extracts, some amino acids and amino acid derivatives have been demonstrated to have anxiolytic activities and suppressive effect against age-related memory loss not only in animal models but also in human trials. Now some amino acids and derivatives are formulated for dietary supplement and functional food ingredients. As described in this book, some food-derived peptides can improve physical health and prevent diseases. Recently, based on animal experiments, some peptides and enzymatic digests have been demonstrated to have anxiolytic and neuroprotective activities by ingestion. However, only few peptides have been evaluated by human trials for their anxiolytic activity. In the present chapter, recent advances in the effect of food-derived peptides on mental health are introduced and their potency discussed.

11.2 Amino Acids and Amino Acid-Related Compounds

As is well known, some amino acids act as neurotransmitters in the central nervous system. Therefore, such amino acids and amino acid-related compounds have the potential to reduce anxiety, relieve stress, enhance attention and performance of task, improve sleep, and so on. The present chapter does not aim to review comprehensively the effects of amino acids and related compounds on the nervous system. The possible effects of these compounds on mental health by oral administration is briefly introduced. γ-Aminobutyric acid (GABA), glycine (Gly), β-alanine (β-Ala), and taurine (Tau) are inhibitory neurotransmitters and serve as stress relievers in the brain.

Among them, GABA plays a chief role in the reduction of nerve impulse transmission through binding to the GABA receptor in the central nervous system. Low GABA levels in the brain are associated with depression, anxiety, and poor mood (Krystal, Sanacora, and Blumberg, 2002; Gottesmann, 2002; Nemeroff, 2003; Kendell, Krystal, and Sanacora, 2005). However, the blood–brain barrier restricts crossing of these amino acids from blood to brain under normal condition. Therefore, it has been assumed that these orally administered amino acids do not have significant anxiolytic activity. Thus, much effort has been focused on developing a drug that increases GABA level in brain and has agonist activity to GABAa receptors.

On the other hand, some researchers demonstrated that dietary GABA has anxiolytic activity to healthy volunteers (Abdou et al., 2006). After 30 min of oral administration of GABA at 100 mg, alpha waves increased and beta waves decreased in the electroencephalographs in comparison to water control. Furthermore, they checked immunoglobulin A (IgA) levels in saliva, a good stress biomarker, of the healthy volunteers who were subjected to stress by crossing a suspended bridge over a deep valley. They demonstrated that

GABA can significantly suppress stress-induced decrease of IgA level and concluded that GABA works as a natural relaxant by ingestion. However, the mechanism for anxiolytic activity by dietary GABA remains unclear. Now, GABA is produced from glutamate by fermentation with lactic acid bacteria (Komatsuzaki et al., 2005) and in Japan and other countries it is used as a food ingredient aiming to relieve stress.

Theanine (γ-glutamyl-ethylamide) is specifically distributed in Japanese green tea and has relaxation and calming effects in humans which were demonstrated by an increase of alpha waves in electroencephalographs (Lu, Gray, and Oliver, 2004; Kimura et al., 2007; Nobre, Rao, and Owen, 2008). It has been demonstrated that theanine increases brain GABA level in rat by ingestion, which suggests that theanine might act as a GABAergic compound (Nathan et al., 2006; Yamada, Tershima, and Kawano, 2009). Theanine also has a neuroprotective effect in the rat model, which might be associated with improved memory and cognitive function (Nathan et al., 2006; Yamada et al., 2008). Now, L-theanine can be synthesized from L-glutamine and ethyl-amine by transglutaminase-induced coupling reaction (Tachiki et al., 2008).

Serotonin is a neurotransmitter and synthesized from tryptophan in pre-synaptic neurons (Figure 11.1). Serotonin levels in brain are associated with a good mood, whereas low serotonin levels have been associated with depression (Young and Leyton, 2002; Ruhe, Mason, and Schene, 2007; Young, 2007).

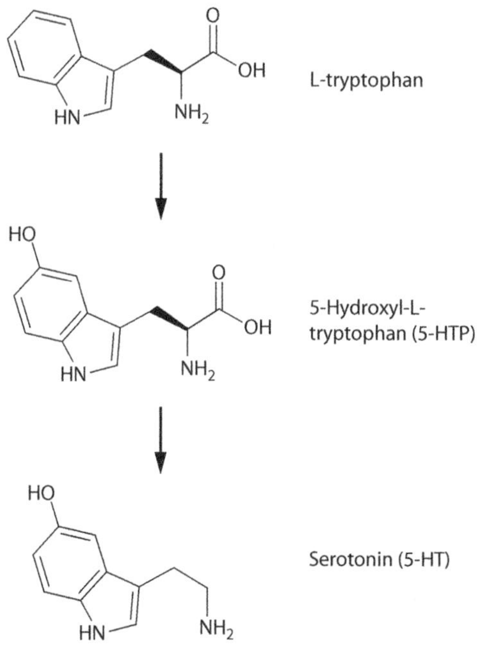

FIGURE 11.1
Synthesis of serotonin from tryptophan.

As tryptophan and 5-HTP can cross the blood–brain barrier, supplementations with tryptophan and 5-HTP elevate the synaptic serotonin level, which has been suggested to enhance mood and a sense of well being.

It has been suggested that supplementation with pyroglutamate, which is produced from glutamine, might improve age-associated memory impairment in rats (Drago et al., 1988) and humans (Grili et al., 1990). Recently, supplementation with glycine (3 g) has been suggested to improve quality of sleep and lessen daytime sleep (Inagawa et al., 2006; Yamadera et al., 2007).

As described here, supplementation of some amino acids and related components can improve mental health and quality of sleeping. However, further studies are necessary to elucidate delivery of these components into the central nervous system and underlying mechanism.

11.3 Casein Hydrolysate

Milk proteins, especially caseins, are the only proteins synthesized by mammals for feeding animal and human infants. It has been demonstrated that enzymatic cleavage of casein produces many peptides with various biological activities beyond its nitrogen source: opioid and opioid-antagonist, antihypertensive, immunostimulative, antithrombotic, antimicrobial, and mineral absorption-enhancing peptides (Meisel, 1997). In addition to this evidence, folk wisdom maintains that milk intake might provide good quality sleep and have a calming effect. To support these episodes, some researchers reported that milk consumption at bedtime provides uninterrupted and longer sleep (Brezinova and Oswald, 1972). On the basis of these findings, a French group has demonstrated that tryptic hydrolysate of α_{s1}-casein has anxiolytic activity both in animal models and human trials. Miclo et al. (2001) demonstrated that intraperitoneal (i.p.) injection of α_{s1}-casein tryptic hydrolysate (3 mg/kg) reduced epileptic symptoms caused by pentylenetetrazole in rats. They also examined anxiolytic activity of the α_{s1}-casein tryptic hydrolysate by using elevated plus-maze and conditioned defensive burying tests.

As shown in Figure 11.2, the elevated plus-maze model is based on the natural aversion of rodents for open spaces. The maze consists of two open arms and two enclosed arms that extend from a central platform. The whole apparatus is located at a height of 70 cm above floor level. The percentage of entries to open and closed arms and periods of time spent in open arm, closed arm, and central platform were examined. Increased activities in the open arms of the maze after administration of test components positively correlated with anxiolytic activity. Head dippings (exploratory movements of head over the sides of the maze) positively correlated with the exploration level of the animals. The conditioned defensive burying test is based

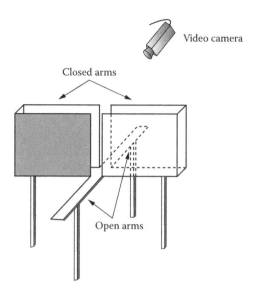

FIGURE 11.2
A plus-maze model apparatus for animal experiments.

on the burying activity of rat. The rat is placed in a chamber with an electric shock probe, which delivers an electric shock to the rat at the first touch by its forepaws. After the electric shock, the rat buries the probe with bed materials. The duration of probe-burying, number of head stretchings, and approaches toward the probe, and number of retreats away from the probe were recorded, and the sum of the ranking of 3 parameters was defined as the anxiety global score. Latency of the first contact with the probe after the electrical shock was also recorded, which is associated with the noninhibiting effect. These ethological studies demonstrated that i.p. injection of the hydrolysate shows anxiolytic activity in rats.

It has been demonstrated that benzodiazepines provide anxiolytic action by binding to a GABAa receptor. These properties make benzodiazepines a useful tool to explore the mechanism of anxiolytic activity of natural compounds. On the basis of displacement studies of [methyl-^{3}H]-flunitrazepam, one of the benzodiazepines, Miclo et al. (2001) revealed that the tryptic hydrolysate of α_{s1}-casein interacts with the benzodiazepine site of the GABAa receptor with an IC50 of 72 μM, suggesting that enhancement of the GABAergic transmission might be involved in the anxiolytic activity of the tryptic hydrolysate. They identified a deca-peptide, Tyr-Leu-Gly-Tyr-Leu-Glu-Gln-Leu-Leu-Arg corresponding to α_{s1}-CN-f (91-100), with affinity for the GABAa receptor, referred to as α-casozepine. The α-casozepine amino acid sequence could be related to the carboxy-terminal sequence of the polypeptide, referred to as the diazepam binding inhibitor, an endogenous ligand of the central GABAa and peripheral-type benzodiazepine receptors.

The authors also confirmed that i.p. injection of the α-casozepine in a small dose (0.4 mg/kg) shows anxiolytic activity in the conditioned defensive-burying rat model. On the basis of these findings, Miclo et al. (2001) proposed that α-casozepine might be an exogenous ligand of the GABAa receptor and might be involved in the traditional calming properties of milk. However, the authors also pointed out an inconsistency that the α-casozepine has 10,000 less affinity for the benzodiazepine site of the GABAa receptor than diazepam, although it was tenfold more efficient than diazepam by i.p. injection to rat. The difference observed between the *in vitro* and *in vivo* activities of α-casozepine could not have been explained by an action via the peripheral-type benzodiazepine receptor as α-casozepine had no affinity for this receptor.

The study by Miclo et al. (2001) was followed by the French groups Violle et al. (2006) and Messaoudi et al. (2009). They examined the anxiolytic and disinhibiting effects of bovine α_{s1}-casein hydrolysate by oral ingestion in the elevated plus-maze and conditioned defensive-burying tests using the male Wistar rat. They found that 15 mg/kg of α_{s1}-casein hydrolysate has significant anxiolytic activity corresponding to 3 mg/kg of diazepam on the basis of the results by the elevated plus-maze and conditioned defensive-burying tests. In their studies, diazepam stimulated total arm entries and head dipping behavior in the elevated plus-maze test and shortened latency of the first contact toward the electric probe after the shock in the conditioned defensive-burying test. These facts imply that diazepam induces not only an anxiolytic state but also a disinhibition state, which could be related to benzodiazepine-induced risk-taking behavior observed in humans. However, α_{s1}-casein hydrolysate did not show such side effects.

It has been speculated that the GABAa receptor agonist activity of α-casozepine might be involved in the anxiolytic activity. However, α-casozepine has lower affinity to the GABAa receptor, and it shows higher anxiolytic activity than diazepam. In addition, α-casozepine does not induce a disinhibition state in rat, which has been observed by administration of diazepam. Therefore, it could be concluded that the α_{s1}-casein hydrolysate may exert anxiolytic activity by the different mechanism of diazepam. There is a possibility that smaller peptides derived from α-casozepine, which can be induced during *in vivo* digestion, might be responsible for anxiolytic activity in the different mechanism for diazepam. To solve these problems, bioavailability and metabolic fate of the peptides in the α_{s1}-casein hydrolysate should be examined.

Safety of α_{s1}-casein hydrolysate (mutagenicity, acute and subacute toxicities, behavioral toxicity, and teraogenicity) has been assessed. Then, α_{s1}-casein hydrolysate has been prepared on an industrial scale and formulated. The formulated α_{s1}-casein hydrolysate has been checked for its anxiolytic activity in human trials. The anxiolytic activity of α_{s1}-casein hydrolysate on humans was first evaluated by Stroop and cold pressor tests (Messaoudi et al., 2005). Healthy male volunteers ingested two capsules containing 200 mg α_{s1}-casein hydrolysate or skim milk as placebo two times in one day before the test. Two hours before the test, subjects ingested a third dose of two 200 mg-capsules. Mental stress was loaded by the Stroop test. During the Stroop test, fast-paced words representing colors appeared on a computer screen. The words were written in a color that did not always match the word. The subjects were asked to identify whether the color matched the word or not by hitting the appropriate key on the computer keyboard. Errors were signaled by an aggressive sound. Physical stress was loaded by putting the right hand into slushy ice water (5°C) for five minutes.

After successive loading of mental and physical stress, blood pressure, heart rate, and blood cortisol level were recorded. Ingestion of the α_{s1}-casein hydrolysate significantly suppressed the stress-induced increase of blood pressure and blood cortisol level. On the other hand, no changes in reactivity, arousal, heart rate, and ambulatory blood pressure were observed between the α_{s1}-casein hydrolysate and placebo groups. On the basis of these data, Messaoudi et al. (2005) concluded that the α_{s1}-casein hydrolysate has antistress efficacy in humans by oral ingestion but is not a β-blocker nor a hypotensive agent.

Subjective improvement of mental health was also examined by double-blind placebo controlled cross-over human study in ambulatory conditions through anxiety questionnaires (see pamphlet for lactium written by Santure). The questionnaire included following areas potentially affected by stress: physical and physiological areas (troubles in digestive tract, respiratory system, cardiovascular system, locomotion system, other physical symptoms), psychological area (troubles in intellectual and emotional functions), and social life (troubles in social relations). Sixty-three women volunteers showing at least one sign of stress received the formulated α_{s1}-casein hydrolysate or placebo (skimmed milk) at 150 mg/day for 30 days. The subjects answered the questionnaire on the 1st, 15th, and 30th days of each treatment period. Although significant improvement of the stress-related symptoms was observed in both groups, the effect of α_{s1}-casein hydrolysate was significantly superior to the placebo effect in five areas: digestive troubles, cardiovascular troubles, intellectual troubles, emotional problems, and social troubles. These data indicate that α_{s1}-casein hydrolysate in solution has anxiolytic activity not only in the animal model but also in healthy and stressed persons.

11.4 Other Peptides

A number of bioactive peptides have been isolated and identified from plant and animal proteins. Ohinata and associates have checked anxiolytic-like activity of the previously identified active peptides and their analogues by using the elevated plus-maze test. As shown in Table 11.1, they found some δ

TABLE 11.1

Peptides Exerting Anxiolytic Activity by Ingestion in Animal Model

Sequence	Origin	Dose for Anxiolytic Activity (mg/kg)	Other Function	Reference
YPLDLF	Spinach Rubisco (rubiscolin-6)	100	δ opioid	Hirata et al. 2007
YPFVV	β-conglycinin (soymorphin-5)	10–30	μ opioid	Ohinata et al. 2007
MRW	Rubisco (rubimetide)	1	hypotensive	Zhao et al. 2007
IY	Mushroom	30	hypotensive	Kanegawa et al., 2009
YL	Analogue peptide	0.3–3	(unknown)	Kanegawa et al., 2009

(Hirata et al., 2007) and μ opioid (Ohinata, Agui, and Yoshikawa, 2007) agonist peptides, and hypotensive peptides (Zhao et al., 2007; Kanegawa et al. 2009) have anxiolytic activity by oral administration. These peptides, especially YL, show anxiolytic activity by oral administration in animal models at low dose as shown in Table 11.1. Ohinata and associates have investigated the mechanism underlying the anxiolytic-like activity of these peptides. Soymorphin-5 (YPFVV), which is derived from soy β-conglycinin, shows anxiolytic activity and acts as a μ opioid agonist. The anxiolytic activity of the soymorphin-5 can be blocked by naloxone, μ opioid antagonist. Then they concluded that the μ opioid agonist activity plays a significant role in the anxiolytic activity of soymorphin-5 (Ohinata et al., 2007).

Analogously, rubiscolin-6 (YPLDLF), derived from spinach Rubisco, shows anxiolytic activity via δ opioid agonist activity (Hirata et al., 2007). They also demonstrated that the anxiolytic activity of the rubiscolin-6 was blocked by antagonists for dopamine D_1 receptor and σ_1 receptors in the central nervous system, whereas the rubiscolin-6 has no affinity to these receptors. Then they speculated that the rubiscolin-6 binds δ opioid receptor, which may successively increase dopamine and σ_1 ligand, and activate D_1 receptor and σ_1 receptor, respectively. In addition, they demonstrated a novel anxiolytic pathway via activation of the proglastandine D_2 system (Zhao et al., 2007). Rubimetide (MRW) is derived from spinach Rubisco and shows hypotensive and anxiolytic activities. The anxiolytic activity of rubimetide can be blocked by an antagonist to the DP_1 receptor, a subtype of the proglastandine D_2 receptor, whereas the rubimetide has no affinity to the DP_1 receptor. On the basis of these data, they concluded that the rubimetide increases proglastandine D_2, followed by activation of the DP_1 receptor and successively downstream adenosine A_{2A} and GABA receptors, consequently inducing anxiolytic activity.

As the hypotensive peptide such as rubimetide has anxiolytic activity, the same group has checked other hypotensive peptides, especially dipeptides for anxiolytic activity. They found that IY, a hypotensive peptide shows anxiolytic activity at 30 mg/kg. By using synthetic peptides analogous to IY, they found a strong anxiolytic dipeptide, YL, which shows the anxiolytic activity at 0.3–3 mg/kg, a dose equivalent to diazepam. YL successively activates serotonin 5-HT$_{1A}$ receptor, dopamine D$_1$ receptor, and GABAa receptor, whereas YL has no affinity to these receptors. On the other hand, YL does not bind δ and μ opioid receptors and DP1 receptor. On the basis of these findings, they concluded that YL might increase the serotonin level, which triggers receptor activations and anxiolytic activity.

11.5 Conclusion and Future Prospects

It is well known that some amino acids and amino acid-related components show anxiolytic activity and provide good sleep. As described above, α$_{s1}$-casein hydrolysate shows anxiolytic activity not only in animal models but also in healthy and stressed human subjects by oral ingestion. In those human studies, skimmed milk, which has similar amino acid composition to the hydrolysate, was used as placebo. Although the placebo also showed significant improvement in mental health, significantly greater improvement was observed by ingestion of the α$_{s1}$-casein casein hydrolysate in comparison to the placebo. In addition, Ohinata and associates have demonstrated that small doses of purified peptides, YL (Kanegawa et al., 2009), and others as shown in Table 11.1, show significant anxiolytic activity in animal models. These facts clearly indicate that some peptides have anxiolytic activity beyond the amino acid source. The series of pharmacological studies by Ohinata and associates using antagonists to receptors associated with opioid and nerve systems have hit on the mechanism of anxiolytic activity. They successfully demonstrated that some food-derived peptides exert anxiolytic activity by activating the proglastandine D$_2$ receptor, serotonin 5-HT$_{1A}$ receptor, dopamine D$_1$ receptor, and GABAa receptor. On the other hand, they found the anxiolytic peptides have no affinity to these receptors. They suggested that the peptide might exert anxiolytic activity by increasing endogenous ligands for these receptors. Detailed mechanisms underlying the increase of the ligands still remain unclear.

The studies mentioned above demonstrate food casein hydrolysate in relatively low dosage can relieve stress in humans and the hydrolysate has been formulated and made commercially available. There is a possibility that other food protein hydrolysates also have stronger anxiolytic activity. Alternatively, a peptide fraction with higher anxiolytic activity could be prepared by a large-scale fractionation procedure as described in Chapter 10 of Sato and Hashimoto (2012). The active peptide fraction from food protein hydrolysate could be easily applied to humans in comparison with the chemically synthesized one. However, it might be difficult to identify really active peptides by ingestion, as most of the peptides in the hydrolysate are degraded during digestion.

The sequences of YL and other anxiolytic peptides listed in Table 11.1 can be found in many food proteins. There is a possibility that ingestion of these food proteins and their partial hydrolysates shows anxiolytic activity by releasing of YL and other anxiolytic peptides. However, there are no available data on the occurrence of these peptides in the human blood system after ingestion of food proteins and hydrolysates. Alternatively, synthesized YL could be used as an anxiolytic medicine or food ingredient/ additive after regulatory clearance. YL can be easily synthesized by chemical methods, however, extensive studies are necessary to demonstrate purity and safety of the chemically synthesized YL before application to humans. Recently, a method synthesized dipeptide from nonprotected amino acids by using L-amino acid ligase (Yagasaki and Hashimoto, 2008). If YL can be synthesized by this approach, it could be easily applied to human trials.

References

Abdou, A.M., Higasgiguchi, S., Horie, K., Kim, M., Hatta, H., and Yokogoshi, H. 2006. Relaxation and immunity enhancement effects of gamma-aminobutyric acid (GABA) administration in humans. *Biofactors* 26: 201–208.

Brezinova, V. and Oswald, I. 1972. Sleep after a bedtime beverage. *Br. Med. J.* 2: 431–433.

Drago, F., Valerio, C., D'Agata, V., Astuto, C., Spadaro, F., Continella, G., and Scapagnini, U. 1988. Pyroglutamic acid improves learning and memory capacities in old rats. *Funct. Neurol.* 3: 137–143.

Gottesmann, C. 2002. GABA mechanisms and sleep. *Neuroscience* 111: 231–239.

Grioli, S., Lomeo, C., Quattropani, M.C., Spignoli, G., and Villardita, C. 1990. Pyroglutamic acid improves the age associated memory impairment. *Fundam. Clin. Pharmacol.* 4: 169–173.

Hirata, H., Sonoda, S., Agui, S., Yoshida, M., Ohinata, K., and Yoshikawa, M. 2007. Rubiscolin-6, a δ opioid peptide derived from spinach Rubisco, has anxiolytic effect via activating σ1 and dopamine D1 receptors. *Peptides* 28: 1998–2003.

Inagawa, K., Hiraoka, T., Kohda, T., Yamadera, W., and Takahashi, M. 2006. Subjective effects of glycine ingestion before bedtime on sleep quality. *Sleep Biol. Rhythms* 4: 75–77.

Jacob, T.C., Moss, S.J., and Jurd, R. 2008. GABAA receptor trafficking and its role in the dynamic modulation of neuronal inhibition. *Nat. Rev. Neurosci.* 9: 331–343.

Kanegawa, N., Suzuki, C., and Ohinata, K. 2009. Dipeptide Tyr-Leu (YL) exhibits anxiolytic-like activity after oral administration via activating serotonin 5-HT1A, dopamine D1 and GABAa receptors in mice. *FEBS Lett.* 584: 599–604.

Kendell, S.F., Krystal, J.H., and Sanacora, G. 2005. GABA and glutamate systems as therapeutic targets in depression and mood disorders. *Expert Opin. Ther. Targets* 9: 153–168.

Kimura, K., Ozeki, M., Juneja, L.R., and Ohira, H. 2007. L-Theanine reduces psychological and ohysiological stress responses. *Biol. Psychol.* 74: 39–45.

Komatsuzaki, J., Shima, J., Kawamoto, S., Momose, H., and Kimura, T. 2005. Production of γ-aminobutyric acid (GABA) by *Lactobacillus paracasei* isolated from traditional fermentated foods. *Food Microbiol.* 22: 497–504.

Krystal, J.H., Sanacora, G., Blumberg, H., Anand, A., Charney, D.S., Marek, G., Epperson, C.N., Goddard, A., and Mason, G.F. 2002. Glutamate and GABA systems as targets for novel antidepressant and mood-stabilizing treatments. *Mol. Psychiatry* 7: S71–80.

Leonardo, E.D. and Hen, R. 2008. Anxiety as a developmental disorder. *Neuropsychopharmacology* 33: 134–140.

Lu, K., Gray, M.A., and Oliver, C. 2004. The acute effects of L-theanine in comparison with alprazolam on anticipatory anxiety in humans. *Hum. Psychopharmacol.* 19: 457–465.

Meisel, H. 1997. Biochemical properties of regulatory peptides derived from milk proteins. *Biopolymers* 43: 119–128.

Messaoudi, M., Lalonde, R., Schroeder, H., and Desor, D. 2009. Anxiolytic-like effects and safety profile of a tryptic hydrolysate from bovine alpha s1-casein in rats. *Fundm. Clin. Pharmacol.* 23: 323–330.

Messaoudi, M., Lefranc-Millot, C., Desor, D., Demagny, B., and Bourdon, L. 2005. Effects of a tryptic hydrolysate from bovine milk α_{s1}-casein on hemodynamic responses in healthy human volunteers facing successive mental and physiological stress situations. *Eur. J. Nutri.* 44: 128–132.

Miclo, L., Perrin, E., Driou, A., Papadopoulos, V., Boujad, N., Vandersse, R., Boudier, J.F., Desor D., Linden, G., and Gaillard, J.-L. 2001. Characterization of αs1-casein with benzodiazepine-like activity. *FASEB J.* Published online June 8. 1096/fj.00-685fje.

Nathan, P.J., Lu, K., Gray, M., and Oliver, C. 2006. The neuropharmacology of L-theanine (N-ethyl-L-glutamine): A possible neuroprotective and cognitive enhancing agent. *J. Herb Pharmacother.* 6: 21–30.

Nemeroff, C.B. 2003. The role of GABA in the pathophysiology and treatment of anxiety disorders. *Psycholpharmacol. Bull.* 37: 133–1346.

Nobre, A.C., Rao, A., and Owen, G.N. 2008. L-Theanine, a natural constituent in tea, and its effect on mental state. *Asia Pac. J. Clin. Nutr.* 17: 167–168.

Ohinata, K., Agui, S., and Yoshikawa, M. 2007. Soymorphins, novel opioid peptides derived from soy β-conglucinin β-subunit have anxiolytic activities. *Biosci. Biotechnol. Biochem.* 71: 2618–2621.

Rot, M., Mathew, S.J., and Channey, D.S. 2009. Links neurobiological mechanisms in major depressive disorder. *CMAJ* 180: 305–313.

Ruhe, H.G., Mason, N.S., and Schene, A.H. 2007. Mood is indirectly related to serotonin, norepinephrine and dopamine levels in humans: A meta-analysis of monoamine depletion studies. *Mol. Psychiatry* 12: 331–359.

Santure, M. Anti-stress efficacy of lactium® on stressed women. Retrieved July 19, 2010 from http://www.lactiumusa.com/pdf/restudy/anti-stess-efficacy.pdf.

Sato, K. and Hashimoto, K. 2012. Large-scale fractionation of biopeptides. In *Food proteins and peptides: Chemistry, functionality, interaction, and commercialization.* N.S. Hettiarachchy, K. Sato, and M. Marshall, eds. Boca Raton, FL: CRC Press.

Short, J.L., Ledent, C., Drago, J., and Lawrence, A.J. 2006. Receptor crosstalk: Characterization of mice deficient in dopamine D1 and adenosine A2A receptors. *Neuropsychopharmacology* 31: 525–535.

Tachiki, T., Okada, Y., Ozeki, M., Okubo, T., Juneja, L.R., and Yamazaki, N. 2008. Process for producing theanine. United States Patent US 7,335,497 B2.

Violle, N., Messaoudi, M., Lefranc-Millot, C., Desor, D., Nejdi, A., Demagny B., and Schroeder, H. 2006. Ethological comparison of the effects of a bovine alpha s1-casein tryptic hydrolysate and diazepam on the behaviour of rats in two models of anxiety. *Pharmacol. Biochem. Behav.* 84: 517–523.

Weeks, B.S. 2009. Formulations of dietary supplements and herbal extracts for relaxation and anxiolytic action. *Med. Sci. Monit.* 15: 256–262.

Yagasaki, M. and Hashimoto, S. 2008. Synthesis and application of dipeptides: Current status and perspectives. *Appl. Microbiol. Biotechnol.* 81: 13–22.

Yamada, T., Tershima, T., and Honma, H., Nagata, S., Okubo, T., Juneja, L.R., and Yokogoshi, H. 2008. Effects of theanine, a unique amino acid in tea leaves, on memory in a rat behavioral test. *Biosci. Biotechnol. Biochem.* 72, 1356–1359.

Yamada, T., Tershima, T., and Kawano, S. 2009. Theanine, gamma-glutamyl-ethyl-amide, a unique amino acid in tea leaves, modulaters neurotransmitter concentrations in the brain striatum interestitium in conscious rats. *Amino Acids* 36: 21–27.

Yamadera, W., Imagawa, K., Chiba, S., Bannai, M., Takahashi, M., and Nakayama, K. 2007. Glycine ingestion improves sleep quality in human volunteers, correlating polysomnographic changes. *Sleep Biol. Rhythms* 5: 126–131.

Young, S.N. 2007. How to increase serotonin in the human brain without drugs. *J. Psychiatry Neurosci.* 32: 394–399.

Young, S.N. and Leyton, M. 2002. The role of serotonin in human model and social interaction. Ingestion from altered tryptophan level. *Pharmacol. Biochem. Behav.* 71: 857–865.

Zhao, H., Ushi, H., Ohinata, K., and Yoshikawa, M. 2007. Rubimetide (Met-Arg-Trp) derived from Rubisco exhibits anxiolytic activity via the DP1 receptor in male ddY mice. *Peptides* 29: 629–632.

12

Proteins and Peptides as Anticancer Agents

Harekrushna Panda, Aruna S. Jaiswal, and Satya Narayan

CONTENTS

12.1 Introduction

Cancer is a disease manifested by uncontrolled cell division that presents over 100 distinct clinical pathologies (Kufe et al., 2003). The development of an effective therapy for such a broad spectrum of disease states represents a unique scientific challenge. The conventional cytotoxic therapies such as chemotherapeutic agents and radiation, which cause DNA damage in actively dividing cells, were intended to selectively kill cancer cells while having limited effect on normal cells. Unfortunately, these cytotoxic agents, although effective in managing certain types of cancer, were limited in their utility due to their toxicity to normal dividing cell populations resulting in adverse side effects. Radiation therapy is relatively precise and is used to achieve local control, whereas chemotherapy exerts a systemic effect and is used in a broad array of cancer treatments. However, both therapies have low therapeutic indices and are often highly toxic with a broad spectrum of severe side effects. Another major limitation of successful cancer treatment with both therapies is the development of resistance, which is currently an important clinical problem. Patients who have a tumor relapse usually present with tumors that are more resistant to chemotherapy than the primary tumor. An alternative to cytotoxic therapies is immunotherapy, which aims to manipulate the immune system to create a hostile environment for cancer cells in the body. This is generally achieved through the delivery of biological response modifiers either by genetic modifications of immune cells or by their adoptive transfer, which provide the host with activation signals missing in cancer cells or with activated cells that have antitumor activity. However, all these therapeutic approaches have so far been proved to be inefficient. Moreover, immunotherapies are associated with other common problems, such as adverse toxicity, reversible autoimmunity, tissue penetration, and easy clearance. The associated toxicities and the limited success of traditional cancer treatments in maximizing cure rates has urged the development of new innovative therapeutic strategies based on the emerging knowledge in tumor biology and host–tumor interaction.

Tumor cells are characterized by specific behaviors such as a limitless replicative potential, self-sufficiency with signals, avoidance of cell death, and sustained angiogenesis along with their ability to evade the attack by the immune system. For developing a successful cancer therapy the above-said characteristic features of the tumor cells are to be exploited. In terms of combating tumor growth, attention has recently shifted to the development of new anticancer strategies such as peptide-based therapies which utilize bioactive peptides that have the potential to be nongenotoxic, genotype-specific alternatives, or adjuvant to the current regimen of cancer therapies. Now peptides have become the prime candidate for cancer treatment due to the possession of certain key advantages over alternative chemotherapy molecules. In contrast to most small molecule drugs, peptides have high affinity, strong specificity for targets, and low toxicity, whereas in contrast to chemotherapeutics antibodies, they have good penetration of tissues because of their small size (Kaplan, Morpurgo, and Linlal, 2007; Mader and Hoskin, 2006; Ruegg et al., 2006; Janin, 2003). It has been reported that some peptides show promising anticancer effects by direct injection into the animal and cell culture system. On the other hand, there is limited information on the anticancer effect of peptides by ingestion.

This chapter compiles the therapeutic prospects and possible mechanisms of action for several natural anticancer peptides such as necrotic peptides, apoptotic peptides, function-blocking peptides, antiangiogenic peptides, and immunostimulating peptides in the context of their ability to induce tumor regression in order to stimulate peptide-based anticancer food.

12.2 Source and Mechanism of Action of Anticancer Peptides

The sources and target for anticancer peptides are diverse and can be categorized into the following main categories (see Figure 12.1).

12.2.1 Necrotic Peptides

Necrosis-inducing peptides are expressed in a wide variety of species including insects, fish, amphibians, and also mammals. A group of necrotic peptides called magainins (Zasloff, 1987; Baker et al., 1993; Cruciani et al., 1991) have been identified from the skin of Australian frogs and toads (Papo et al., 2003; Rozek et al., 2000; Papo et al., 2002). These peptides have anticancer properties

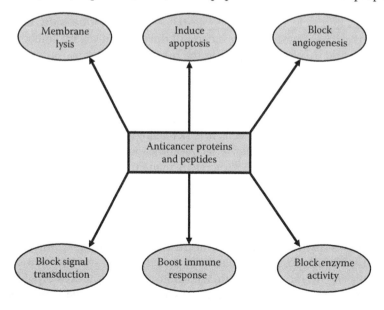

FIGURE 12.1
A generalized schematic representation of modes of action of protein- and peptide-based anticancer drugs.

against all classes of human cancers, including leukemias, melanomas, and cancer of the lung, colon, central nervous system (CNS), ovary, kidney, prostate, and breast (Apponyi et al., 2004). Similarly, cercopins A and B are peptides found in mammals and various insects, for example, in the hemolymph of the giant silk moth, *Hyalophora Ceropia*, and are effective against several lymphomas and leukemias *in vitro* (Moore, Devine, and Bibby, 1994). Another necrosis-inducing peptide group includes the bee venom melitin (Habermann and Jentsch, 1967), tachyplesin II isolated from the horseshoe crab (Dimarcq et al., 1998; Mai et al., 2001), human neutrophil defensins (Lichtenstein et al., 1986; Andreu and Rivas, 1998), and the human LL-37 (Frohm et al., 1997; Johansson et al., 1998). Most of the above studies are based on *in vitro* (cell culture-based) experiments and their effect on animal studies are lacking.

Necrosis-inducing peptides are groups of cell membrane lytic peptides that target the cell membrane and cause its lysis, thereby killing the cell. Generally the outer membranes of the cancer cells overexpress negatively charged phosphatidylserine and O-glycosylated mucins which make them carry a slightly greater net negative charge in comparison to that on normal eukaryotic cells (Mader and Hoskin, 2006). This favors their selective attack by the positively charged necrosis-inducing peptides, which are usually short, cationic, form amphipathic structure in nonpolar solvents, and show very little sequence similarity among themselves. When these peptides bind to the plasma membrane they disrupt its integrity either by micellization or through pore formation resulting in cell lysis and death of cancer cells (Mader and Hoskin, 2006; Papo and Shai, 2005). Necrotic peptides have a higher degree of selectivity for neoplastic cells than traditionally used chemotherapy drugs. These peptides also kill the chemoresistant cells by membrane lysis, and do not depend upon the proliferative status of the target.

12.2.2 Apoptotic Peptides

Cancer cells have an elevated apoptotic threshold and peptides that are able to induce apoptosis in tumor cells are increasingly seen as promising candidates for the development of new effective anticancer therapeutics. Several cationic peptides such as bovine lactoferricin, magainin2, $hCAP_{109-135}$ (C-terminal domain of human cationic antibacterial protein of 18 kDa, CAP18), and BMAP-28 from bovine myeloid cathelicidin, have been found to induce mitochondria-dependent apoptosis (Mader et al., 2007; Lehmann et al., 2006; Okumura et al., 2004; Risso et al., 2002). After passing through the plasma membrane these peptides disturb the integrity of the negatively charged mitochondrial membranes, which results in release of cytochrome c into the cytosol. The release of cytochrome c from damaged mitochondria induces Apaf-1 oligomerization, caspase-9 activation, and subsequently the conversion of procaspase-3 to caspase-3, which ultimately causes the apoptosis (Cory and Adams, 1998). Several apoptotic peptides have been identified; for example, Ellerby and colleagues have reported on a cationic membrane-active antimicrobial peptide (KLAKLAKKLAKLAK) fused to a CNGRC homing domain that exhibited antitumor activity by targeting mitochondria and triggering apoptosis (Ellerby et al., 1999). Similarly, granulysin, a cytolytic molecule released by cytotoxic T lymphocyte (CTL) via granule-mediated exocytosis, was shown to depolarize the mitochondrial membrane potential and to induce tumoral cell death via the mitochondrial pathway of apoptosis (Pardo et al., 2001). The heptadeca cationic antimicrobial peptide tachyplesin, which is conjugated to RGD, an integrin homing domain (Chen et al., 2001; Park et al., 1992), interacts with mitochondrial membranes of cancer cells and induces apoptosis in tumor cells. Likewise, peptides from several species in the Euphobiaceae family and *Abrus* agglutinin induce mitochondrial apoptosis in cultured cells (Bhutia et al., 2008; Uthaisang et al., 2004). However, animal-based studies demonstrating the effect of these apoptotic peptides are lacking.

12.2.3 Function-Blocking Peptides

Many peptides have been identified that block other cancer-related processes such as receptor interaction, cell adhesion, and metastasis.

12.2.3.1 Receptor-Interacting Peptides

Several tumor-cell receptors have been targeted by small peptides for development of anticancer therapy. The important hormone receptors in this regard are: somatostatin receptors, gastrin-releasing peptide (GRP) receptor, and the luteinizing hormone-releasing hormone (LHR) receptor. Hormone receptor analogues exhibit strong anticancer activity, both *in vitro* (in cultured cells) and *in vivo* (given to animal as injections), on a wide range of tumor models, and some are being tested in clinical trials (Schally and Nagy, 2003). The hormone bradykinin (BK) stimulates cancer-cell growth directly and also promotes cancer migration and invasion by stimulating the activity of matrix metalloprotease (MMP) enzymes. The peptide antagonist of bradykinin has been developed which blocks the growth, migration, and metastasis of cancerous cells, thereby offering the advantage of combination therapy in a single compound (Stewart et al., 2005). The aminopeptidase N (CD13), a cell-surface protease, which is expressed in a variety of cells has an important role in the process of angiogenesis. It is upregulated in endothelial cells in mouse and human tumors. A tripeptidic sequence, NGR, which was identified by *in vivo* selection of phage-display libraries as the tumor-targeting peptide of aminopeptidase N, inhibits angiogenesis (Pasqualini et al., 2000). Vascular endothelial growth factor (VEGF) is a well-characterized positive regulator of angiogenesis and, as such, constitutes another major target for tumor therapy. By screening a peptide library, several members of a group of peptides have been identified which exhibited antitumor properties by blocking interaction between VEGF and the VEGF receptor (An et al., 2004; Starzec et al., 2006).

12.2.3.2 Antiangiogenic Peptides

Angiogenesis, new blood vessel formation from established veins, is critical for neoplastic growth and metastasis but is essentially quiescent in adults (Fidler and Ellis, 1994; Folkman, 1995). New therapies in cancer treatment are focusing on multifaceted approaches to starve and kill the tumors by blocking the angiogenesis process. Several endogenous inhibitors of angiogenesis have been discovered, many of which are fragments of naturally occurring components of extracellular matrix (ECM) and hematostasis pathways (Staton and Lewis, 2005; Nyberg, Xei, and Kalluri, 2005). Inhibitors of angiogenesis can interfere with any of several steps in the formation of new blood vessels, including the degradation of basement membrane of the endothelium, endothelial cell proliferation, endothelial cell migration, and formation of new blood vessels. The majority of endogenous inhibitors of angiogenesis are large complex fragments that are too big for tissue penetration and easy clearance and too costly to be produced in sufficient quantities for therapeutic use. Therefore researchers are now focusing on identifying small peptides that mimic the effects of the larger endogenous proteins. For example, a heptapeptide from the N-terminal domain of thrombospondin-1 (TSP-1) has been identified to be responsible for its antiangiogenic activity and a derivative of this peptide, designated ABT-510, is currently undergoing phase II clinical trials (Ebbinghaus et al., 2007). Similarly, recently it has been shown that a 27-residue peptide corresponding to the N-terminal end of endostatin mimics the entire set of antitumor, antimigration, and antipermeability activities exhibited by endostatin (Tjin Tham Sjin et al., 2005). Alphastatin™ a 24-residue peptide derived from the E-fragment of fibrinogen (FgnE) mimics many antiangiogenic activities of FgnE (Staton et al., 2004). Antineoplastic urinary protein (ANUP), a 32-kDa protein normally secreted in human urine, has both antiproliferative and antiangiogenic activities and a synthetic peptide from the N-terminus end of this protein exhibits the same effects as does a full-length protein (Folkman, 1995; Hehir et al., 2004).

12.2.3.3 Peptides That Bind to Cell-Adhesion Proteins

Inhibitors of cell adhesion alter the properties of extracellular matrix (ECM) proteins, which play a central role in metastasis and angiogenesis (Ruoslahti and Pierschbacher, 1987; Hehir et al., 2004). For example, a tripeptidic RGD integrin-recognition motif was able to inhibit $\alpha V \beta 3$ integrin-mediated cell adhesion to ECM ligands, thereby blocking tumor invasion and angiogenesis. Furthermore, cilengitide, a cyclic form of RGD, has shown improved potency compared with that of the noncyclic peptide and is currently undergoing clinical trial (Hariharan et al., 2007) and it is worth mentioning that these anti-cell-adhesion peptides were effective when injected into the experimental animals. Similarly to target integrin $\alpha 5 \beta 1$, ATN-161 (Ac-PHSCN) was designed as a modified version of the PHSRN so-called "synergy region" of fibronectin, a region known to potentiate the binding of fibronectin to α5β1. ATN-161 acts by interfering with this interaction (Livant et al., 2000).

A second example concerns the laminin receptor, which is a high-affinity laminin-binding protein that is overexpressed on the tumor cell surface in a variety of cancers. Peptide YIGSR blocks the laminin-1-binding domain on the laminin-1 receptor, thereby inhibiting the adherence of cancer cells to the extracellular stroma and, consequently, tumor migration and evasion (Ardini et al., 2002). Finally, neural (N)-cadherins, which comprise a family of trans-membrane glycoproteins, mediate homophilic cell–cell interactions through a conserved His-Ala-Val (HAV) sequence. Linear peptides harboring this motif are capable of inhibiting a variety of cadherin-dependent processes. For example, a cyclic peptide N-Ac-CHAVC-NH$_2$ named ADH-1 (Exherin) perturbs the N-cadherin-mediated endothelial cell interactions necessary for angiogenesis and results in progressive apoptotic death (Erez et al., 2004).

12.2.3.4 *Antimitotic Peptides (Tubulin or Actin Function Interfering Peptides)*

Microtubules and actin filaments are the cytoskeletal protein polymers involved in cell growth and division. The polymerization of α-tubulin dimers in microtubules is the target of many important antitumor drugs. It is the drug-induced alteration of tubulin polymerization or depolymerization dynamics which hampers the correct occurrence of microtubules and is the underlying mechanism of the antitumor action (Correia, 1991; Jordan and Wilson, 1998). Similarly other compounds owe their cytotoxicity to the disorganization of actin filaments that are made from the polymerization of actin monomers. Several peptides such as moroidin, celogentin, tubulysin D, and vitilevuamide have been identified as potent inhibitors of microtubule dynamics. The linear peptides dolastatin 10 (Bai, Pettit, and Hamel, 1990) and dolastatin 15 (Bai et al., 1992) were found to act on tubulin polymerization. Their structures led to the synthesis of cematodin (de Arruda et al., 1995; Jordan et al., 1998) and TZT-1027 (Kobayashi et al., 1997) which are today, along with dolastatin 10, at different stages of clinical development (Madden, Tran, and Beck, 2000; Supko, Tran, and Beck, 2000; Vaishampayan et al., 2000).

The phallotoxins, such as phalloidin, are toxic peptides isolated from the poisonous mushroom *Amanita phalloides* (Wieland, 1968) that also interact with actin by preventing the depolymerization of F-actin into G-actin (Miyamoto et al., 1986). Several other peptides such as jasplakinolide, chrondamide, and dolastatin 11 have been reported to disrupt the proper function of actin. For example, dolastatin 11 induces hyperpolymerization of purified actin (Bai et al., 2001). However, the pharmaceutical development of this class of compounds was discontinued as pulmonary toxicity was found for jasplakinolide.

12.3 Enzyme Inhibitors

Several peptides have been identified that block the functions of key enzymes thereby preventing carcinogenesis.

12.3.1 Blocking Histone Deacetylase

Histone deacetylase catalyzes the removal of the acetyl group from the ε-amine of lysine residues present near the amino terminus of nucleosomal histones. Inhibition of this process was shown to be a target for cytotoxic compounds, possibly by the prevention of DNA unwinding from around the histone prior to its transcription (Yoshida et al., 2001). The antitumor peptide FR901228 (or FK 228) was isolated from *Chromobacterium violaceum* (Ueda et al., 1994a,b) and was later found to be also an inhibitor of histone deacetylase (Nakajima et al., 1998). Its pharmacological and antiangiogenic properties made it a good candidate for clinical development (Piekarz et al., 2001). Moreover, lunasin, a remarkable soybean-extracted 43-mer peptide, or fragments as short as the 17-mer Cys-Glu-Lys-His-Ile-Met-Glu-Lys-Ile-Gln-Gly-Arg-Gly-Asp-Asp-Asp-Asp, were recently found to bind deacetylated histones and thus block cell proliferation by preventing their acetylation (Galvez et al., 2001). On the basis of this mechanism, lunasin significantly suppresses foci formation of mouse fibroblasts cells C3H10T1/2 induced by the chemical carcinogens; DMBA and MCA (Galvez et al., 2001), and also viral oncogenes (Lam, Galvez, and de Lumen, 2003). The effect of lunasin on cancer prevention has been also been demonstrated using an *in vivo* mouse model. In this model, dermal application of lunasin at 250 μg/week reduced skin tumor incidence in SENCAR mice treated with DMBA and TPA by approximately 70% compared with the untreated control (Galvez et al., 2001). This treatment also reduced the tumor multiplicity (tumor/mouse), and delayed by two weeks the appearance of papilloma in the mice relative to the untreated control.

The affinity of lunasin for hypoacetylated chromatin and its inhibitory effect on histone acetylation is relevant to the proposed epigenetic mechanism of action using the E1A-Rb-HDAC model (de Lumen, 2005). See Chapter 13 by Hsieh and colleagues for detailed structure and action of lunasin.

12.3.2 Blocking Protein Kinases

Protein kinases are commonly activated in cancer cells and have become attractive targets for anticancer therapy. The kinase family can be divided into those enzymes specific to tyrosine phosphorylation (tyrosine kinases) and those specific to serine or threonine phosphorylation (serine/threonine kinases). Small peptides have been developed to inhibit the interaction of kinases with their substrates (Mendoza et al., 2005). For example, a peptide named P15 that abrogates casein kinase 2 (CK2)-mediated phosphorylation elicits a strong antitumor effect without any toxic effect when systemically injected in both human and murine tumors implanted in mice (Perera et al., 2008).

12.3.3 Blocking Functions of Proteases

Tumors possess lytic machinery comprising different proteolytic enzymes such as cathepsins, urokinase-type plasminogen activator (uPA), and matrix metalloproteinases (MMPs), which facilitate tumor invasion and metastasis. Cyclic peptides containing the sequence His-Trp-Gly-Phe (HWGF) selectively inhibit metalloproteases MMP-2 and MMP-9 and have been shown to suppress migration and invasion of tumors (Koivunen et al., 1999). Urokinase-type plasminogen activator (uPA) binds with high affinity to its specific cell-surface receptor (uPAR, also known as CD87) through a well-defined sequence within the N-terminal region of uPA (uPA_{19-31}). The high-affinity synthetic uPA-derived cyclic peptides WX-360 and WX-360-Nle competitively interfere with the uPA–uPAR interaction and have been reported to generate an effective reduction of tumor growth in a human ovarian cancer xenograft model (Sato et al., 2002). Finally, a peptidic derivative Z-Phe-Gly-NHO-Bz is an inhibitor of the cysteine protease cathepsins and induces apoptosis in human cancer cells (Zhu and Uckun, 2000). Interestingly, soybean has the Bowman–Birk protease inhibitor (BBI) which inhibits both trypsin and chymotrypsin and has been found to inhibit carcinogenesis in the colon, esophagus, liver, lungs, and also in the oral cavity (Losso, 2008).

12.4 Immunostimulatory Peptides

In most of the cancers, the host immune response is very weak and often suppressed. The tumor microenvironment contains a wide array of immune-suppressive molecules that prevent efficient eradication of the cancerous cells by the immune defense system (Zitvogel, Tesniere, and Kroemer, 2006). The host response to cancer therapies might therefore be stimulated by the simultaneous administration of immunoadjuvants, which activate the type-1 innate and acquired immune system (Dredge et al., 2002). Many small peptides of microbial origin possess a nonspecific immunostimulatory response (Dutta, 2002). Muramyl dipeptide (MDP) from mycobacterium has long been known to act as a vaccine adjuvant and to have potential in cancer therapy. This antitumor effect of MDP was attributed to the release of proinflammatory molecules by activation of macrophages, NK cells, and lymphocytes (Kotani et al., 1983). Moreover, MDP analogues and MDP derivatives have been synthesized in efforts to solicit improved activity. Recently, a synthetic

MDP analogue named romurtide has shown tumoricidal properties upon oral administration (Ueda and Yamazaki, 2001).

Other well-studied immuno-enhancing peptides, such as bestatin and FK-565 from Streptomyces sp., are also known to possess antitumor properties (Schorlemmer, Bosslet, and Sedlacek, 1983; Wang et al., 1989). Potent immunostimulatory peptides from plasma proteins such as globulin have been reported. For example, the naturally occurring tetrapeptide tuftsin from the Fc region of leukophilic immunoglobulin G (IgG) binds specifically to macrophages, monocytes, and polymorphonuclear leukocytes, and stimulates their immunomodulatory activity in tumor-suppressed hosts (Nishioka et al., 1981). Similarly, peptide RS-83277 derived from C-reactive protein has been found to activate tumoricidal properties of murine and human monocytes and macrophages *in vitro* and to produce antitumor effects in experimental tumor models (Barna et al., 1993). The antitumor activities were found to be mediated through production of proinflammatory cytokines and monocyte chemoattractant protein by macrophages (Barna et al., 1996).

Several other immunoactive peptides have been isolated from a variety of sources. For example, tyroserleutide (YSL) is a tripeptide degradation product from porcine spleen that shows macrophage-mediated antitumor properties in hepatocarcinoma. It also inhibits the growth, invasion, and adhesion of tumors (Yao et al., 2006). YSL is currently undergoing clinical trial as a novel drug for the treatment of liver cancer (Yao et al., 2005). Another peptide, myelopeptide-2 (MP-2) from porcine bone marrow, has been found to stimulate T cells. MP-2 inhibits tumor growth by means of its ability to restore the functional activity of T cells that are suppressed in tumors, and this represents a promising approach for an antitumor therapy (Strelkov, 1996). Similarly, a cytokine-like material obtained from the hemolymph of the insect Calliphora vicina challenged with bacteria has been found to stimulate natural killer cells to produce interferon-γ (INF-γ), resulting in enhanced antitumor activity in mice (Chernysh et al., 2002).

12.5 Cancer-Preventive Peptides from Natural Food Sources

Dietary proteins are known to carry a wide range of nutritional, functional, and biological properties. The specific biological properties of some of the proteins make them potential ingredients in functional or health-promoting food. Many of these specific biological properties are attributed to physiologically active peptides encrypted in protein molecules. The richest sources of such bioactive peptides are milk and eggs. However, they are also found in meat and plant products. These peptides are inactive within the sequence of the parent protein and are released during gastrointestinal digestion by digestive enzymes such as trypsin or even by the microbial enzymes of the gut flora, and also during food processing by fermentation and protease treatment. These peptides usually consist of 3–20 amino acid residues and depending on the amino acid sequences, these bioactive peptides exhibit different functions *in vivo*. Bioactive peptides may be liberated from food proteins throughout the whole intestine, and they can exhibit their function in the small and large intestine. It is well known that di- and tripeptides are easily absorbed in the intestine (Adibi, 1976), however, little is known about the absorption of bioactive peptides with larger molecular weights. Most known bioactive peptides are not absorbed in the intestinal tract. Hence, many proteins and peptides may either act directly in the intestinal tract or via receptor and cell signaling in the gut. Several bioactive peptides from milk have been identified with antihypertensive, antithrombic, antimicrobial, mineral-binding, and immunomodulatory effects (Meisel, 1997). Similarly many bioactive peptides having antihypertensive and antioxidative properties have been identified from protease hydrolyzed meat from chicken, bovine, and porcine (Arihara and Ohata, 2008). Proteins isolated from different plant sources including wheat germ (Borel et al., 1989) and wheat flour (Tani, Ohishi, and Watanabe, 1994) have shown lipase inhibitory action *in vitro* whereas proteins from water-extracted defatted rice bran have been reported to reduce the plasma triglyceride levels in rats when fed orally (Tsutsumi et al., 2000).

Although a substantial amount of literature information is available on anticancer peptides for medicinal application, limited information has been published on food-based anticancer peptides and their potential use as nutraceuticals and pharmaceuticals. This is perhaps due to the fact that the food-based anticancer proteins and peptides go through a series of processes such as digestion and assimilation and we do not know what the final fate is of these peptides after these processes (see Chapter 6). The active compound may be degraded during digestion or not attain the appropriate concentrations in blood and target tissues that are required for acting significantly. Certain processing procedures, especially heating, may influence the bioactivity and may also lead to the formation of undesired toxic, allergic, or carcinogenic substances. In addition, the bioactivity may be reduced through molecular alteration during food processing or interaction with other food ingredients. The bitter taste of protein hydrolysates prevents their use as food additives. The challenge for food technologists will be to develop functional foods and nutraceuticals without the undesired side effects of the added peptides and to retain the stability of the added peptides within the shelf-life of the product. It is hoped that in the future researchers will address these fundamental problems and we can have more food-based natural antidisease agents delivered at low cost and high efficacy.

Recently, few researchers have reported anticarcinogenic activities of rice bran (Nam et al., 2005; Kannan et al., 2008), soy peptide, lunasin, and so on. Researchers have demonstrated the feasibility of producing large quantities of natural lunasin from soybeans for animal and human studies (Jeong et al., 2003). So we can hope that this will open up the opportunity to develop nutraceuticals and pharmaceuticals based on lunasin technology which can be used as a potent weapon against the ever-increasing threat of cancers. Identification and characterization of edible bioactive peptides from natural food sources may generate the possibility of supplementing our diets with them for fighting cancer and other fatal diseases.

12.6 Structural Features of Anticancer Peptides

The anticancer peptides are diverse in their structure and can be grouped in two main categories, that is, linear peptides and cyclic peptides. The linear peptides are relatively simple in structure, and can be as short as 3 amino acid residues (tripeptide) or can be as long as 20–50 amino acid residues. However most of them are 5–10 amino acid residues long. A large number of these naturally occurring and synthetic linear peptides contain pyruglutamate as their first amino acid residue such as pGlu-Glu-Asp-Ser-Gly (Jensen et al., 1990), pGlu-Phe-Gly-NH2 (Gembistky et al., 2000), pGlu-Glu-Gly-Ser-Asp or pGlu-Glu-Gly-Ser-Asn (Paulsen et al., 1992), pGlu-Glu-Asp-Cys-Lys (Foa et al., 1987) or pGlu-His-Gly (Paulsen, 1993), and are reported for their cellular growth inhibitory properties. Similarly the N-acetylated glutamate derivative Ac-Glu-Ser-Gly-NH2 has also been found to inhibit lymphocyte growth (Liu et al., 2000). However more elaborate linear cytotoxic peptides have also been characterized such as the two linear peptides, majusculamide D and deoxymajusculamide D (Moore and Entzeroth, 1988).

The cyclic peptides are more complex in their structure. The cyclodepsipeptide dolastatin 14 containing a 14 carbon long lipophilic hydroxy acid was isolated from sea hare *Dolabella auricularia* and was found to be very active (Pettit et al., 1990). Papuamides are depsipeptides displaying a lipophilic decanedienoic side chain that have been isolated from the sponges *Theonella mirabilis* and *Thoenella swonhoei* and display powerful anti-HIV and cytotoxic properties (Ford et al., 1999). Although it is 6-hydroxylated, an homoproline residue is also present in a number of much less cytotoxic depsipeptides such as microcystilide (Tsukamoto et al., 1993), dolastatin 13 (Pettit et al., 1989), symplostatin 2 (Harrigan et al., 1999), or somamides (Nogle, Williamson, and Gerwick, 2001). Many cytotoxic thiazole-containing cyclopeptidic compounds have been isolated such as trunkamide A (Caba et al., 2001; Wipf and Uto, 1999) and apratoxin A (Luesch et al., 2002) which are notable in terms of cytotoxicity. Several depsipeptides, featuring β-amino acid, were reported such as cyclic octapeptides made of Asn-Tyr-Asn-Gln-Pro-Asn-Ser and various fatty β-amino acid, dolastatin D (Sone et al., 1993) and dolastatin 17 (Pettit et al., 1998). A remarkable series of cyclo-β-tripeptides, bearing lipophilic side chains, have been recently reported (Gademann and Seebach, 2001).

Moreover the cytotoxic kailiuns (Harrigan et al., 1997) and pitipeptolides (Luesch et al., 2001) display a fatty side chain as well as lipophilic residues. Also reported for their cytotoxicity are the lipophilic cyclic peptides axinellins (Randazzo, Dal Piaz, and Orru, 1998), cycloleonuripeptides (Morita et al., 1996), mollamide (Carroll et al., 1994), and onchidins (Fernandez et al., 1996).

12.7 Characterization of Anticancer Proteins and Peptides

Several different kinds of anticancer proteins and peptides have been identified and characterized from diverse sources both from plants and animals. The traditional method of characterization involves crude isolation of these proteins or peptides from their sources, their purification by different chromatographic methods such as high-performance liquid chromatography (HPLC), ion exchange chromatography, and immunoaffinity column chromatography followed by their identification by matrix-assisted laser ionization (MALDI) analysis. The next step involves the functional characterization of these proteins or peptides by colony forming assay or by MTT assay (Mosmann, 1983) and also by *in vivo* studies in a variety of tumor systems.

Recently, the phage display system is being widely used to characterize bioactive proteins and peptides (An et al., 2004; Doorbar and Winter, 1994). Phage display, first introduced by G. Smith in 1985, is a very effective way of producing large numbers of diverse peptides and proteins and isolating molecules that perform specific functions (Ladner, 2000; Hoogenboom et al., 1998). This technique can also be used to study protein–ligand interactions (Cesareni, 1992), receptor and antibody-binding sites (Griffiths, 1993; Winter et al., 1994), and to improve or modify the affinity of proteins for their binding partners (Burton, 1995; Neri, Petrul, and Roncucci, 1995). Phage display involves the expression of proteins, including antibodies, or peptides on the surface of filamentous phage. DNA sequences of interest are inserted into a location in the genome of filamentous bacteriophages such that the encoded protein is expressed or "displayed" on the filamentous phage surface as a fusion product to one of the phage coat proteins. Therefore, instead of having to genetically engineer proteins or peptide variants one by one and then express, purify, and analyze each variant, phage display libraries containing several billion variants can be constructed simultaneously. These libraries can then be easily used to select and purify specific phage particles bearing sequences with desired binding specificities from the nonbinding variants. A significant aspect of phage display lies in linking the phenotype of a bacteriophage-displayed peptide or protein with the genotype encoding that molecule, packaged within the same virion. This permits the selection and amplification of specific phage clones representing desired binding sequences from pools of billions of phage clones. In the case of filamentous phage, amplification is simply accomplished by infecting male *E. coli*. The genotype–phenotype linkage also permits the rapid determination of the amino acid sequence of the specific binding peptide or protein molecule by DNA sequencing of the specific insert in the phage genome.

12.8 Delivery System for Peptide-Based Drugs

The use of protein and peptides as therapeutic agents is hampered by their rapid elimination from the circulation through renal filtration, enzymatic degradation, uptake by the reticuloendothelial system (RES), and accumulation in nontargeted organs and tissues. Many anticancer drugs are designed just to kill cancer cells, often in a semi-specific fashion; the distribution of anticancer drugs in healthy organs or tissues is especially undesirable because of the potential for severe side effects. Because of the rapid elimination and widespread distribution of drugs into nontargeted organs and tissues the drug needs to be administered in large quantities, which is often not economical and is sometimes complicated by nonspecific toxicity. Therefore novel approaches, such as conjugating the peptides with various soluble polymers, incorporation into microparticulate drug carriers, and drug association with serum albumen are being used to improve drug stability and longevity, as well as to generate enhanced permeability and retention in the body (Barna et al., 1996; Torchilin and Lukyanov, 2003). Several polymers have been used for protein stabilization with varying degrees of success.

Currently, polyethylene glycol (PEG) is the most widely used polymer for the modification of proteins with therapeutic potential. Oncospar® (PEG-modified L-asparaginase) is one such drug that has been used in the United States for lymphoma and leukemia treatments since 1995. PEGylated interferon α-2b, which was recently approved in the United States by the FDA for treating chronic hepatitis C also shows potential as an anticancer agent because it can stimulate the immune system and is now undergoing clinical trials (Bukowski et al., 2002). Similarly few peptide-based drugs have been conjugated to poly(styrene-co-maleic acid anhydride; SMA) which increases the circulation time, protects them from degradation, and decreases their immunogenicity (Maeda, 2001). The neocarzinostatin–SMA conjugate, is one such drug currently approved in Japan for treating hepatoma (Maeda, Sawa, and Konno, 2001). Among particulate drug carriers, liposomes are the most extensively studied and possess the most suitable characteristics for peptide and protein encapsulation and delivery to the target cells. Currently, liposomal forms of at least two conventional anticancer drugs, doxorubicin (Gabizon, 2001) and daunorubicin (FDA, 1996) are in therapeutic use. Furthermore, new approaches to intracellular drug delivery, including the use of transduction peptides such as Antennapedia (Antp) from *Drosophila*

(Derossi et al., 1994) and Tat from human immunodeficiency virus (HIV; Schwarze et al., 1999), are being developed for efficient delivery of peptide-based anticancer drugs to the cells (Perera et al., 2008).

12.9 Production, Design, and Commercialization of Anticancer Proteins and Peptides

Many peptides and proteins have some kind of biological activity which makes them potential candidates for anticancer therapeutics. Several peptides such as somatostatin analogues octreotide, lanreotide, and vapreotide are being used to treat pituitary and gastrointestinal tumors (Froidevaux and Eberle, 2002). Endostatin and several other peptide inhibitors of angiogenesis are in different stages of clinical trial and are promising candidates for anticancer therapy (Figg et al., 2002; Kerbel and Folkman, 2002). Recently, identification of depsipeptides in several microorganisms and marine invertebrates has revealed a group of potential anticancer agents (Ballard, Yu, and Wang, 2002). The enzyme L-asparaginase, that inhibits the growth of tumors by breaking down the aminoacid asparagine which is required in much higher amounts by some type of tumor cells than the normal cells is routinely being used for the treatment of leukemia (Asselin, 1999). Nowadays, antibodies against certain cancer-specific ligands have been used as anticancer drugs. Trastuzumab and Rituxan® developed by Genentech Inc. are examples of two such drugs approved by the FDA. Trastuzumab is an antibody that has been raised against HER2/neu, the extracellular domain of the HER-2 protein expressed in certain types of breast cancer and it acts by inducing the immune-mediated responses, downregulates the HER-2 receptor, and promotes the production of cell cycle inhibitors (Baselga and Albanell, 2001). Similarly Rituxan is raised against the B-cell specific antigen CD20 and is effective against follicular lymphoma (Marshall, 2001).

The enormous untapped potential of peptides to serve as anticancer drugs has enabled the scientific community to come up with novel and better ways to combat tumors. The effectiveness of any targeted therapeutics mostly depends upon whether the appropriate drug can be delivered to the desired location in sufficient quantities and in a timely manner. Despite many technological hurdles, such as weak efficiency *in vivo* and adverse toxicity, the peptide-based anticancer drugs are promising candidates because of their smaller size, specificity for the tumor cells, lower immunogenicity, and relative ease to synthesize and modify. To increase the efficacy of the peptide drugs, new strategies such as amino acid substitution, incorporation of D isomers of amino acids into native molecules, use of peptomometics, and peptide engineering are being developed. The AFPep, D-K_6L_9, BT-510, hunter-killer peptides, and Anginex™ are a few examples of such modified peptide-based anticancer drugs (Ebbinghaus et al., 2007; Bennett et al., 2006; Papo et al., 2006; Griffioen et al., 2001; Ellerby et al., 1999). Researchers are also working on efficient peptide delivery methods to further enhance the potential of these anticancer peptides. Furthermore, combination therapies are also being developed that take advantage of multifunctional peptides or place different active peptides together, representing a novel therapeutic approach to address the problem of drug resistance. Certain tumor types overexpress specific receptors and some peptide-based receptor antagonists are being developed which are conjugated with standard anticancer drugs for efficient tumor-targeted therapy (Schally and Nagy, 2003). Advances in solid-phase peptide synthesis, recombinant DNA, and hybridoma technologies have enabled us to produce large quantities of clinical grade peptides and proteins. Our current knowledge provides some promising approaches on how to deliver peptide- and protein-based anticancer drugs into tumor and tumor cells. These advancements may lead to the development of new peptide-based anticancer drugs and treatment in the near future. Furthermore, information from the human genome project, together with advances in proteomics, promises further progress in the identification and development of peptide- and protein-based anticancer drugs. Similarly, identification and characterization of edible bioactive peptides from natural food sources along with the advancement in food processing techniques may generate the possibility of supplementing our diets with them for fighting cancer and other fatal diseases. More distant future approaches might also include various methods of gene or DNA delivery to tumor cells where the biosynthesis of proteins or peptides can be initiated that can serve as potent anticancer agents or act as vaccines against cancer.

12.10 Conclusion

Peptide-based therapies have the potential to be nongenotoxic, genotype-specific, and can be used along with the current range of traditional cancer treatments. A patient-tailored cancer-cell-directed therapeutic approach is feasible for peptide-based therapies which would have fewer side effects and could well be more effective than the current drug- or combination-based regimens. The ability of novel synthetic as well as natural anticancer peptides such as necrotic peptides, apoptotic peptides, function-blocking peptides, antiangiogenic peptides, and immunostimulatory peptides in the context of their ability to block or inhibit several important pathways crucial for tumor growth and thus induce tumor regression is a key factor in their successful use as anticancer agents. The potential of peptide and protein anticancer agents has yet to be fully explored owing to several unresolved problems concerning their delivery to the site of a tumor and also into the tumor cells. Our current understanding of the mechanisms underlying the biodistribution and the fate of protein- and peptide-based drugs has advanced to the stage where this knowhow can be exploited for effective delivery of these drugs into the cancer cells. Current approaches to improve the stability, longevity, and targeting of peptides and proteins in the body, such as their modification with various soluble polymers, incorporation into microparticular drug carriers, and enhanced permeability using transduction proteins and peptides are being developed. It is hoped that these developments along with advances in solid-phase peptide synthesis, recombinant DNA, and hybridoma technologies will enable us to produce large quantities of clinical grade peptide- and protein-based anticancer drugs for our use.

Acknowledgments

The research work in our lab is currently supported by NCI-NIH (CA-097031, CA-100247) and for Satya Narayan, Flight Attendant Medical Research Institute, Miami, Florida. We thank Mary Wall for proofreading the manuscript.

References

Adibi, S.A. 1976. Intestinal phase of protein assimilation in man. *Am. J. Clin. Nutr.* 29: 205–215.

An, P., Lei, H., Zhang, J. et al. 2004. Suppression of tumor growth and metastasis by a VEGFR-1 antagonizing peptide identified from a phage display library. *Int. J. Cancer* 111: 65–173.

Andreu, D. and Rivas, L. 1998. Animal antimicrobial peptides: An overview. *Biopolymers* 47: 415–433.

Apponyi, M.A., Pukala, T.L., Brinkworth, C.S. et al. 2004. Host-defence peptides of Australian anurans: Structure, mechanism of action and evolutionary significance. *Peptides* 25: 1035–1054.

Ardini, E., Sporchia, B., Pollegioni, L. et al. 2002. Identification of a novel function for 67-kDa laminin receptor: increase in laminin degradation rate and release of motility fragments. *Cancer Res.* 62: 1321–1325.

Arihara, K. and Ohata, M. 2008. Bioactive compounds in meat. In *Meat Biotechnology* (F. Toldra, Ed.), New York: Springer, pp. 231–249.

Asselin, B.L. 1999. The three asparaginases. Comparative pharmacology and optimal use in childhood leukemia. *Adv. Exp. Med. Biol.* 457: 621–629.

Bai, R., Friedman, S.J., Pettit, G.R., and Hamel, E. 1992. Dolastatin 15, a potent antimitotic depsipeptide derived from Dolabella auricularia. Interaction with tubulin and effects of cellular microtubules. *Biochem. Pharmacol.* 43: 2637–2645.

Bai, R., Pettit, G.R., and Hamel, E. 1990. Dolastatin 10, a powerful cytostatic peptide derived from a marine animal. Inhibition of tubulin polymerization mediated through the vinca alkaloid binding domain. *Biochem. Pharmacol.* 12: 1941–1949.

Bai, R., Verdier-Pinard, P., Gangwar, S. et al. 2001. Dolastatin 11, a marine depsipeptide, arrests cells at cytokinesis and induces hyperpolymerization of purified actin. *Mol. Pharmacol.* 59: 462–469.

Baker, M.A., Maloy, W.L., Zasloff, M., and Jacob, L.S. 1993. Anticancer efficacy of Magainin2 and analogue peptides. *Cancer Res.* 53: 3052–3057.

Ballard, C.E., Yu, H., and Wang, B. 2002. Recent developments in depsipeptide research. *Curr. Med. Chem.* 9: 471–498.

Barna B.P., Eppstein, D.A., Thomassen, M.J. et al. 1993. Therapeutic effects of a synthetic peptide of C-reactive protein in pre-clinical tumor models. *Cancer Immunol. Immunother.* 36: 171–176.

Barna, B.P., Thomassen, M.J., Zhou, P., Pettay, J., Singh-Burgess, S., and Deodhar, S.D. 1996. Activation of alveolar macrophage TNF and MCP-1 expression *in vivo* by a synthetic peptide of C-reactive protein. *J. Leukoc. Biol.* 59: 397–402.

Baselga, J. and Albanell, J. 2001. Mechanism of action of anti-HER2 monoclonal antibodies. *Ann. Oncol.* 12: S35–S41.

Bennett, J.A., DeFreest, L., Anaka, I. et al. 2006. AFPep: An anti-breast cancer peptide that is orally active. *Breast Cancer Res. Treat.* 98: 133–141.

Bhutia, S.K., Mallick, S.K., Stevens, S.M., Prokai, L., Vishwanatha, J.K., and Maiti, T. 2008. Induction of mitochondria-dependent apoptosis by *Abrus* agglutinin derived peptides in human cervical cancer cell. *Toxicol. In Vitro* 22: 344–351.

Borel, P., Liron, D., Termine, E., Gratoroli, R., and Lafont, H. 1989. Isolation and properties of lipolysis inhibitory proteins from wheat germ and wheat bran. *Plant Foods Hum. Nutr.* 39: 339–348.

Bukowski, R., Ernstoff, M.S., Gore, M.E. et al. 2002. Pegylated interferon alfa-2b treatment for patients with solid tumors: a phase I/II study. *J. Clin. Oncol.* 20: 3841–3849.

Burton, D.R. 1995. Phage display. *Immunotechnology* 1: 87–94.

Caba, J.M., Rodriguez, I.M., Manzanares, I., Giralt, E., and Albericio, F. 2001. Solid-phase synthesis of trunkamide A. *J. Org. Chem.* 66: 7568–7574.

Carroll, A.R., Bowden, B.F., Coll, J.C., Hockless, C.R., Skelton, B.W., and White A.H. 1994. Studies of Australian ascidians. Mollamide, a cytotoxic cyclic heptapeptide from the compound ascidian *Didemnum molle*. *Austral. J. Chem.* 47: 61–69.

Cesareni, G. 1992. Peptide display on filamentous phage capsids. A new powerful tool to study protein-ligand interaction. *FEBS Lett.* 307: 66–70.

Chen, Y., Xu, X., Hong, S. et al. 2001. RGD-tachyplesin inhibits tumor growth. *Cancer Res.* 61: 2434–2438.

Chernysh, S., Kim, S.I., Bekker, G. et al. 2002. Antiviral and antitumor peptides from insects. *Proc. Natl. Acad. Sci. USA* 99: 12628–12632.

Correia J.J. 1991. Effects of antimitotic agents on tubulin-nucleotide interactions. *Pharmacol. Ther.* 52: 127–147.

Cory, S. and Adams, J.M. 1998. Matters of life and death: Programmed cell death at Cold Spring Harbor. *Biochim. Biophys. Acta* 1377: R25–R44.

Cruciani, R.A., Barker, J.L., Zasloff, M., Chen, H.C., and Colamonici, O. 1991. Antibiotic magainins exert cytolytic activity against transformed cell lines through channel formation. *Proc. Natl. Acad. Sci. USA* 88: 3792–3796.

de Arruda, M., Cocchiaro, C.A., Nelson, C.M. et al. 1995. LU103793 (NSC D-669356): A synthetic peptide that interacts with microtubules and inhibits mitosis. *Cancer Res.* 55: 3085–3092.

de Lumen, B.O. 2005. Lunasin: A cancer-preventive soy peptide. *Nutr. Rev.* 63: 16–21.

Derossi, D., Joliot, A.H., Chassaing, G., and Prochiantz, A. et al. 1994. The third helix of the Antennapedia homeodomain translocates through biological membranes. *J. Biol. Chem.* 269: 10444–10450.

Dimarcq, J.L., Bulet, P., Hetru, C., and Hoffmann, J. 1998. Cysteine-rich antimicrobial peptides in invertebrates. *Biopolymers* 47: 465–477.

Doorbar, J. and Winter, G. 1994. Isolation of a peptide antagonist to the thrombin receptor using phage display. *J. Mol. Biol.* 244: 361–369.

Dredge, K., Marriott, G.B., Todryk, S.M., and Dalgleish, A.G. 2002. Adjuvants and the promotion of Th1-type cytokines in tumor immunotherapy. *Cancer Immunol. Immunother.* 51: 521–531.

Dutta, R.C. 2002. Peptide immunomodulators versus infection; An analysis. *Immunol. Lett.* 83: 153–161.

Ebbinghaus, S., Hussain, M., Tannir, N. et al. 2007. Phase 2 Study of ABT-510 in patients with previously untreated advanced renal cell carcinoma. *Clin. Cancer Res.* 13: 6689–6695.

Ellerby, H.M., Arap, W., Ellerby, L.M. et al. 1999. Anti-cancer activity of targeted pro-apoptotic peptides. *Nat. Med.* 5: 1032–1038.

Erez, N., Zamir, E., Gour, B.J., Blashuck, O.W., and Geiger, B. 2004. Induction of apoptosis in cultured endothelial cells by a cadherin antagonist peptide: Involvement of fibroblast growth factor receptor-mediated signaling. *Exp. Cell Res.* 294: 366–378.

Fernandez, R., Rodriguez, J., Quinoa, E. et al. 1996. Onchidin B: A new cyclodepsipeptide from the mollusc *Onchidinium* sp. *J. Am. Chem. Soc.* 118: 11635–11643.

Fidler, I.J., and Ellis, L.M. 1994. The implications of angiogenesis for the biology and therapy of cancer metastasis. *Cell* 79: 185–188.

Figg, W.D., Kruger, E.A., Price, D.K., Kim, S., and Dahut, W.D. 2002. Inhibition of angiogenesis: Treatment options for patients with metastatic prostate cancer. *Invest. New Drugs* 20: 183–194.

Foa, P., Chillemi, F., Lombardi, L., Lonati, S., Miaolo, A.T., and Polli, E.E. 1987. Inhibitory activity of synthetic pentapeptide on leukemic myelopoiesis both *in vitro* and *in vivo* in rats. *Eur. J. Haematol.* 39: 399–403.

Folkman, J. 1995. Angiogenesis in cancer, vascular, rheumatoid and other disease. *Nat. Med.* 1: 27–31.

Food and Drug Administration. 1996. FDA approves DaunoXome as first-line therapy for Kaposi's sarcoma. *J. Int. Assoc. Physicians AIDS Care* 2: 50–51.

Ford, P.W, Gustafson, K.R., McKee, T.C. et al. 1999. Papuamides A-D, HIV-inhibitory and cytotoxic depsipeptides from the sponges *Theonella mirabilis* and *Theonella swinhoei* collected in Papua New Guinea. *J. Am. Chem. Soc.* 121: 5899–5909.

Frohm, M., Agerberth, B., Ahangari, G. et al. 1997. The expression of the gene coding for the antibacterial peptide LL-37 is induced in human keratinocytes during inflammatory disorders. *J. Biol. Chem.* 272: 15258–15263.

Froidevaux, S. and Eberle, A.N. 2002. Somatostatin analogs and radiopeptides in cancer therapy. *Biopolymers 66:* 161–183.

Gabizon, A.A. 2001. Pegylated liposomal doxorubicin: Metamorphosis of an old drug into a new form of chemotherapy. *Cancer Invest.* 19: 424–436.

Gademann, K. and Seebach, D. 2001. Synthesis of cyclo-β-tripeptides and their biologial *in vitro* evaluation as antiproliferative against the growth of human cancer cell lines. *Helv. Chim. Acta.* 84: 2924–2937.

Galvez, A.F., Chen, N., Macsaieb, J., and de Lumen, B.O. 2001. Chemopreventive property of a soybean peptide (lunasin) that binds to deacetylated histones and inhibits their acetylation. *Cancer Res.* 61: 7473–7478.

Gembistky, D.S., De Angelis, P.M., Reichelt, K.L., and Elgjo, K. 2000. An endogenous melanocyte-inhibiting tripeptide pyroGlu-Phe-GlyNH2 delays *in vivo* growth of monoclonal experimental melanoma. *Cell Prolif.* 33: 91–99.

Griffioen, A.W., van der Schaft D., Barendsz-Janson, A.F. et al. 2001. Anginex, a designed peptide that inhibits angiogenesis. *Biochem. J.* 354: 233–242.

Griffiths, A.D. 1993. Production of human antibodies using bacteriophage. *Curr. Opin. Immunol.* 5: 263–267.

Habermann, E. and Jentsch, J. 1967. Sequence analysis of melittin from tryptic and peptic degradation products. *Hoppe Seyler's Z. Physiol. Chem.* 348: 37–50.

Hariharan, S., Gustafson, D., Holden, S. et al. 2007. Assessment of the biological and pharmacological effects of the $\alpha_v\beta_3$ and $\alpha_v\beta_3$ integrin receptor antagonist, cilengitide (EMD 121974), in patients with advanced solid tumors. *Ann. Oncol.* 18: 1400–1407.

Harrigan, G.G., Harrigan, B.L., and Davidson, B.S. 1997. Kailuins A-D, new cyclic acyldepsipeptides from cultures of a marine-derived bacterium. *Tetrahedron* 53: 1577–1582.

Harrigan, G.G., Luesch, H., Yoshida, W.Y., Moore, R.E., Nagle, D.G., and Paul, V.J. 1999. Symplostatin 2: A dolastatin 13 analogue from marine cyanobacterium *Symploca hydnoides. J. Nat. Prod.* 62: 655–658.

Hehir, K.M., Baguisi, A., Pennington, S.E., Bates, J.M., and Di Tullio, P.A. 2004. A potential antitumor peptide therapeutic derived from antineoplastic urinary protein. *Peptides* 25: 543–549.

Hoogenboom, H.R., de Bruine A.P., Hufton, S.E., Hoet, R.M., Arends, J.W., and Roovers, R.C. 1998. Antibody phage display technology and its applications. *Immunotechnology* 4: 1–20.

Janin, Y.L. 2003. Peptides with anticancer use or potential. *Amino Acids* 25: 1–40.

Jensen, P.K.A., Elgjo, K., Laerum, O.D., and Bolund, L. 1990. Synthetic epidermal pentapeptide and related growth regulatory peptides inhibit proliferation and enhance differentiation in primary and regenerating cultures of human epidermal keratinocytes. *J. Cell Sci.* 97: 51–58.

Jeong, H.J., Park, J.H., Lam, Y., De Lumen, B.O. 2003. Characterization of lunasin isolated from soybean. *J. Agric. Food Chem.* 51, 7901–7906.

Johansson, J., Gudmundsson, G.H., Rottenberg, M.E., Berndt, K.D., and Agerberth, B. 1998. Conformation-dependent antibacterial activity of the naturally-occuring human peptide LL-37. *J. Biol. Chem.* 273: 3718–3724.

Jordan, M.A., Walker, D., De Arruda, M., Barlozzari, T., and Panda, D. 1998. Suppression of microtubule dynamics by binding of cemadotin to tubulin: possible mechanism for its antitumor action. *Biochemistry* 37: 17571–17578.

Jordan, M.A. and Wilson, L. 1998. Microtubules and actin filaments: dynamic targets for cancer chemotherapy. *Curr. Opin. Cell Biol.* 10: 123–130.

Kannan, A., Hettiarachchy, N., Johnson, M.G., and Nannapaneni, R. 2008. Human colon and liver cancer proliferation inhibition by peptide hydrolysates derived from heat stabilized defatted rice bran. *J. Agric. Food Chem.* 56: 11643–11647.

Kaplan, N., Morpurgo, N., and Linlal, M. 2007. Novel families of toxin-like peptides in insects and mammals: A computational approach. *J. Mol. Biol* 369: 553–566.

Kerbel, R. and Folkman, J. 2002. Clinical translation of angiogenesis inhibitors. *Nat. Rev. Cancer* 2: 727–739.

Kobayashi, M., Natsume, T., Tamaoki, S. et al. 1997. Antitumor activity of TZT-1027, a novel dolastatin 10 derivative. *Jpn. J. Cancer Res.* 88: 316–327.

Koivunen, E., Arap, W., Valtanen, H. et al. 1999. Tumor targeting with a selective gelatinase inhibitor. *Nat. Biotechnol.* 17: 768–774.

Kotani, S., Azuma, I., Takada, H., Tsujimoto, M., and Yamamura, Y. 1983. Muramyl dipeptides: prospect for cancer treatments and immunostimulation. *Adv. Exp. Med. Biol.* 166: 117–158.

Kufe, D.W., Pollock, R.E., Weichselbaum, R.R., Bast, R.C., Jr., Gansier, T.S., and Holland, J.F. 2003. *Cancer Medicine*, 6th edition. Hamilton, Ontario: American Cancer Society Inc. and B.C. Decker.

Ladner, R.C. 2000. Phage display and pharmacogenomics. *Pharmacogenomics* 1: 199–202.

Lam, Y., Galvez, A., and de Lumen, B.O. 2003. Lunasin suppresses E1A-mediated transformation of mammalian cells but does not inhibit growth of immortalized and established cancer cell lines. *Nutr. Cancer,* 47: 88–94.

Lehmann, J., Retz, M., Sidhu S.S. et al. 2006. Antitumor activity of the antimicrobial peptide magainin II against bladder cancer cell lines. *Eur. Urol.* 50: 141–147.

Lichtenstein, A., Ganz, T., Selsted, M.E., and Lehrer, R.I. 1986. *In vitro* tumor cell cytolysis mediated by peptide defensins of human and rabbit granulocytes. *Blood* 68: 1407–1410.

Liu, Y., Elgjo, K., Wright, M., and Reichelt, K.L. 2000. Novel lymphocyte growth-inhibiting tripeptide: N-acetyl-Glu-Ser-Gly-NH2. *Biochem. Biophys. Res. Commun.* 277: 562–567.

Livant, D.L., Brabec, R.K., Pienta, K.J. et al. 2000. Anti-invasive, antitumorigenic, and antimetastatic activities of the PHSCN sequence in prostate carcinoma. *Cancer Res.* 60: 309–320.

Losso, J.N. 2008. The biochemical and functional food properties of the Bowman–Birk inhibitor. *Crit. Rev. Food Sci. Nutr.* 48: 94–118.

Luesch, H., Pangilinan, R., Yoshida, W.Y., Moore, R.E., and Paul, V.J. 2001. Pitipeptolides A and B, new cyclodepsipeptides from the marine cyanobacterium *Lyngya majuscula. J. Nat. Prod.* 64: 304–307.

Luesch, H., Yoshida, W.Y., Moore, R.E., and Paul, V.J. 2002. New apratoxins of marine cyanobacterial origin from Guam and Palau. *Bioorg. Med. Chem.* 10: 1973–1978.

Madden, T., Tran, H.T., and Beck, D. 2000. Novel marine-derived anticancer agents: a phase I clinical, pharmacological, and pharmacodynamic study of dolastatin 10 (NSC 376128) in patients with advanced solid tumors. *Clin. Cancer Res.* 6: 1293–1301.

Mader, J.S. and Hoskin, D.W. 2006. Cationic antimicrobial peptides as novel cytotoxic agents for cancer treatment. *Expert Opin. Investig. Drugs* 15: 933–946.

Mader, J.S., Richardson, A., Salsman, J. et al. 2007. Bovine lactoferricin causes apoptosis in Jurkat T-leukemia cells by sequential permeabilization of the cell membrane and targeting of mitochondria. *Exp. Cell Res.* 313: 2634–2650.

Maeda, H. 2001. SMANCS and polymer-conjugated macromolecular drugs: advantages in cancer chemotherapy. *Adv. Drug Deliv. Rev.* 46: 169–185.

Maeda, H., Sawa, T., and Konno, T. 2001. Mechanism of tumor-targeted delivery of macromolecular drugs, including the EPR effect in solid tumor and clinical overview of the prototype polymeric drug SMANCS. *J. Control Release.* 74: 47–61.

Mai, J.C., Mi, Z., Kim, S.H., Ng, B., and Robbins, P.D. 2001. A proapoptotic peptide for the treatment of solid tumors. *Cancer Res.* 61: 7709–7712.

Marshall, H. 2001. Anti-CD20 antibody therapy is highly effective in the treatment of follicular lymphoma. *Trends Immunol.* 22: 183–184.

Meisel, H. 1997. Biochemical properties of regulatory peptides derived from milk proteins. *Peptide Sci.* 43: 119–128.

Mendoza, F.J., Espino, P.S., Cann, K.L., Bristow, N., McCrea, K., and Los, M. et al. 2005. Anti-tumor chemotherapy utilizing peptide-based approaches: Apoptotic pathways, kinases, and proteasome as targets. *Arch. Immunol. Ther. Exp. (Warsz).* 53: 47–60.

Miyamoto, Y., Kuroda, M., Munekata, E., and Masaki, T. 1986. Stoichiometry of actin and phalloidin binding: One molecule of the toxin dominates two actin subunits. *J. Biochem.* 100: 1677–1680.

Moore, A.J., Devine, D.A., and Bibby, M.C. 1994. Preliminary experimental anticancer activity of cecropins. *Pept. Res.* 7: 265–269.

Moore, R.E. and Entzeroth, M. 1988. Majusculamide D and deoxymajusculamide D, two cytotoxins from *Lyngbya majuscula. Phytochemistry* 27: 3101–3103.

Morita, H., Gonda, A., Takeya, K., and Itokawa, H. 1996. Cycloleonuripeptide A, B and C, three new proline-rich cyclic nonapeptides from *Leonurus heterophyllus. Bioorg. Med. Chem. Lett.* 6: 767–770.

Mosmann, T. 1983. Rapid colorimetric assay for cellular growth and survival: Application to proliferation and cytotoxicity assays. *J. Immunol. Meth.* 65: 55–63.

Nakajima, H., Kim, Y.B., Terano, H., Yoshida, M., and Horinouchi, S. 1998. FR901228, a potent antitumor antibiotic, is a novel histone deacetylase inhibitor. *Exp. Cell Res.* 241: 126–133.

Nam, S.H., Choi, S.P., Kang, M.Y., Kozukue, N., and Friedman, M. 2005. Antioxidative, antimutagenic, and anticarcinigenic activities of rice bran extracts in chemical and cell assays. *J. Agric. Food Chem.* 53: 816–822.

Neri, D., Petrul, H., and Roncucci, G. 1995. Engineering recombinant antibodies for immunotherapy. *Cell Biophys.* 27: 47–61.

Nishioka, K., Babcock, G.F., Phillips, J.H., and Noyes, R.D. 1981. Antitumor effect of tuftsin. *Mol. Cell. Biochem.* 41: 13–18.

Nogle, L.M., Williamson, R.T., and Gerwick, W.H. 2001. Somamides A and B, two new depsipeptide analogues of dolastatin 13 from a Fijian cyanobacterial assemblage of *Lyngbya majuscula* and *Schizothrix* species. *J. Nat. Prod.* 64: 716–719.

Nomura, Y., Oohashi, K., Watanabe, M., and Kasugai, S. 2005. Increase in bone mineral density through oral administration of shark gelatin to ovariectomized rats. *Nutrition* 21: 1120–1126.

Nyberg, P., Xei, L., and Kalluri, R. 2005. Endogenous inhibitors of angiogenesis. *Cancer Res.* 65: 3967–3979.

Okumura, K., Itoh, A., Isogai, E. et al. 2004. C-terminal domain of human CAP18 anti-microbial peptide induces apoptosis in oral squamous cell carcinoma SAS-H1 cells. *Cancer Lett.* 212: 185–194.

Papo, N., Oren, Z., Pag, U., Sahl, H.G., and Shai, Y. 2002. The consequence of sequence alteration of an amphipathic alpha-helical antimicrobial peptide and its diaste-reomers. *J. Biol. Chem.* 277: 33913–33921.

Papo, N., Seger, D., Makovitzki, A. et al. 2006. Inhibition of tumor growth and elimi-nation of multiple metastases in human prostate and breast xenografts by sys-temic inoculation of a host defense-like lytic peptide. *Cancer Res.* 66: 5371–5378.

Papo, N., Shahar, M., Eisenbach, L., and Shai, Y. 2003. A novel lytic peptide composed of D, L amino acids selectively kills cancer cells in culture and in mice. *J. Biol. Chem.* 278: 21018–21023.

Papo, N. and Shai, Y. 2005. Host defense peptides as new weapons in cancer treat-ment. *Cell. Mol. Life Sci.* 62: 784–790.

Pardo, J., Perez-Galan, P., Gamen, S. et al. 2001. A role of the mitochondrial apoptosis inducing factor in granulysin-induced apoptosis. *J. Immunol.* 167: 1222–1229.

Park, N.G., Lee, S., Oishi, O. et al. 1992. Conformation of tachyplesin I from *Tachypleus tridentatus* when interacting with lipid matrices. *Biochemistry* 31: 12241–12247.

Pasqualini, R., Koivunen, E., Kain, R. et al. 2000. Aminopeptidase N is a receptor for tumor-homing peptides and a target for inhibiting angiogenesis, *Cancer Res.* 60: 722–727.

Paulsen, J.E. 1993. The synthetic colon peptide pyroGlu-His-GlyOH inhibits growth of human colon carcinoma cells (HT-29) transplanted subcutaneously into athy-mic mice. *Carcinogenesis* 14: 1719–1721.

Paulsen, J.E., Hall, K.S., Rugstad, H.E., Reichelt, K.L., and Elgo, K. 1992. The syn-thetic hepatic peptides Pyr-Glu-Gly-Ser-Asp and Pyr-Glu-Glu-Gly-Ser-Asp acid inhibit growth of MH1C1 rat hepatoma cells transplanted into buffalo rats or athymic mice. *Cancer Res.* 53: 1218–1221.

Perera, Y., Farina, H.G., Hernandez, I. et al. 2008. Systemic administration of a pep-tide that impairs the protein kinase (CK2) phosphorylation reduces solid tumor growth in mice. *Int. J. Cancer* 122: 57–62.

Pettit, G.R., Kamano, Y., Herald, C.L. et al. 1989. Isolation and structure of the cyto-static depsipeptide dolastatin 13 from the sea hare *Dolabella auricularia*. *J. Am. Chem. Soc.* 111: 5015–5017.

Pettit, G.R., Kamano, Y., Herald, C.L. et al. 1990. Antineoplastic agents. 190. Isolation and structure of the cyclodepsipeptide dolastatin 14. *J. Org. Chem.* 55: 2989–2990.

Pettit, G.R., Xu, J.P., Hogan, F., and Cerny, R.L. 1998. Isolation and structure of dolas-tatin 17. *Heterocycles* 47: 491–496.

Piekarz, R.L., Robey, R., Sandor, V. et al. 2001. Inhibitor of histone deacetylation, dep-sipeptide (FR901228), in the treatment of peripheral and cutaneous T-cell lym-phoma: A case report. *Blood* 98: 2865–2868.

Randazzo, A., Dal Piaz, F., and Orru, S. 1998. Axinellins A and B: New prolinecon-taining antiproliferative cyclopeptides from the Vanuatu sponge *Axinella carteri*. *Eur. J. Org. Chem.* 1998: 2659–2665.

Risso, A., Braidot, E., Sordano, M.C. et al. 2002. BMAP-28, an antibiotic peptide of innate immunity, induces cell death through opening of the mitochondrial per-meability transition pore. *Mol. Cell. Biol.* 22: 1926–1935.

Rozek, T., Wegener, K.L., Bowie, J.H. et al. 2000. The antibiotic and anticancer active aurein peptides from the Australian Bell Frogs *Litoria aurea* and *Litoria raniformis* the solution structure of aurein 1.2. *Eur. J. Biochem.* 267: 5330–5341.

Ruegg, C., Hasmim, M., Lejeune, F.J., and Alghisi, G.C. 2006. Antiangiogenic peptides and proteins: from experimental tools to clinical drugs. *Biochim. Biophys. Acta.* 1765: 155–177.

Ruoslahti, E. and Pierschbacher, M.D. 1987. New perspectives in cell adhesion: RGD and integrins. *Science* 238: 491–497.

Sato, S., Kopitz, C., Schmalix, W.A. et al. 2002. High-affinity urokinase-derived cyclic peptides inhibiting urokinase/urokinase receptor-interaction: Effects on tumor growth and spread. *FEBS Lett.* 528: 212–216.

Schally, A.V. and Nagy, A. 2003. New approaches to treatment of various cancers based on cytotoxic analogs of LHRH, somatostatin and bombesin. *Life Sci.* 72: 2305–2320.

Schorlemmer, H.U., Bosslet, K., and Sedlacek, H.H. 1983. Ability of the immunomodulating dipeptide bestatin to activate cytotoxic mononuclear phagocytes. *Cancer Res.* 43: 4148–4153.

Schwarze, S.R., Ho, A., Vocero-Akbani, A., and Dowdy, S.F. 1999. *In vivo* protein transduction: Delivery of a biologically active protein into the mouse. *Science* 285: 1569–1572.

Smith, G.P. 1985. Filamentous fusion phage: Novel expression vectors that display cloned antigens on the virion surface. *Science* 228: 1315–1357.

Sone, H., Nemoto, T., Ishiwata, H., Ojika, M., and Yamada, K. 1993. Isolation, structures and synthesis of dolastatin D, a cytotoxic cyclic depsipeptide from the sea hare *Dolabella auricularia*. *Tetrahedron Lett.* 34: 8449–8452.

Starzec, A., Vassy, R., Martin, A. et al. 2006. Antiangiogenic and antitumor activities of peptide inhibiting the vascular endothelial growth factor binding to neuropilin-1. *Life Sci.* 79: 2370–2381.

Staton, C.A., Brown, N.J., Rodgers, G.R. et al. 2004. Alphastatin, a 24-amino acid fragment of human fibrinogen, is a potent new inhibitor of activated endothelial cells *in vitro* and *in vivo*. *Blood* 103: 601–606.

Staton, C.A. and Lewis, C.E. 2005. Angiogenesis inhibitors found within the haemostasis pathway. *J. Cell. Mol. Med.* 9: 286–302.

Stewart, J.M., Gera, L., Chan, D.C. et al. 2005. Combination cancer chemotherapy with one compound: Pluripotent bradykinin antagonists. *Peptides* 26: 1288–1291.

Strelkov, L.A., Mikhailova, A.A., Sapozhnikov, A.M., Fonina, L.A., and Petrov, R.V. 1996. The bone marrow peptide (myelopeptide-2) abolishes induced by human leukemia HL-60 cell suppression of T lymphocytes. *Immunol. Lett.* 50: 143–147.

Supko, J.G., Lynch, T.J., and Clark, J.W. 2000. A phase I clinical and pharmacokinetic study of the dolastatin analogue cemadotin administered as a 5-day continuous intravenous infusion. *Cancer Chemother. Pharmacol.* 46: 319–328.

Tani, H., Ohishi, H., and Watanabe, K. 1994. Purification and characterization of proteinous inhibitor of lipase from wheat flour. *J. Agric. Food Chem.* 42: 2382–2385.

Tjin Tham Sjin, R.M, Satchi-Fainaro, R., Birsner, A.E., Ramanujam, V.M., Folkman, J., and Javaherian, K. 2005. A 27-amino-acid synthetic peptide corresponding to the NH2-terminal zinc-binding domain of Endostatin is responsible for its antitumor activity. *Cancer Res.* 65: 3656–3663.

Torchilin, V.P. and Lukyanov, A.N. 2003. Peptide and protein drug delivery to and into tumors: Challenges and solutions. *Drug Discov. Today* 8: 259–266.

Tsukamoto, S., Painuly, P., Young, K.A., Yang, X., and Shimizu, Y. 1993. Microcystilide A: A novel cell-differentiation-promoting depsipeptide from *Microcystis aeruginosa* NO-15-1840. *J. Am. Chem. Soc.* 115: 11046–11047.

Tsutsumi, K., Kawauchi, Y., Kondo, Y. et al. 2000. Water extract of defatted rice barn auppresses visceral fat accumulation in rats. *J. Agric. Food Chem.* 48: 1653–1656.

Ueda, H., Manda, T., Matsumoto, S. et al. 1994a. FR901228, a novel antitumor bicyclic depsipeptide produced by *Chromobacterium violaceum* No. 968 III. Antitumor activities on experimental tumors in mice. *J. Antibiot.* 47: 315–323.

Ueda, H., Nakajima, H., Hori, Y. et al. 1994b. FR901228, a novel antitumor bicyclic depsipeptide produced by *Chromobacterium violaceum* No. 968 I. Taxonomy, fermentation, isolation, physicochemical and biological properties and antitumor activity. *J. Antibiot.* 47: 301–310.

Ueda, H. and Yamazaki, M. 2001. Induction of tumor necrosis factor-alpha in solid tumor region by the orally administered synthetic muramyl dipeptide analogue, romurtide. *Int. Immunopharmacol.* 1: 97–104.

Uthaisang, W., Reutracul, V., Krachangchaeng, C., Wilairat, P., and Fadeel, B. 2004. VR-3848, a novel peptide derived from Euphobiaceae, induces mitochondria-dependent apoptosis in human leukemia cells. *Cancer Lett.* 208: 171–178.

Vaishampayan, U., Glode, M., Du, W. et al. 2000. Phase II study of dolastatin-10 in patients with hormone-refractory metastatic prostate adenocarcinoma. *Clin. Cancer Res.* 6: 4205–4208.

Wang Y.L., Kaplan, S., Whiteside, T., and Herberman, R.B. 1989. *In vitro* effects of an acyltripeptide, FK565, on antitumor effector activities and on metabolic activities of human monocytes and granulocytes. *Immunopharmacology* 18: 213–222.

Wieland, T. 1968. Poisonous principles of mushrooms of the genus Amanita. Four-carbon amines acting on the central nervous system and cell-destroying cyclic peptides are produced. *Science* 159: 946–952

Winter, G., Griffiths A.D., Hawkins, R.E., and Hoogenboom, H.R. 1994. Making antibodies by phage display technology. *Annu. Rev. Immunol.* 12: 433–55.

Wipf, P. and Uto, Y. 1999. Total synthesis of the putative structure of the marine metabolite trunkamide A. *Tetrahedron Lett.* 40: 5165–5169.

Yao, Z., Lu, R., Jia, J. et al. 2006. The effect of tripeptide tyroserleutide (YSL) on animal models of hepatocarcinoma. *Peptides* 27: 1167–1172.

Yao, Z., Qui, S., Wang, L. et al. 2005. Tripeptide tyroserleutide enhances the antitumor effects of macrophages and stimulates macrophage secretion of IL-1β, TNF-α and NO *in vitro*. *Cancer Immunol. Immunother.* 55: 56–60.

Yoshida, M., Furumai, R., Nishiyama, M., Komatsu.Y., Nishino, N., and Horinouchi, S. 2001. Histone deacetylase as a new target for cancer chemotherapy. *Cancer Chemother. Pharmacol.* 48: S20–S26.

Zasloff, M. 1987. Magainins, a class of antimicrobial peptides from *Xenopus* skin: Isolation, characterization of two active forms and partial cDNA sequence of a precursor. *Proc. Natl. Acad. Sci. USA* 84: 5449–5453.

Zhu, D.M. and Uckun, F.M. 2000. Z-Phe-Gly-NHO-Bz, an inhibitor of cysteine cathepsins, induces apoptosis in human cancer cells. *Clin. Cancer Res.* 6: 2064–2069.

Zitvogel, L., Tesniere, A., and Kroemer, G. 2006. Cancer despite immunosurveillance: Immunoselection and immunosubversion. *Nat. Rev. Immunol.* 6: 715–727.

13

Lunasin: A Novel Seed Peptide with Cancer Preventive Properties

Chia-Chien Hsieh, Blanca Hernández-Ledesma, and Ben O. de Lumen

CONTENTS

13.1 Introduction: Diet and Cancer

Last century was a time of enormous progress in human nutrition. Essential minerals, vitamins, amino acids, and fatty acids were identified, metabolic pathways were described, and the effect of genetic variants on metabolism was studied. Recent evidence has shown that the large majority of chronic disorders, such as cardiovascular disease, diabetes, and cancer can be prevented by modifications of nutritional and lifestyle habits. In 2007, 12.3 million new cancer cases and 7.6 million deaths from cancer worldwide were reported by the American Cancer Society (Garcia et al., 2007). It has been estimated that as many as 35% of these cases may be related to dietary factors (Manson, 2003). The rising worldwide prevalence of cancer and the corresponding rise in healthcare costs is propelling interest among researchers and consumers for the multiple health benefits of food compounds, including reduction in cancer risk and modification of tumor behavior (Béliveau and Gingras, 2007; Kaefer and Milner, 2008). Epidemiological evidence, cell culture, and animal tumor model studies have demonstrated that a large number of natural compounds present in the diet could lower cancer risk and even sensitize tumor cells in anticancer therapies (Ban, Lee, and Guyatt, 2008; Chen and Dou, 2008; de Kok, van Breda, and Manson, 2008; Fimognari, Lenzi, and Hrelia, 2008; Kaefer and Milner, 2008; Ramos, 2008).

Carcinogenesis is characterized by a complex process that involves a series of individual steps in which distinct molecular and cellular alterations occur. Tumor development includes three different steps: initiation, promotion, and progression (Pitot, 1993). During the initiation phase, the carcinogenic agent interacts with target cell DNA causing damage (Cohen, 1995). Blocking this genotoxic damage at early stages of carcinogenesis has been considered one of the most effective ways for preventing cancer (Ramos, 2008). It can be achieved by scavenging the reactive oxygen species or by inducing the phase-II conjugating enzymes (Welss et al., 2003). During the promotion phase, those mechanisms stopping or slowing down cell division, such as induction of cell cycle arrest or apoptosis, could be potentially beneficial to restore the lost balance between cell proliferation and programmed cell death (Schwartz and Shah, 2005). At the last phase of progression, the interruption of angiogenesis or the prevention of malignant cells to escape from the original location and invade other tissues (metastasis) could also be potentially useful (Bhat and Sing, 2008).

Phytochemicals are compounds present in plant foods with the capacity to affect all the carcinogenesis stages. They can regulate multiple key proteins involved in diverse signal transduction pathways such as regulation of cellular proliferation, differentiation, apoptosis, angiogenesis, or metastasis, resulting in a potential beneficial effect (Fimongnari et al., 2008; Ramos, 2008; van Breda, de Kok, and van Delft, 2008). Numerous studies using cell culture and cancer mouse models have demonstrated the chemopreventive and chemotherapeutic properties against human cancers of different phytochemicals such as epigallocatechin gallate [(-)-EGCG] (green tea polyphenol; Chen and Dou, 2008; Yang et al., 2008), genistein (soybean), apigenin (celery, parsley), isothiocyanates (broccoli), anthocyanins (berries; Fimognari et al., 2008), quercetin (onions), kaempferol (broccoli, grapefruits; Ramos, 2008), curcumin (turmeric; Chen and Dou, 2008), diallyl trisulfide (garlic; Seki et al., 2008), and lycopene (tomatoes; Singh and Goyal, 2008). Recently, growing epidemiological and preclinical evidence has shown the important cancer preventive role of culinary herbs and spices (Kaefer and Milner, 2008). The bioactive compounds present in these herbs have been reported by *in vitro* studies to inhibit different pathways involved in regulation of cell division, cell proliferation, and detoxification (Kaefer and Milner, 2008).

Daily intake of food rich in anticancer molecules could be compared to a preventive, nontoxic version of chemotherapy that is harmless to the physiology of normal tissue and stops microtumors (Béliveau and Gingras, 2007). These molecules have become an invaluable treasure in cancer prevention and chemotherapy, and further research is currently under way to explore their properties and mechanisms of action.

13.2 Soy and Cancer

Soybean (*Glycine max*) is an ancient legume consumed worldwide, but most commonly in Asian countries, such as China, Japan, Korea, Taiwan, and Indonesia. It contains a high concentration of proteins (40–50%), lipids (20–30%), and carbohydrates (26–30%). In traditional Asian diets, soy is consumed in many forms, including soybeans, soybean sprouts, toasted soy protein flours, soy milk, tofu, and fermented soy products, such as tempeh, miso, natto, soybean paste, and soy sauce (Coward et al., 1993; Wang and Murphy, 1994). Asians consume an average of 20 to 80 g of traditional soy foods daily (Coward et al., 1993; Messina and Flickinger, 2002), which equates to a daily intake of between 25 and 100 mg total isoflavones (Messina, McCaskill-Stevens, and Lampe, 2006) and between 8 and 50 g soy protein (Erdman, Jr. et al., 2004). Western populations consume much less soy, only about 1 to 3 g daily, and this is mostly in processed forms, such as soy drinks, breakfast cereals, energy bars, and soy "burgers" (Fournier, Erdman, Jr., and Gordon, 1998).

Soybeans, soy-derived foods, and soy food supplements have recently generated a great deal of interest due to their potential health effects, particularly with respect to cardiovascular health and cancer prevention. A number of epidemiological studies have demonstrated an association between the consumption of soybeans and improved health, particularly as a reduced risk for cardiovascular diseases (Anderson, Smith, and Washnock, 1999) and cancer, such as breast (Wu et al., 1996; Zheng, Dai, and Custer, 1999), prostate (Jacobsen, Knutsen, and Fraser, 1998; Kolonel et al., 2000), endometrial (Goodman et al., 1997), lung (Swanson et al., 1992), and bladder cancer (Sun et al., 2002). In addition, a number of animal studies support anticancer properties of soy, soy foods, and soy constituents that have been shown to suppress tumor growth in a variety of tissues including skin, bladder, mammary, and prostate (Messina and Flickinger, 2002). Some clinical studies also support these soy properties, but the issue is complicated by the lack of knowledge of the particular components of soy that may give rise to the putative anticancer effects (Kerwin, 2004). Efforts to identify these components have led to the identification of a number of constituents that may play a role in the protective effects of soy foods. An array of biologically active compounds or phytochemicals with cancer preventive effects have been isolated and identified in soy foods. These include isoflavones, saponins, phenolic acids, phytosterols, protease inhibitors, and other soy proteins and peptides, such as lectins and lunasin (see reviews: Messina and Flickinger, 2002; Losso, 2008; Hernández-Ledesma, Hsieh, and de Lumen, 2009).

13.2.1 Isoflavonoids

The health-promoting effects of soybean consumption have more recently been linked to the biological activities of a specific group of phenolic compounds known as isoflavonoids (McCue and Shetty, 2004). Genistein, daidzein, and glycitein are the three major isoflavonoids found in soybean and soy products whose properties have been extensively studied (Park and Surh, 2004; Khan, Afaq, and Mukhtar, 2008). Large bodies of epidemiological studies have shown people consuming high amounts of these soy isoflavonoids in their diets have lower rates of several cancers, including breast, prostate, and endometrial cancer (Lof and Weiderpass, 2006). In animal models, these compounds have been reported to inhibit the development of different types of tumors (Barnes et al., 1990; Li et al., 1999), but the results are not completely conclusive.

A number of mechanisms of action have been suggested for these compounds. They are structurally and functionally similar to 17β-estradiol, the most potent mammalian estrogen (Setchell et al., 2002). Because of this similarity, isoflavonoids are sometimes called phytoestrogens (Wuttke et al., 2003). Their affinity and competition with endogenous estrogens in binding with estrogen receptors have been reported as the main mechanism by which isoflavones may influence cancer development (Zheng and Zhu, 1999). Soy isoflavones exert both estrogenic and antiestrogenic effects, depending upon the compound, assay, and tissues in which they are acting (Wiseman et al., 2000; Alekel et al., 2000; Messina and Loprinzi, 2001). This has led to their being referred to as "nature's selective estrogen receptor modulators" (Setchell, 2001). The hormonal action of isoflavones has been postulated to be through a number of pathways, including the ability to inhibit many tyrosine kinases involved in regulation of cell growth (Akiyama et al., 1987), to augment transformation growth factor-β which inhibits the cell cycle progression (Zhou and Lee, 1998), as well as to influence the transcription factors that are involved in the expression of stress response-related genes involved in programmed cell death (Zhou and Lee, 1998).

Other nonhormonal mechanisms by which isoflavones are believed to exert their anticarcinogenic effects are via their antioxidant, antiproliferative, antiangiogenic, and anti-inflammatory properties (Gilani and Anderson, 2002). The chemical structure of soy isoflavones are responsible of their antioxidative effects, which results in a decrease in lipid peroxidation (Wiseman et al.,

2000) and oxidative DNA damage (Djuric et al., 2001), both important factors for carcinogenesis.

Isoflavones exert both hormonal and nonhormonal action in the prevention of cancer, but the mechanism by which these compounds exert these chemo-preventive properties is not yet clear and is currently a hot topic for research.

13.2.2 Bowman–Birk Protease Inhibitor

Recently, soybean proteins, as major ingredients of soybean, are receiving more and more attention. Soy protein itself, which is lower in sulfur amino acid content than animal protein, has been shown to inhibit the development of carcinogen-induced tumors in animals (Koski, 2006). Soybean proteins also can be a source of bioactive peptides with diverse and unique health benefits that can be used in the prevention of age-related chronic disorders, such as cardiovascular disease, obesity, decreased immune function, and cancer.

Bowman–Birk protease inhibitor (BBI) is a polypeptide with 71 amino acids and molecular weight of 7975 Da (Birk, 1985). It has the ability to inhibit serine proteases through competition with substrates for access to the active site of the enzyme. BBI has two inhibitory domains and forms a stable stei-chiometric complex with the digestive enzymes trypsin and chymotrypsin. The trypsin inhibitory site of BBI has been associated with negative effects on bioavailability of dietary proteins, whereas the chymotrypsin inhibitory site of BBI has been implicated in cancer chemopreventive effects (Kennedy et al., 1993; Clemente et al., 2005).

Kennedy has demonstrated that soybean BBI has significant cancer che-mopreventive activity in both *in vitro* and *in vivo* bioassay systems (Kennedy, 1998a,b). *In vitro* studies have shown that BBI can inhibit malignant transfor-mation of C3H/10T1/2 cells by X-rays (Yavelow et al., 1983) or 3-methylcho-lanthrene (MCA; St. Clair, 1991). Similarly, BBI suppresses the enhancement of radiation-induced transformation by the tumor promoter tetradecanoylphor-bol-13-acetate (TPA) (Su, Toscano, and Kennedy, 1991) and suppresses the *in vitro* growth of human small cell lung cancer cell lines (Clark et al., 1993), and a number of human prostate cancer cell lines (Kennedy and Wan, 2002). Prevention of transformation in cultured cells by BBI can be achieved at nanomolar levels (Kennedy, 1998a,b).

In animal models, the BBI or the BBI concentrate (BBIC, a preparation largely containing BBI that was developed due to the cost of producing pure BBI) exerts a protective effect in dimethylhydrazine (DMH)-treated animals, reducing the incidence and frequency of colon tumors in mice and rats, without any adverse side effects documented for animal growth or organ physiology (St. Clair, Billings, and Kennedy, 1990). Other animal models have shown that BBI or the BBIC exerts a suppressive effect on 7,12-dimethylbenza(a)anthracene (DMBA)-induced oral cancer (Kennedy et al., 1993), MCA-induced lung cancer (Witschi and Kennedy, 1989), methyl-benzylnitrosamine-induced esophageal cancer (von Hofe, Newberne, and

Kennedy, 1991), and radiation-induced lymphosarcomas (Evans et al., 1992). BBIC also demonstrates activity against established neoplastic lesions, as it inhibits the growth of human prostate cancer xenografts in nude mice (Wan, Ware, and Zhang, 1999), and prevents the formation of pulmonary metastases in mice after subcutaneous injections of tumor cells (Kobayashi et al., 2004).

As a result of this evidence, BBI acquired the status of "investigational new drug" from the FDA in 1992 and currently is being evaluated in large-scale human trials as an anticarcinogenic agent in the form of BBIC. The results of phase I and II clinical trials have shown that BBIC has a substantial positive clinical effect in patients with oral leukoplakia, a preneoplastic lesion in the oral cavity (Armstrong et al., 2000, 2003; Meyskens, 2001). At this time, a phase IIb randomized placebo-controlled clinical trial to determine the clinical effectiveness of BBIC is under way. BBIC is also being used to investigate its efficacy in the treatment of benign prostatic hyperplasia and ulcerative colitis (Kennedy, 2006).

Although many studies have been performed to determine the mechanisms for the anticarcinogenic effects of protease inhibitors, they are still unknown. Several different hypotheses have been discussed, and several different mechanisms may exist. Many of these hypotheses are related to the fact that these agents prevent the release of the superoxide anion radical and hydrogen peroxide from polymorphonuclear leukocytes and other cell types stimulated with tumor-promoting agents (Goldstein et al., 1979; Frenkel et al., 1987). BBI can keep free radicals from being produced in cells and thereby decrease the amount of oxidative damage (Frenkel et al., 1987). It is assumed that the ability to prevent the release of oxygen free radicals is also related to the potent anti-inflammatory activity of BBI. Because inflammation is closely associated with carcinogenesis, the anti-inflammatory activity of BBI could be the major mechanism by which BBI prevents cancer (Kennedy, 1998a,b). Other mechanisms contributing to the anti-inflammatory activity of BBI are the direct and potent inhibitory effects on the catalytic activities of major proteases involved in inflammatory processes, such as cathepsin G, elastase, and chymase (Larionova et al., 1993; Ware et al., 1997). It has also been hypothesized that BBI suppresses carcinogenesis by affecting certain types of proteolytic activities (Messadi et al., 1986; Billings et al., 1987, 1988; Billings, Habres, and Kennedy, 1990, 1991; Billings and Habres, 1992; Billings, 1993) or the expression of certain proto-oncogenes, both of which are thought to play important roles in carcinogenesis.

13.2.3 Other Proteins and Peptides

Recently, there has been increased interest in the potential health benefits of other bioactive polypeptides and proteins from soybean, including lectins and lunasin. Soy lectins are a significant group of biologically active glycoproteins that have been shown to possess cancer chemopreventive activity *in vitro, in vivo,* and in human case studies (González de Mejia, Bradford, and Hasler, 2003). The suggested mechanisms of action for lectins include their effect on tumoral cell membranes, the reduction in cell proliferation, the induction of tumor-specific cytotoxicity of macrophages, and the induction of apoptosis. Another suggestion is that lectins could have a strong effect on the immune system by altering the production of various interleukins (González de Mejia and Prisecaru, 2005).

There is still much to learn about the effects of soybean lectins on cancer risk. However, they are currently being used as therapeutics agents in cancer treatment studies and this area of research holds considerable potential.

13.3 Lunasin

13.3.1 Discovery and Sequence of Lunasin

Lunasin was initially identified in the soybean cotyledon when a cDNA encoding a post-translationally processed 2S-albumin (Gm2S-1) was cloned from mid-maturation soybean seed (Galvez, Revilleza, and de Lumen, 1997). Gm2S-1 coded not only for the methionine-rich protein that was sought by these authors but also for other three proteins, a signal peptide, a linker peptide, and a small subunit. The subunit was termed lunasin from the Tagalog word *lunas* for cure.

Lunasin is a peptide composed of 43 amino acid residues with a MW of 5.5 kDa. This peptide, whose sequence is SKWQHQQDSCRKQKQGVNLTPC-EKHIMEKIQG-*RGD*-**DDDDDDDD,** contains a poly-aspartyl (D) carboxyl end (bold) preceded by a cell adhesion motif Arg-Gly-Asp (*RGD*) (italics) and a predicted helix region with structural homology to a conserved region of chromatin-binding proteins (underlined; Galvez et al., 2001). In 1999, Galvez and de Lumen discovered that transfection and constitutive expression of the lunasin gene into mammalian cells (murine hepatome, breast cancer cells, and murine fibroblasts) disrupted mitosis and induced chromosomal fragmentation and apoptosis (Galvez and de Lumen, 1999). The killing effect of the lunasin gene, however, affects both normal and cancer cells. The authors showed that in the lysed cells, lunasin adheres to the fragmented chromosomes preceded by the asymmetric distribution of metaphase chromosomes and elongated spindle fibers that remain unattached to the kinetochores. Therefore, Galvez and de Lumen proposed a model for the mechanism of action of the lunasin gene. They attributed the antimitotic effect of lunasin to the binding of its negatively charged poly-D carboxyl end to the highly basic histones found within the nucleosomes of condensed chromosomes, probably to regions that contain more positively charged deacetylated histones, such as the hypoacetylated chromatin found in telomeres and centromeres. The displacement by lunasin of the kinetochore proteins normally bound to the centromeres could lead to the failure of spindle fiber attachment and eventually to mitotic arrest and cell death (Galvez and de Lumen, 1999). *In vitro* binding studies have demonstrated that lunasin binds specifically to deacetylated core histones but not to acetylated histones. The poly-D is required for this binding whereas the helical region may play a role targeting lunasin to core histones due to its similarity to a number of chromosome binding proteins (Galvez et al., 2001).

Moreover, the sequence of lunasin contains the motif RGD known to allow tumor cell attachment to the extracellular matrix (Ruoslahti and Pierschbacher, 1986). Peptides containing this motif have been reported to prevent metastasis of tumor cells by competitive adhesion to extracellular matrices (Akiyama, Aota, and Yamada, 1995). It has been demonstrated that this motif is required for internalization of lunasin into the nucleus of C3H10T1/2 cells, but unnecessary for internalization into the nucleus of another cell line (NIH3T3), suggesting that the role of the RGD motif could be cell-line specific (Galvez et al., 2001). All these observations suggest a possible role of the lunasin gene in cancer therapy but the gene has to be coupled with a cancer-cell-specific delivery system; the constitutive expression of the lunasin gene kills both cancer and normal cells.

13.3.2 *In Vitro* Cancer Preventive Properties of Lunasin Peptide and Mechanism of Action

In contrast to the antimitotic effect of the constitutive expression of the lunasin gene, lunasin peptide has been demonstrated to have chemopreventive properties both *in vitro* and *in vivo*. If carcinogenic agents are not present, cell morphology and proliferation do not seem to be affected by this peptide. However, in the foci formation assay, it has been reported that lunasin suppresses transformation by about 62 to 90% relative to the positive control (chemical carcinogens DMBA and MCA alone) at concentrations ranging from 10 nM to 10 µM in a dose-dependent manner (Galvez et al., 2001). Lunasin has been demonstrated to be fourfold more active than the BBIC, another known cancer preventive soybean protein (Section 13.2.2). Lunasin's suppressive effect has also been shown in transformation of mouse fibroblast cells C3H10T1/2 and NIH3T3 caused by oncogenes and genes that inactivate tumor suppressor proteins (Galvez et al., 2001; Lam, Galvez, and de Lumen, 2003; Jeong et al., 2003). Addition of lunasin at 20 nM to cells stably transfected with the viral oncogene E1A has been reported to suppress colony formation (Lam et al., 2003). This oncogene is known to be associated with human tumors due to its capacity to induce cell cycle progression and transforming cells by inactivating the tumor suppressor retinoblastoma protein (Rb; Helt and Galloway, 2003).

Ras-oncogenes are well-studied cancer-related genes due to their frequent activation in human cancers, and play a central role in the ras/mitogen activated protein kinase (MAPK) signaling cascade, which has a pivotal role in cell proliferation, differentiation, survival, and cell death (Barbacid, 1987; Malumbres and Barbacid, 2003). Jeong and coworkers demonstrated that lunasin prevents transformation of NIH3T3 cells transfected with inducible forms of this oncogene (Jeong et al., 2003).

The results obtained by these *in vitro* experiments have suggested the role of lunasin against different carcinogenic pathways that share common mechanisms. The new studies carried out since then have been focused on demonstrating the epigenetic nature of lunasin's mechanism of action. Epigenetic mechanisms are responsible for the organization and compaction of the genome in such a manner that the set of genes needed in a particular cell type is accessible for transcription (Fog, Jensen, and Lund, 2007). These mechanisms influence processes such as cell-cycle control, angiogenesis, migration, and DNA damage responses. In recent years, epigenetic changes have become established as being important molecular signatures of human cancer (Zheng et al., 2008). The two major epigenetic changes are DNA methylation and posttranslational modifications of histone proteins, such as acetylation, phosphorylation, and methylation. Histone acetylation is one of these modifications catalyzed by histone acetyltransferases (HATs), which transfer an acetyl group from acetyl-CoA to lysine residues on histone proteins. Acetylation neutralizes the positive charge of these histones and disrupts the electrostatic interactions between DNA and histone proteins that promote chromatin unfolding which has been associated with transcription and gene expression (Davis and Ross, 2007).

First studies with lunasin have reported that the anticancer potential of this peptide can be attributed to its capability to selectively kill cells that are being transformed or are newly transformed by disrupting the dynamic of histone acetylation-deacetylation. Under steady-state conditions in the cell, the core H3 and H4 histones are mostly deacetylated (repressed state). In the presence of sodium butyrate, a well-known deacetylase inhibitor, lunasin has been found to inhibit histone acetylation in both normal and cancerous mammalian cells (Galvez et al., 2001; Jeong et al., 2003). Moreover, lunasin has been demonstrated to compete with different HATs, such as yGCN5 and PCAF, in binding to deacetylated core histone, inhibiting the acetylation and repressing the cell cycle progression (Jeong et al., 2007a). With these findings, a model of the lunasin's mechanism of action has been proposed. This model stipulates that when tumor suppressor proteins, such as Rb, p53, and pp32, are inactivated by chemical carcinogens or viral oncogenes, HATs enzymes can act by acetylating histones and stimulating gene transcription and cell cycle progression. If lunasin is present in the nucleus, it is acting as a surrogate tumor suppressor by tightly binding to deacetylated core histones and disrupting the balance between acetylation–deacetylation, which is perceived by the cell as abnormal and leads to cell death (de Lumen, 2005).

Rb is an important regulator of cell-cycle progression and differentiation (Riley, Lee, and Lee, 1994; Weinberg, 1995). It is able to suppress inappropriate proliferation by arresting cells in the G1 phase of the cell cycle (Dyson, 1998; Helin, 1998). Inactivation of its growth-suppressing function can be achieved through its phosphorylation in Ser 780 of the protein chain (Bartek, Bartkova, and Lukas, 1997). Cyclin-dependent kinase (CDK)/cyclin complexes are responsible for phosphorylation and inactivation of Rb in cells to allow cell-cycle progression, and they are upregulated in tumor cells (Bartek et al., 1997). A recent study carried out with lunasin has demonstrated its capacity to inhibit Rb phosphorylation induced by cyclin D1 (Jeong et al., 2007a). This might affect the cell cycle control pathway, especially G1/S arrest, resulting in keeping the core histone deacetylated and inhibiting abnormal cell growth. Further research is currently under way to elucidate the complete epigenetic mechanism of action of lunasin.

13.3.3 *In Vivo* Cancer Preventive Properties of Lunasin Peptide

Chemopreventive properties of lunasin have also been demonstrated *in vivo*. The first animal model used to demonstrate these properties was the SENCAR mouse model treated with a chemical carcinogen, a commonly used model of skin cancer (Chen et al., 1995; Galvez et al., 2001). Dermal application with 250 µg of lunasin per week to SENCAR mice treated with chemical DMBA and promoter TPA reduces skin tumor incidence, decreases tumor yield/mouse, and increases the tumor latency period compared with the untreated control (Galvez et al., 2001). New studies using the ^2H_2O labeling method allows sensitive detection of changes in epidermal cell proliferation within a much shorter time period than the traditional methods. This has been reported to be very useful for evaluating the efficacy of chemopreventive agents with antipromotional and antiproliferative actions, for example, lunasin. Lunasin delayed the appearance of papilloma because of its capacity of slowing down keratinocyte proliferation in mouse skin by DMBA challenge (Hsieh et al., 2004).

Because oral administration is an important characteristic of an ideal cancer preventive agent, new studies are being conducted to demonstrate the chemopreventive properties of lunasin after oral ingestion. Preliminary bioavailability studies carried out in mice and rats fed lunasin-enriched soy protein have found that 35% of ingested lunasin reaches the target tissues and organs in an intact and active form (Jeong et al., 2007a,b). It has also been reported that soy naturally occurring protease inhibitors, such as BBI and Kunitz trypsin inhibitor, provide a combined protection against digestion of lunasin by gastrointestinal enzymes (Park, Jeong, and de Lumen, 2007). This protection plays a major role in making lunasin available in soy protein. The capacity of lunasin to survive degradation by gastrointestinal and serum proteinases and peptidases reaching blood and other organs in a bioactive form, as well as *in vitro* chemopreventive properties make lunasin a perfect candidate to exert a potent *in vivo* cancer preventive activity. The experiments currently under way will be very useful in proving this activity.

13.3.4 Lunasin Concentration in Soy and Other Seeds

Initially, lunasin peptide was identified in the soybean cotyledon, proposing an intriguing role of this peptide in seed development. The angiosperm seed development consists of three stages. During the first two stages, a rapid cell division and differentiation, followed by cessation of cell division in the central parenchyma cells of the cotyledon or endosperm, and enlargement of the cells accompanied by biosynthesis of storage forms of carbohydrates, proteins, lipids, and nucleic acids for the germinating seeds occur (Spencer and Higgins, 1981). Finally, during the last stage, the seed dehydrates. The second stage is considered unique to the angiosperm seeds in endoreduplication of DNA without cell division, allowing DNA accumulation for purposes of storage (Hellerstein, 1999). We have proposed that lunasin is a molecule that allows arrest of cell division and initiates the second stage of seed development. Thus, in theory, all angiosperm seeds should contain lunasin. The first studies demonstrated the presence of this peptide in soybean, in a concentration ranging from 4.4 to 70.5 mg lunasin/g protein (González de Mejia et al., 2004; Jeong et al., 2007b). These variations indicate that the soybean genotype has a significant effect on lunasin concentration and suggest the possibility of selecting and breeding varieties of soy with higher lunasin content (González de Mejia et al., 2004). The stages of seed development have also been found to affect lunasin concentration, a notable increase happens during seed maturation (Park et al., 2005). However, sprouting leads to a continuing decrease of lunasin with soaking time. Light and dark conditions do not seem to affect the content of this peptide. Recently, Wang and coworkers have reported the influence of the environmental factors, mainly temperature and soil moisture, and the processing conditions on lunasin concentration (Wang et al., 2008). It has been demonstrated that large-scale processing of soy to produce different protein fractions also influences lunasin concentration. This content varied from 12 to 44 mg lunasin/g of flour when different commercially available soy proteins were analyzed (Jeong et al., 2003; González de Mejia et al., 2004).

In the search for natural sources of lunasin in addition to soybean, a first screening has been carried out using different beans, grains, and herbal plants. Lunasin has been found in barley and wheat, two cereal grains known for its health effects (Jeong et al., 2002, 2007c). Several seeds of oriental herbal and medicinal plants were analyzed, with the finding that lunasin is present in all of the *Solanaceae* family, except *L. chinensis.*, but not in any of the *Phaseolus* beans analyzed (Jeong et al., 2007a). Recently, lunasin has been identified in amaranth, called the "golden grain" by the Aztecs because it provides grains and leaves with high nutritional value and known physiological properties (Silva-Sánchez et al., 2008). A more rigorous and systematic search of lunasin and lunasin homologues in different seeds is currently being carried out in order to establish a relation between the presence of this peptide and the taxonomic properties of the plants.

13.4 Conclusion

Epidemiological evidence has demonstrated an association between the consumption of soybean and improved health, particularly reduced risk for cardiovascular diseases and cancer. *In vitro* as well as *in vivo* studies support the cancer preventive properties of soy and soy compounds responsible for these properties. This review has summarized the chemopreventive activity of isoflavonoids and peptides that contribute to reported cancer preventive effects of soybean.

Lunasin is a novel cancer preventive seed peptide initially identified in soybean and now found in other seeds. Its efficacy has been established in cell culture models and in a skin cancer mouse model against chemical carcinogens and oncogenes. Lunasin has been proposed to selectively kill cells that are being transformed by inhibiting histone acetylation and disrupting the normal dynamics of histone acetylation–deacetylation. Initial studies show that lunasin administered in soy proteins is bioavailable, evidently protected from digestion by naturally occurring protease inhibitors in soy. Research using animal models is currently being conducted to demonstrate lunasin's chemopreventive activity against different types of cancer, such as breast, colon, prostate, and oral cancer. Moreover, studies applying genomics, proteomics, and biochemical tools are being carried out to fully elucidate the epigenetic nature of lunasin's mechanism of action. Other aspects, such as searching for lunasin in other seeds, optimization of techniques to enrich products with this peptide, and studying lunasin's interactions with other food constituents affecting its activity should also be conducted. The results of all these studies would position lunasin as a novel and potent naturally occurring cancer preventive agent.

Acknowledgments

The authors would like to acknowledge the American Institute for Cancer Research (AICR) for research funding. Blanca Hernández-Ledesma is a Marie-Curie post-doctoral fellow from the European Commission coordinated by the Spanish National Research Council.

References

Akiyama, S.K., Aota, S., and Yamada, K.M. 1995. Function and receptor specificity of a minimal 20-Kilodalton cell adhesive fragment of fibronectin. *Cell Adhes. Commun.* 3: 13–25.

Akiyama, T., Ishida, J., Nakagawa, S. et al. 1987. Genistein a specific inhibitor of tyrosine protein kinases. *J. Biol. Chem.* 262: 5592–5595.

Alekel, D.L., Germain, A.S., Peterson, C.T. et al. 2000. Isoflavone-rich soy protein isolate attenuates bone loss in the lumbar spine of perimenopausal women. *Amer. J. Clin. Nutr.* 72: 844–852.

Anderson, J.W., Smith, B.M., and Washnock, C.S. 1999. Cardiovascular and renal benefits of dry bean and soybean intake. *Amer. J. Clin. Nutr.* 70: 464S–474S.

Armstrong, W.B., Kennedy, A.R., Wan, X.S. et al. 2000. Single-dose administration of Bowman–Birk inhibitor concentrate in patients with oral leukoplakia. *Cancer Epidemiol. Biomark. Prevent.* 9: 43–47.

Armstrong, W.B., Wan, X.S., Kennedy, A.R. et al. 2003. Development of the Bowman–Birk inhibitor for oral cancer chemoprevention and analysis of neu immunohistochemical staining intensity with Bowman–Birk inhibitor concentrate treatment. *Laryngoscope* 113: 1687–1702.

Ban, J.-M., Lee, E.J., and Guyatt, G. 2008. Citrus fruit intake and stomach cancer risk: A quantitative systematic review. *Gastric Cancer* 11: 23–32.

Barbacid, M. 1987. Ras genes. *Ann. Rev. Biochem.* 56: 779–827.

Barnes, S., Grubbs, C., Setchell, K.D.R. et al. 1990. Soybeans inhibit mammary tumors in models of breast cancer. In *Progress in Clinical and Biological Research*, Volume 347. *Mutagens and Carcinogens in the Diet* (M.W. Pariza, et al., Eds.), New York, Chichester, UK: Wiley-Liss, pp. 239–254.

Bartek, J., Bartkova, J., and Lukas, J. 1997. The retinoblastoma protein pathway in cell cycle control and cancer. *Exper. Cell Res.* 237: 1–6.

Béliveau, R. and Gingras, D. 2007. Role of nutrition in preventing cancer. *Canad. Family Physician* 53: 1905–1911.

Bhat, T.A. and Singh, R.P. 2008. Tumor angiogenesis—A potential target in cancer chemoprevention. *Food Chem. Toxicol.* 46:1334–1345.

Billings, P.C. 1993. Approaches to studying the target enzymes of anticarcinogenic protease inhibitors. In *Protease Inhibitors as Cancer Chemopreventive Agents* (W. Troll and A.R. Kennedy, Eds.), New York: Plenum Press, pp. 191–198.

Billings, P.C., Carew, J.A., Keller-McGandy, C.E. et al. 1987. A serine protease activity in C3H/10T1/2 cells that is inhibited by anticarcinogenic protease inhibitors. *Proc. Nat. Acad. Sci. USA* 84: 4801–4805.

Billings, P.C. and Habres, J.M. 1992. A growth regulated protease activity which is inhibited by the anticarcinogenic Bowman–Birk protease inhibitor. *Proc. Nat. Acad. Sci.* 89: 3120–3124.

Billings, P.C., Habres, J.M., and Kennedy, A.R. 1990. Inhibition of radiation induced transformation of C3H10T1/2 cells by specific protease substrates. *Carcinogenesis* 11: 329–332.

Billings, P.C., Habres, J.M., Liao, D.C. et al. 1991. A protease activity in human fibroblasts which is inhibited by the anticarcinogenic Bowman–Birk protease inhibitor. *Cancer Res.* 51: 5539–5543.

Billings, P.C., St. Clair, W., Owen, A.J. et al. 1988. Potential intracellular target proteins of the anticarcinogenic Bowman–Birk protease inhibitor identified by affinity chromatography. *Cancer Res.* 48: 1798–802.

Birk, Y. 1985. The Bowman–Birk inhibitor. *Int. J. Peptide Protein Res.* 25: 113–131.

Chen, D. and Dou, Q.P. 2008. Tea polyphenols and their roles in cancer prevention and chemotherapy. *Int. J. Molec. Sci.* 9: 1196–1206.

Chen, L.C., Tarone, R., Huynh, M. et al. 1995. High dietary retinoic acid inhibits tumor promotion and malignant conversion in a two-stage skin carcinogenesis protocol using 7,12-dimethylbenz[a]anthracene as the initiator and mezerein as the tumor promoter in female SENCAR mice. *Cancer Lett.* 95: 113–118.

Clark, D.A., Day, R., Seidah, N. et al. 1993. Protease inhibitors suppress in-vitro growth of human small-cell lung-cancer. *Peptides* 14: 1021–1028.

Clemente, A., Gee, J.M., Johnson, I.T. et al. 2005. Pea (*Pisum sativum* L.) protease inhibitors from the Bowman–Birk class influence the growth of human colorectal adenocarcinoma HT29 cells *in vitro*. *J. Agric. Food Chem.* 53: 8979–8986.

Cohen, S.M. 1995. Cell proliferation in the bladder and implications for cancer risk assessment. *Toxicology* 102: 149–159.

Coward, L., Barnes, N., Setchell, K. et al. 1993. Genistein, daidzein, and their beta-glycoside conjugates: antitumor isoflavones in soybean food from American and Asian diets. *J. Agric. Food Chem.* 41: 1961–1967.

Davis, C.D. and Ross, S.A. 2007. Dietary components impact histone modifications and cancer risk. *Nutr. Rev.* 65: 88–94.

de Kok, T.M., van Breda, S.G., and Manson, M.M. 2008. Mechanisms of combined action of different chemopreventive dietary compounds. *Euro. J. Nutr.* 47: 51–59.

de Lumen, B.O. 2005. Lunasin: A cancer-preventive soy peptide. *Nutr. Rev.* 63: 16–21.

Djuric, Z., Chen, G., Doerge, D.R. et al. 2001. Effect of soy isoflavone supplementation on oxidative stress in men and women. *Cancer Lett.* 172: 1–6.

Dyson, N. 1998. The regulation of E2F by pRB-family proteins. *Genes Dev.* 12: 2245–2262.

Erdman, J. Jr., Badger, T., Lampe, J. et al. 2004. Not all soy products are created equal: Caution needed in interpretation of research results. *J. Nutr.* 134: S1229–1233.

Evans, S.M., Szuhaj, B.F., Van Winkle, T. et al. 1992. Protection against metastasis of radiation induced thymic lymphosarcoma and weight loss in C57Bl/6NCr1BR mice by an autoclave resistant factor present in soybeans. *Radiation Res.* 132: 259–262.

Fimognari, C., Lenzi, M., and Hrelia, P. 2008. Chemoprevention of cancer by isothiocyanates and anthocyanins: Mechanisms of action and structure-activity relationship. *Curr. Med. Chem.* 15: 440–447.

Fog, C.K., Jensen, K.T., and Lund, A.H. 2007. Chromatin-modifying proteins in cancer. *APMIS* 115: 1060–1089.

Fournier, D.B., Erdman, J.W. Jr., and Gordon, G.B. 1998. Soy, its components, and cancer prevention: A review of the *in vitro*, animal, and human data. *Cancer Epidemiol. Biomark. Prev.* 7: 1055–1065.

Frenkel, K., Chrzan, K., Ryan, C.A. et al. 1987. Chymotrypsin specific protease inhibitors decrease H2O2 formation by activated human polymorphonuclear leukocytes. *Carcinogenesis* 8: 1207–1212.

Galvez, A.F., Chen, N., Macasieb, J. et al. 2001. Chemopreventive property of a soybean peptide (Lunasin) that binds to deacetylated histones and inhibits acetylation. *Cancer Res.* 61: 7473–7478.

Galvez, A.F. and de Lumen, B.O. 1999. A soybean cDNA encoding a chromatin-binding peptide inhibits mitosis of mammalian cells. *Nature Biotechnol.* 17: 495–500.

Galvez, A.F., Revilleza, M.J.R., and de Lumen, B.O. 1997. A novel methionine-rich protein from soybean cotyledon: Cloning and characterization of cDNA (accession No. AF005030). Plant Gene Register #PGR97-103. *Plant Physiol.* 114: 1567–1569.

Garcia, M., Jemal, A., Ward, E.M. et al. 2007. *Global Cancer Facts & Figures 2007*. Atlanta, GA: The American Cancer Society.

Gilani, G.S. and Anderson, J.J.B. 2002. *Phytoestrogens and Health*. Champaign, IL: AOCS Press.

Goldstein, B.D., Witz, G., Amoruso, M. et al. 1979. Protease inhibitors antagonize the activation of polymorphonuclear leukocyte oxygen consumption. *Biochem. Biophys. Res. Commun.* 88: 854–860.

González de Mejia, E. and Prisecaru, V.I. 2005. Lectins as bioactive plant proteins: a potential in cancer treatment. *Crit. Rev. Food Sci. Nutr.* 45: 425–455.

González de Mejia, E., Bradford, T., and Hasler, C. 2003. The anticarcinogenic potential of soybean lectin and lunasin. *Nutr. Rev.* 61: 239–246.

González de Mejia, E., Vásconez, M., de Lumen, B.O., and Nelson, R. 2004. Lunasin concentration in different soybean genotypes, commercial soy protein, and isoflavone products. *J. Agric. Food Chem.* 52: 5882–5887.

Goodman, M.T., Wilkens, L.R., Hankin, J.H. et al. 1997. Association of soy and fiber consumption with the risk of endometrial cancer. *Amer. J. Epidemiol.* 146: 294–306.

Helin, K. 1998. Regulation of cell proliferation by the E2F transcription factors. *Curr. Opin. Gen. Dev.* 8: 28–35.

Hellerstein, M. H. 1999. Antimitotic peptide characterized from soybean: Role in protection from cancer? *Nutr. Rev.* 57: 359–61.

Helt, A.M. and Galloway, D.A. 2003. Mechanisms by which DNA tumor virus oncoproteins target the Rb family of pocket proteins. *Carcinogenesis* 24: 159–169.

Hernández-Ledesma, B., Hsieh, C.-C., and de Lumen, B.O. 2009. Lunasin: A novel peptide for cancer prevention. *Perspect. Med. Chem.* 2: 75–80.

Hsieh, E.A., Chai, C.M., de Lumen, B.O. et al. 2004. Dynamics of keratinocytes *in vivo* using ^2H_2O labeling: A sensitive marker of epidermal proliferation state. *J. Invest. Dermatol.* 123: 530–536.

Jacobsen, B.K., Knutsen, S.F., and Fraser, G.E. 1998. Does high soy milk intake reduce prostate cancer incidence? The Adventist health study (United States). *Cancer Causes Control* 9: 553–557.

Jeong, H.J., Jeong, J.B., Kim, D.S. et al. 2007b. Inhibition of core histone acetylation by the cancer preventive peptide lunasin. *J. Agric. Food Chem.* 55: 632–637.

Jeong, H.J., Jeong, J.B., Kim, D.S. et al. 2007c. The cancer preventive peptide lunasin from wheat inhibits core histone acetylation. *Cancer Lett.* 255: 42–48.

Jeong, H.J., Lam, Y., and de Lumen, B.O. 2002. Barley lunasin suppresses *ras*-induced colony formation and inhibits core histone acetylation in mammalian cells. *J. Agric. Food Chem.* 50: 5903–5908.

Jeong, H.J., Park, J.H., Lam, Y. et al. 2003. Characterization of lunasin isolated from soybean. *J. Agric. Food Chem.* 51: 7901–7906.

Jeong, J.B., Jeong, H.J., Park, J.H. et al. 2007a. Cancer-preventive peptide lunasin from *Solanum nigrum* L. inhibits acetylation of core histones H3 and H4 and phosphorylation of retinblastoma protein (Rb). *J. Agric. Food Chem.* 55: 10707–10713.

Kaefer, C.M. and Milner, J.A. 2008. The role of herbs and spices in cancer prevention. *J. Nutr. Biochem.* 19: 347–361.

Kennedy, A.R. 1998a. Chemopreventive agents: Protease inhibitors. *Pharmacol. Therapeut.* 78: 167–209.

Kennedy, A.R. 1998b. The Bowman–Birk inhibitor from soybeans as an anticarcinogenic agent. *Amer. J. Clin. Nutr.* 68 (suppl): 1406S–1412S.

Kennedy, A.R. 2006. The status of human trials utilizing Bowman–Birk inhibitor concentrate from soybeans. In *Soy in Health and Disease Prevention* (M. Sugano, Ed.), Boca Raton, FL: CRC Press.

Kennedy, A.R., Billings, P.C., Maki, P.A. et al. 1993. Effects of various preparations of dietary protease inhibitors on oral carcinogenesis in hamsters induced by DMBA. *Nutr. Cancer* 19: 191–202.

Kennedy, A.R. and Wan, X.S. 2002. Effects of the Bowman–Birk inhibitor on growth, invasion, and clonogenic survival of human prostate epithelial cells and prostate cancer cells. *Prostate* 50: 125–133.

Kerwin, S.M. 2004. Soy saponins and the anticancer effects of soybeans and soy-based foods. *Curr. Med. Chem.* 4: 263–272.

Khan, N., Afaq, F., and Mukhtar, H. 2008. Cancer chemoprevention through dietary antioxidants: progress and promise. *Antiox. Redox Signal.* 10: 475–510.

Kobayashi, H., Fukuda, Y., Yoshida, R. et al. 2004. Suppressing effects of dietary supplementation of soybean trypsin inhibitor on spontaneous, experimental and peritoneal disseminated metastasis in mouse model. *Int. J. Cancer* 112: 519–524.

Kolonel, L.N., Hankin, J.H., Whittemore, A.S. et al. 2000. Vegetables, fruits, legumes and prostate cancer: A multiethnic case-control study. *Cancer Epidemiol. Biomark. Prev.* 9: 795–804.

Koski, S.L. 2006. Soy protein ingredients and their health benefits. *Agro Food Industry Hi-Tech.* 17: 33–35.

Lam, Y., Galvez, A.F., and de Lumen, B.O. 2003. Lunasin suppresses E1A-mediated transformation of mammalian cells but does not inhibit growth of immortalized and established cancer cell lines. *Nutr. Cancer* 47: 88–94.

Larionova, N.I., Gladysheva, I.P., Tikhonova, T.V. et al. 1993. Inhibition of cathepsin G and human granulocyte elastase by multiple forms of Bowman–Birk type soybean inhibitor. *Biokhimiia* 58: 1437–1444.

Li, D.H., Yee, J.A., Mcguire, M.H. et al. 1999. Soybean isoflavones reduce experimental metastasis in mice. *J. Nutr.* 129: 1075–1078.

Lof, M. and Weiderpass, E. 2006. Epidemiological evidence suggests that dietary phytoestrogens intake is associated with reduced risk of breast, endometrial and prostate cancers. *Nutr. Res.* 26: 609–619.

Losso, J.N. 2008. The biochemical and functional food properties of the Bowman–Birk Inhibitor. *Crit. Rev. Food Sci. Nutr.* 48: 94–118.

Malumbres, M. and Barbacid, M. 2003. RAS oncogenes: The first 30 years. *Nature Rev. Cancer* 3: 459–465.

Manson, M. 2003. Cancer prevention – The potential for diet to modulate molecular signaling. *Trends Molec. Med.* 9: 11–18.

McCue, P. and Shetty, K. 2004. Health benefits of soy isoflavonoids and strategies for enhancement: A review. *Crit. Rev. Food Sci. Nutr.* 44: 361–367.

Messadi, P.V., Billings, P., Shklar, G. et al. 1986. Inhibition of oral carcinogenesis by a protease inhibitor. *J. Nat. Cancer Inst.* 76: 447–452.

Messina, M. and Flickinger, B. 2002. Hypothesized anticancer effects of soy: Evidence points to isoflavones as the primary anticarcinogens. *Pharmaceut. Biol.* 40: S6–S23.

Messina, M., McCaskill-Stevens, W., and Lampe, J.W. 2006. Addressing the soy and breast cancer relationship: Review, commentary, and workshop proceedings. *J. Nat. Cancer Inst.* 98: 1275–1284.

Messina, M.J. and Loprinzi, C.L. 2001. Soy for breast cancer survivors: A critical review of the literature. *J. Nutr.* 131(suppl 11): 3095S–3108S.

Meyskens, F.L. 2001. Development of Bowman–Birk inhibitor for chemoprevention of oral head and neck cancer. *Cancer Prevent.* 952: 116–123.

Park, J.H., Jeong, H.J., and de Lumen, B.O. 2005. Contents and bioactivities of lunasin, Bowman–Birk inhibitor, and isoflavones in soybean seed. *J. Agric. Food Chem.* 53: 7686–7690.

Park, J.H., Jeong, H.J., and de Lumen, B.O. 2007. *In vitro* digestibility of the cancer-preventive soy peptides lunasin and BBI. *J. Agric. Food Chem.* 55: 10703–10706.

Park, O.J. and Surh, Y.-H. 2004. Chemopreventive potential of epigallocatechin gallate and genistein: evidence from epidemiological and laboratory studies. *Toxicol. Lett.* 150: 43–56.

Pitot, H.C. 1993. The molecular biology of carcinogenesis. *Cancer* 72: 962–970.

Ramos, S. 2008. Cancer chemoprevention and chemotherapy: Dietary polyphenols and signaling pathways. *Molec. Nutr. Food Res.* 52: 507–526.

Riley, D.J., Lee, E.Y., and Lee, W.H. 1994. The retinoblastoma protein: More than a tumor suppressor. *Ann. Rev. Cell . Dev. Biol.* 10: 1–29.

Ruoslahti, E. and Pierschbacher, M.D. 1986. Arg-Gly-Asp: A versatile cell recognition signal. *Cell* 44: 517–518.

Schwartz, G.K. and Shah, M.A. 2005. Targeting the cell cycle: A new approach to cancer therapy. *J. Clin. Oncol.* 23: 9408–9421.

Seki, T., Hosono, T., Hosono-Fukao, T. et al. 2008. Anticancer effects of diallyl trisulfide derived from garlic. *Asia Pacific J. Clin. Nutr.* 17: 249–252.

Setchell, K.D. 2001. Soy isoflavones: Benefits and risks from nature's selective estrogen receptor modulators (SERMs). *J. Amer. Coll. Nutr.* 20(suppl 5): 354S–362S.

Setchell, K.D., Brown, N.M., Zimmer-Nechemias, L. et al. 2002. Evidence for lack of absorption of soy isoflavone glycosides in humans, supporting the crucial role of intestinal metabolism for bioavailability. *Amer. J. Clin. Nutr.* 76: 447–453.

Silva-Sánchez, C., Barba de la Rosa, A.P., León-Galván, M.F. et al. 2008. Bioactive peptides in amaranth (*Amaranthus hypochondriacus*) seed. *J. Agric. Food Chem.* 56: 1233–1240.

Singh, P. and Goyal, G.K. 2008. Dietary lycopene: Its properties and anticarcinogenic effects. *Comp. Rev. Food Sci. Food Safety* 7: 255–270.

Spencer, D. and Higgins, T.J.V. 1981. Molecular aspects of seed protein biosynthesis. In *Commentaries in Plant Science*, Volume 2 (H. Smith, Ed.), New York: Pergamon Press, pp. 175–189.

St. Clair, W.H. 1991. Suppression of 3-methylcholanthrene induced cellular transformation by timed administration of the Bowman–Birk protease inhibitor. *Carcinogenesis* 12: 935–938.

St. Clair, W.H., Billings, P.C., and Kennedy, A.R. 1990. The effects of the Bowman–Birk protease inhibitor on c-*myc* expression and cell proliferation in the unirradiated and irradiated mouse colon. *Cancer Lett.* 52: 145–152.

Su, L.N., Toscano, W.A., and Kennedy, A.R. 1991. Suppression of phorbol ester-enhanced radiation-induced malignancy invitro by protease inhibitors is independent of protein-kinase-C. *Biochim. Biophys. Res. Commun.* 176: 18–24.

Sun, C.L., Yuan, J.M., Arakawa, K. et al. 2002. Dietary soy and increased risk of bladder cancer: The Singapore Chinese health study. *Cancer Epidemiol. Biomark. Prev.* 11: 1674–1677.

Swanson, C.A., Mao, B.L., Li, J.Y. et al. 1992. Dietary determinants of lung-cancer risk - Results from a case-control study in Yunnan province, China. *Int. J. Cancer* 50: 876–880.

van Breda, S.G.J., de Kok, T.M.C.M., and van Delft, J.H.M. 2008. Mechanisms of colorectal and lung cancer prevention by vegetables: A genomic approach. *J. Nutr. Biochem.* 19: 139–157.

von Hofe, E., Newberne, P.M., and Kennedy, A.R. 1991. Inhibition of N-nitroso-methylbenzylamine induced esophageal neoplasms by the Bowman–Birk protease inhibitor. *Carcinogenesis* 12: 2147–2150.

Wan, X.S., Ware, J.H., and Zhang, L.L. 1999. Treatment with soybean-derived Bowman Birk inhibitor increases serum prostate-specific antigen concentration while suppressing growth of human prostate cancer xenografts in nude mice. *Prostate* 41: 243–252.

Wang, H. and Murphy, P. 1994. Isoflavone content in commercial soybean foods. *J. Agric. Food Chem.* 42: 666–673.

Wang, W., Dia, V.P., Vasconez, M. et al. 2008. Analysis of soybean protein-derived peptides and the effect of cultivar, environmental conditions, and processing on lunasin concentration in soybean and soy products. *J. AOAC Int.* 91: 936–946.

Ware, J.H., Wan, X.S., Rubin, H. et al. 1997. Soybean Bowman–Birk protease inhibitor is a highly effective inhibitor of human mast cell chymase. *Arch. Biochem. Biophysics* 344: 133–138.

Weinberg, R.A. 1995. The retinoblastoma protein and cell cycle control. *Cell* 81: 323–330.

Welss, T., Papoutsaki, M., Michel, G. et al. 2003. Molecular basis of basal cell carcinoma: analysis of differential gene expression by differential display PCR and expression array. *Int. J. Cancer* 104: 66–72.

Wiseman, H., O'Reilly, J.D., and Adlercreutz, H. et al. 2000. Isoflavone phytoestrogens consumed in soy decrease F2-isoprostane concentrations and increase resistance of low-density lipoprotein to oxidation in humans. *Amer. J. Clin. Nutr.* 72: 395–400.

Witschi, H. and Kennedy, A.R. 1989. Modulation of lung tumor development in mice with the soybean-derived Bowman–Birk protease inhibitor. *Carcinogenesis* 10: 2275–2277.

Wu, A.H., Ziegler, R.G., Horn-Ross, P.L. et al. 1996. Tofu and risk of breast cancer in Asian-Americans. *Cancer Epidemiol. Biomark. Prev.* 5: 901–906.

Wuttke, W., Jarry, H., Becker, T. et al. 2003. Phytoestrogens: Endocrine disrupters or replacement for hormone replacement therapy? *Maturitas* 44: S9–S20.

Yang, C.S.Y., Lu, G., Xiao, H. et al. 2008. Cancer prevention by tea and tea polyphenols. *Asia Pacific J. Clin. Nutr.* 17: 245–248.

Yavelow, J., Finlay, T.H., Kennedy A.R. et al. 1983. Bowman–Birk soybean protease inhibitor as an anticarcinogen. *Cancer Res.* 43: 2454–2459.

Zheng, G. and Zhu, S. 1999. Antioxidant effects of soybean isoflavones. In *Antioxidants in Human Health* (T.K. Basu, N.J. Temple, and M.L. Garg, Eds.), Oxon, UK: CAB, pp. 123–130.

Zheng, W., Dai, Q., and Custer, L.J. 1999. Urinary excretion of isoflavonoids and the risk of breast cancer. *Cancer Epidemiol. Biomark. Prevent.* 8: 35–40.

Zheng, Y.G., Wu, J., Chen, Z. et al. 2008. Chemical regulation of epigenetic modifications: Opportunities for new cancer therapy. *Med. Res. Rev.* 28: 645–687.

Zhou, Y. and Lee, A.S. 1998. Mechanism for the suppression of the mammalian stress response by genistein, an anticancer phytoestrogen from soy. *J. Nat. Cancer Inst.* 90: 381–388.

14

Computer-Aided Optimization of Peptide Sequences and Integrated Delivery of Selected Peptides to Targets: A Case Study on Blood Pressure of Oldest-Old Patients

Linping Wu, Wasaporn Chanput, Rotimi E. Aluko, Jianping Wu, Yasumi Horimoto, and Shuryo Nakai

CONTENTS

14.1 Introduction

The quantitative structure–activity relationship (QSAR) is the most fundamental approach in identifying a chemical compound, in order to elucidate
its functionality based on the chemical structure. In the case of protein, the
four structure categories, primary, secondary, tertiary, and quarterly, are
involved. Above all, the sequence data, in other words "primary structure,"
is the basis of all other structure data. Recently, the importance of the function of oligopeptide, especially tripeptide, has been emphasized in explaining the mechanism of functionality of the entire molecule of protein (Nakai
et al., 2005a). Similarly, the critical role of peptide sequence to construct the
best function in peptide/protein has been recognized as the final as well as
eternal objective of protein chemistry.

In the meantime, the mathematical approach for finding the best function
that could be the minimum or maximum function depending on the objective of interest is called "optimization study" in order to accomplish single
to multifactor investigations. We have developed "random-centroid optimization" (RCO) for general processing purposes (Nakai et al., 1998a); as well,
RCG (for genetic projects) was then derived from the RCO for optimizing
peptide functions (Nakai et al., 1998b). To achieve the objectives, the appropriate physical as well as chemical properties of peptides should be identified to
evaluate response values in the RCG computation; those peptide properties
are, thus, included as the subprogram by saving them in the RCG software.

Computer-assisted technologies for experimental design and response-
surface methodology were compared in a handbook (Nakai et al., 2007).
Our optimization techniques of RCO and RCG were recently included in the
book entitled *Optimization in Food Engineering* (Nakai et al., 2008). In the case
of angiotensin I-converting enzyme (ACE)-inhibitory peptides, it was found
that tripeptides with proper distribution of hydrophobicity-patches in the
peptide sequences played a critical role (Wu, Aluko, and Nakai, 2006).

It is well known that ACE inhibitory activity is indispensable in controlling blood pressure (BP; Williams et al., 2004). Figures 14.1–14.3 show the
results of our study on BP changes under different medications. The combination of metoprolol ($C_{15}H_{25}NO_3$, β-blocker) and amilodipine ($C_{20}H_{25}CIN_2O_5\cdot$
$C_6H_6O_3S$, calcium channel blocker) decreased the uncontrolled systolic BP
of 190–200 mmHg to 120–130 mmHg after 8 weeks. Discontinuing Norvasc
alone increased BP to 160 mmHg in the 12th week. Oat digest and Calpis (a
lactic fermented dairy drink sold in Japan) were not as effective as that of
Norvasc as shown in Figure 14.1 during the 10th–12th weeks.

In the functional classification of bioactive peptide/protein, such as enzymes
and hormones, the importance of sequence similarity among them has been
recognized, thereby resulting in creation of homology similarity analysis
(HSA) software (Nakai et al., 2003). Thereafter, its modified version of homology similarity search (HSS) was developed on the basis of HSA for searching

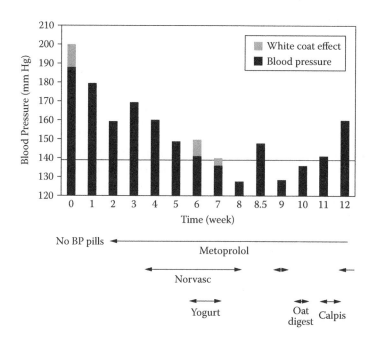

FIGURE 14.1
Blood pressure changes under different medications.

for the similarity in peptide/protein in anticipation of a similar function that should belong to the same category of enzymes, hormones, or functional proteins (Nakai, 2004; Nakai et al., 2004, 2005a). A computer-aided process of the combined RCG-HSS is one of our new strategies as proposed in the peptide QSAR study to be able to accomplish the best functions for the molecules.

Cardiovascular diseases include diseases of the heart and blood vessels, as well as stroke (American Heart Association, Learn and Live, 2008). High blood pressure (hypertension) is their most common sign. For cardiovascular diseases, different aspects of some peptides, that is, antihypertensive and antioxidative activities both play key roles (Korhonen and Pihlanto, 2006). The latter is deeply involved in aging. Thus, in order to discuss high blood pressure of oldest-old patients, it is essential to take into consideration these two aspects of peptides.

The purpose of this chapter is to describe the results of our study using RCO/RCG along with HSA/HSS or separately. Recently, the interest of the majority of researchers is focused on health and medical applications of food proteins and we are not an exception (Nakai et al., 2005b). Among the many different medical applications areas, we have chosen BP of hypertensive patients, especially in elderly people. In many studies utilizing food proteins as biomarkers, attention is required in interpreting results due to the extremely complicated BP mechanism within the human body. Recent arguments in that area are also reiterated in this chapter.

14.2 Antihypertensive Activity of Food Protein Digests

There is a quantitative *in silico* analysis, mostly in the case of meta-analysis, to combine the results of critical trials in terms of bioactivity, for example, ACE inhibitory activity that can exert an antihypertensive effect (Vermeissen et al., 2004; Korhonen and Pihlanto, 2006). This activity against the peptide size was compared along with experimental results. The ACE inhibitory peptides which were released during *in vitro* gastrointestinal digestion were estimated in *in silico* digestion using the database server BIOPEP as reported by Dziuba and Iwaniak (2006). Despite the fact that we have applied the same BIOPEP, it was frequently conducted for the purpose of selecting ACE inhibitory food to search for appropriate peptides. Interestingly, the crystal structure of ACE-lisinopril complex was reported by Natesh et al. (2003), in which lisinopril had a 3D structure similar to bioactive tripeptides reported in the literature. Importance of annotation of oligopeptides for interpreting the protein function was also reported by Burke and Deane (2001) and Polacco and Babbitt (2006).

In vitro inhibition of ACE by 141 tripeptides (Table 14.1), as we have previously reported (Wu et al., 2006), was employed in principal component similarity analysis for classifying their sequences (SPCS; Vodovotz, Arteaga, and Nakai, 1993). This approach replaced conventional principal component analysis (PCA) because of the multifunctional nature of hypertensive BP as well as complications in preventing the possible recurrence of stroke (O'Rourke et al., 2004). Inasmuch as the values of log IC_{50} were extensively varied from -0.68 to 4.0 (or IC_{50} of 0.2 μM to 10 mM; Wu et al., 2006), log IC_{50} < 0.52 (Table 14.2) was arbitrarily chosen to replace <4.0 as the upper bound (Table 14.1). This modification provided greater emphasis on the antihypertensive activity of oligopeptides, which were included in the food protein digests. Meanwhile, Wu et al. (2006) evaluated the activity of dipeptides in addition to that of tripeptides. However, the smallest log IC_{50} value of dipeptides was 0.15 ($IC_{50} = 1.4$ μM) and only few dipeptides were included in the total of 168 peptides having log IC_{50} < 0.52 (3.3 μM). Furthermore, many of the dipeptides consisted of part of the sequences of 141 tripeptides, thus the QSAR study of tripeptides could simultaneously contain dipeptide QSAR.

The HSS using pattern similarity constants of peptide segments versus reference peptide based on side chain properties was introduced in search of active as well as binding sites in the sequences of proteins/peptides (Nakai et al., 2005a). The same approach of the HSS was successfully applied again to elucidate the mechanism of peptide emulsions (Nakai et al., 2004). In addition to the HSS, the BIOPEP program (Dziuba and Iwaniak, 2006) was also simultaneously employed in this study as stated above (Figure 14.2), in conjunction with the prediction of resistant peptides derived from food proteins after the systemic digestion.

TABLE 14.1

Inhibitory Tripeptides of Angiotensin Converting Enzyme[a]

Number	Peptide	Log(IC$_{50}$)	Number	Peptide	Log(IC$_{50}$)	Number	Peptide	Log(IC$_{50}$)	Number	Peptide	Log(IC$_{50}$)
1	FEP	1.08		MNP	1.82		GPM	1.23		YEY	0.6
	IKP	**0.23**		NPP	2.46		**GKP**	**2.55**		PPY	1.2
	LNY	1.91	40	PPK	3		**IPA**	**2.15**	110 (B)	LGI	1.46
	HQG	2.87		ITT	2.83		**VYP**	**2.46**		ITF	1.69
	HHT	2.9		TTN	2.83		VMY	0.97		**IPP**	**1.4**
	ALP	2.38		TNP	2.32		FYN	1.26		**IAP**	**2.61**
	LKP	0.2		**GQP**	**0.51**		YGG	4		**QPP**	**1.52**
	LYP	0.82		GPH	1.51		GGY	0.11		FAP	0.62
	DYG	3.43		RML	3.01		YPR	1.22		**PVY**	**3.23**
10	AQK	3.26		YVA	0.15	80	PRY	0.4		LEK	2.9
	IEP	**0.2**	50	GKV	0.59		YGL	2.61	120 (C)	GVY	2.6
	IKY	**−0.68**		PVY	0.91		VFK	3.01		IRP	0.26
	LAP	**0.54**		FFL	1.57		**IKW**	**−0.68**		**LPP**	**0.98**
	LKP	**−0.49**		IFL	1.65		**IKP**	**0.84**		LVL	1.09
	GRP	1.3		LPF	1.6		**IWH**	**0.54**		LRP	−0.57
	RFH	2.52		GPP	1.25		VAP	0.3		LPP	0.23
	AKK	0.5		AGP	2.75		FAP	0.58		LEP	0.28
	RVY	2.31		VIY	0.88		**PLW**	**1.56**		VPP	1
	LKL	2.27		**RIY**	**1.45**		FGK	2.2		LEP	1.63
20	HIR	2.98		AFL	1.8	90	AVP	2.53		LNP	1.76
	HHL	1.73		FAL	1.42		PYP	2.34		LLP	1.2
	HLL	1.76		IAQ	1.54		LVR	1.15		VLP	0.59
	HHL	0.73		VVF	1.55		TAP	0.54		LAY	1.4

(continued)

TABLE 14.1 (continued)

Inhibitory Tripeptides of Angiotensin Converting Enzyme[a]

Number	Peptide	Log(IC$_{50}$)	Number	Peptide	Log(IC$_{50}$)	Number	Peptide	Log(IC$_{50}$)	Number	Peptide	Log(IC$_{50}$)
	LIY	-0.09		IVQ	1.98		VRP	0.34	130 (D)	IRA	1.11
60	LAY	0.59		VQV	0.94		MPP	0.98		LAA	1.11
	LLP	1.76		AQL	1.76		LKP	0.6		LEE	2
	LEE	2		LVQ	1.15		TVY	1.18		MKY	0.86
	FNE	2.53	100 (A)	FDK	2.59		**IVY**	**-0.32**		LRY	0.7
	GPL	0.35		IVY	0.38		IMY	0.26		VPW	1.37
30	GPV	0.67		VIP	1.91		DGL	0.33		LWA	1.1
	IPP	0.7		VLY	1.49		TKY	0.36		VTR	2.13
	GPL	0.41		ILP	1.51		LTF	0.44		IKW	-0.27
	DLP	0.68		**VPP**	**0.95**		AGP	1.95	140 (E)	VGP	1.42
	GLY	0.95	70	RPP	1.78		FNF	0.84		**MRW**	**-0.22**
	LLF	1.9		RPK	3.27		AVL	0.85		IAY	1.1
	LAP	**0.43**									

Source: Wu, J., Aluko, R.E., and Nakai, S. *J. Agric. Food Chem.* 54, 732–738, 2006.

[a] Peptides in bold have been tested *in vivo* for antihypertensive activity.

Repeated peptides indicate results from different sources.

IC$_{50}$ is the concentration of ACE inhibitor required to inhibit the ACE activity.

TABLE 14.2

Tripeptides with Log IC_{50} Lower than 0.52

Group I				Group II			
Number	Sequence	Log IC_{50}	Alias	Number	Sequence	Log IC_{50}	Alias
2	IKP	**0.23**		17	AKK	0.50	
11	IEP	**0.20**		29	GPL	0.35	32 (0.41)
12	IKY	−0.68		43	GQP	0.51	
14	LKP	−0.49	7 (0.20)	46	YVA	0.15	
24	LIY	−0.09		79	GGY	0.11	
36	IAP	**0.43**		81	PRY	0.11	
65	IVY	0.38	99 (−0.32)	87	VAP	0.30	
84	IKW	−0.68	D8 −0.27)	95	VRP	0.34	
A0	IMY	0.26		A1	DGL	0.33	
A3	LTF	0.44		A2	TKY	0.36	
B8	IRP	0.26		E0	**MRW**	**−0.22**	
C1	LRP	−0.57					
C2	LSP	0.23					
C3	LEP	0.28					

Source: Wu, J., Aluko, R.E., and Nakai, S. *J. Agric. Food Chem.* 54, 732–738, 2006.
Bold phase of sequences show *in vivo* assayed.
Tripeptide numbers are designed to be two digits as shown in Table 14.1.

14.3 Biopeptides Effective in BP Reduction

Because of complication in the mechanism, BP should not be focused on alone (Williams et al., 2004); rather BP is one of the most important risk factors for cardiovascular diseases including stroke. In other words, BP reduction plays an important role in maintaining a healthy lifespan.

When antioxidative activities of protein in salmon, milk, soy, cotton, canola, barley, buckwheat, rice, rye, sunflower, oat, wheat, and eggs were compared according to the linoleic acid peroxide assay (Mitsuda, Yasumoto, and Iwami, 1966), barley hordein was found superior to other food proteins. Table 14.3 is the result of ferric reducing activity (Benzie and Strain, 1996) that measures a similar but different aspect of antioxidative activity. By applying the HSS program to barley holdein molecules, 49, 18, and 27 antioxidative tripeptides were found in C, B, and D hordein. However, their distribution was 10–20, 70–80, and 2–4%, respectively, in the total barley protein. Therefore, it is economically reasonable to use the hordein fraction as a whole by ignoring the effect of minor D hordein. The above antioxidative activity did not change after acid-solubilization, thus no modification in the procedure was required during WPN (Trp-Pro-Asn) → WPD (Trp-Pro-Asp) variation along with the regular HSS computation.

We conducted the RCG (Nakai et al., 1998b) and found that tripeptides WPN showed the most potent antioxidative activity (Table 14.4). Thus,

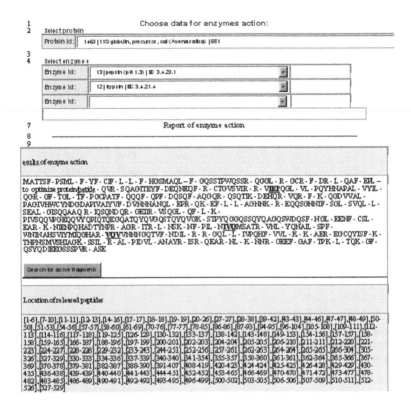

FIGURE 14.2
BIOPEP printout of oat globulin digested after a pepsin–trypsin combined action.

the WPN was used thereafter as a reference biomarker of the HSS in our laboratory because there could be no similarity of >1.0 in one HSS run. Unfortunately, this search was not completed after the first cycle, because there is less chance of finding better antihypertensive activity than medication using metoprolol and amilodipine. The medication was more effective than polypeptides (Figure 14.3). The amilodipine–metoprolol combination showed 100–115 mmHg lower than the 5 g/day dose of AmylS, sardine peptide, yogurt, and oat globuline as reported by Li et al. (2004). Furthermore, unreasonable expenses for acquiring the quantity of synthetic tripeptides required for human application prohibited us from further experimentation.

Oolong tea was effective in improving BP due to its chemical nature of polyphenol, mainly because of its antiobesity activities (Han et al., 1999). These effects of oolong tea could be added to the BP reducing activity of peptide nanoparticles, when used together by taking advantage of a different mechanism in the BP control.

TABLE 14.3

Reducing Activity of Crude Hordein Fractions, Rice Bran Protein Fractions, Partially Purified Hordein Fractions, Their Hydrolysates, and Commercial Antioxidants

Extracts	Reducing Activity (Micromolar of Fe^{2+})[a,b]
Barley hordein	
Hordein fraction	
Protein	$1,332.8 \pm 197.9$[m]
Hydrolysate	$1,620.8 \pm 16.9$[l]
Partailly purified B hordein	
Protein	880.8 ± 5.6[n]
Hydrolysate	$1,342.8 \pm 8.5$[m]
Partailly purified C hordein	
Protein	$2,172.8 \pm 16.9$[ij]
Hydrolysate	$4,129.6 \pm 0.0$[f]
Partailly purified D hordein	
Protein	912.8 ± 5.6[n]
Hydrolysate	$1,228.8 \pm 5.6$[m]
Rice bran protein	
Albumin	
Protein	$6,964.0 \pm 28.3$[d]
Hydrolysate	$8,744.0 \pm 2 8.3$[c]
Globulin	
Protein	$2,904.0 \pm 28.3$[g]
Hydrolysate	$5,128.0 \pm 0.0$[e]
Prolamin	
Protein	$2,016.8 \pm 5.6$[jk]
Hydrolysate	$2,444.0 \pm 28.3$[h]
Glutelin	
Protein	$1,808.8 \pm 11.3$[kl]
Hydrolysate	$2,384.0 \pm 0.0$[hi]
BHT	$79,440.0 \pm 282.8$[a]
Vitamin E	$74,840.0 \pm 282.8$[b]

[a] Means $\pm$ SD in the same column with different letters are significantly different ($p \leq 0.05$, $n = 3$).

[b] Relative activity refers to the amount of Fe^{2+} that was produced from the reduction of Fe^{3+} complex by each protein fraction and its hydrolysate.

14.3.1 Cell-Penetrating Peptides

Hypertensive compounds should be able to penetrate cell membranes to exert bioactive effects in the cell. Using amphoteric model peptides, effects of peptide helicity and membrane surface charge on the electrostatic and

TABLE 14.4

Random-Centroid Search of Optimal Tripeptide with
High Antioxidative Activity

	Variables		Antioxidative Activity (TEAC ± SD)
Cycle 1 Random search			
(1)	Ile-His-Phe	IHF	2.72 ± 0.21
(2)	Leu-His-Gln	LHN	6.5 ± 0.34
(3)	Gly-Leu-Tyr	GLY	1.37 ± 0.15
(4)	Trp-Glu-Tyr	WEY	4.28 ± 0.58
(5)	Cys-Ile-Trp	CIW	10.74 ± 1.03
(6)	Leu-Ser-His	LSH	3.62 ± 0.41
(7)	Trp-Phe-Arg	WFR	18.09 ± 1.75
(8)	Pro-Ser-Ala	PSA	4.01 ± 0.15
(9)	His-Phe-Phe	HFF	3.68 ± 0.34
Cycle 1 Centroid search			
(10)	Val-Val-Val	VVV	0.12 ± 0.013
(11)	Trp-Pro-Asn	WPN	27.39 ± 1.91
(12)	Val-VAl-Pro	VVP	3.74 ± 0.23
(13)	Tyr-Val-Pro	YVP	9.57 ± 1.30

hydrophobic interactions with lipid bilayers and biological membranes were compared, resulting in stability of the membrane and penetration of the peptides (Dathe et al., 1996). It was suggested that hydrophobic interactions had induced penetration of the amphipathic peptide structure into the inner membrane region, thus disturbing the arrangement of the lipid acyl chains, thereby causing local disruption.

14.4 Blood Pressure of Elderly Patients

Mattila et al. (1988) reported that high BP did not always increase the risk of death in elderly men aged 85 or more. Mortality was least in subjects with systolic pressure of 160 mmHg or more (e.g., even >200 mmHg) and diastolic pressure of 90 mmHg or more (e.g., >110 mmHg). This paradoxical survival of elderly patients was also discussed by Langer, Ganiats, and Barrett-Connor (1989). Satish et al. (2001) reported that men 85 and older with systolic BP ≥180 mmHg have significantly lower mortality than those with <130 mmHg. Bulpitt et al. (2003) stated that a clinical trial was currently under way to help address this issue of hypertension in very elderly patients; however, they reported that the final results might be years away. Recommendations were recently made that the reductions in intensity of

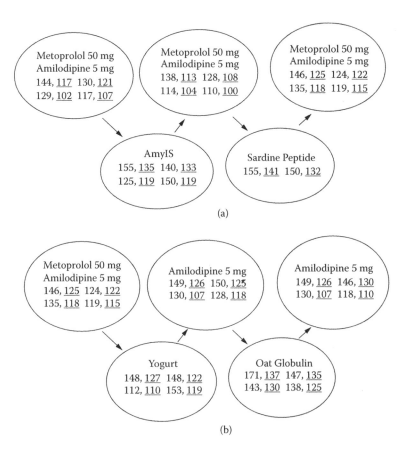

FIGURE 14.3

Comparison of BP upon medication cycling with different polypeptides. Upper and lower BP sets in A and B are in early and after three days monitoring; first value and underlined second value are the first reading and the leveled-off value. First and second two digits were a.m./p.m. readings.

therapy for patients whose BP was below 140/90 should be less aggressive (Pates et al., 2007). It was warned that extrapolation of data of hypertension studies involving younger patients in the very old group might be erroneous and even dangerous. Suggestion was thus made that higher blood pressure might even play a protective role, perhaps through maintenance or improvement of vital organ perfusion.

On the contrary, however, a large-scale investigation (Beckett et al., 2008) with the antihypertensive treatment group (1,933 patients) versus the placebo group (1,912 patients) concluded more recently that in persons 80 years or older, the antihypertensive treatment using medication was beneficial with 15.0/6.1 mmHg lower than in the placebo group. A 30% reduction in the rate of fatal or nonfatal stroke and 39% reduction in the rate of death from stroke were observed.

14.5 Our Experiences of Home Blood Pressure Monitoring of the Elderly Since 2007

Despite the statistical limitations as mentioned earlier, individual BP data of a small group of elderly patients has been stated here in order to derive a more reliable summary under the same circumstances in the future. Because high-normal patients (>140 mmHg) occupy more than two-thirds of the entire population (Oates et al., 2007), information on how to control their BP should be beneficial to the general public although it is difficult to be expressed in a statistically acceptable form to apply to large populations.

For assisting easier interpretation, systolic BP values monitored were classified as A: <130 mmHg, B: 131–140 mmHg, C: 141–150 mmHg, D: 151–160 mmHg, E: 161–170 mmHg, F: <180 mmHg and G: ≥180 mmHg. Life Source UA-767 Plus Digital Blood Pressure Monitor manufactured by A & D Medical, Milpitas, CA was used in this study. The BP monitored were in classes A or B in 2007, and increased to classes B or C by the end of July 2008, when class D started sporadically appearing even by relying on medication using metoprolol and Novasc. Thereafter, the intake of oolong tea and barley hordeins appeared to retain the BP better than class level C without using medications for limited duration. Concurrently, discomfort due to side effects by medications could be relieved (for example, backache, leg-ache, exhaustion) without such suffering. Generally, peripheral edema (feet and ankles) was reported as a frequently observed side effect (in one out of 10 users) of amilodipine in addition to many other possible side effects.

We have partially purified three major hordein fractions of barley (i.e., B, C, and D hordeins) by gel filtration. Albumin, globulin, prolamin, and glutelin fractions of rice bran were also fractioned by the Osborne extraction method. Hydrolysates of these protein fractions were prepared by digesting with pepsin followed by trypsin. Antioxidant properties in terms of antioxidative activity against linoleic acid peroxidation as well as reducing activity without the lipid adjuvant were investigated (Table 14.3). The globulin fraction from rice bran protein revealed the most powerful antioxidative activity throughout the incubation time of seven days ($p \leq 0.05$). Meanwhile, the albumin fraction of rice bran protein showed the highest reducing activity (6964 micromoles of Fe^{2+}) followed by globulin, prolamin, glutelin, and hordein fractions with the activities of 2904, 2017, 1809, and 1333 micromoles of Fe^{2+}, respectively ($p \leq 0.05$). Partially purified C hordein showed the greatest reducing activity compared with B and D hordeins. Protein hydrolysates obtained after digestion with pepsin and trypsin exhibited much greater antioxidative as well as reducing activities than those before digestion.

Many articles regarding rice bran that decreases blood pressure have been reported in the literature (Jensen et al., 2004; Ardiansyah et al., 2006). Hot cereal prepared using 70 g of rice bran containing 10 g protein, 14 g fat, and 19 g dietary fiber reduced BP to about the same extent as that of barley hordeins without medication.

14.6 Discussion

High BP is one of the most readily preventable causes of stroke and other cardiovascular complications (Cappuccio et al., 2004). Probably because of its medical significance from every aspect, the health/medical significance has been estimated on the basis of different studies in the literature. Therefore, the QSAR study was conducted in our laboratory depending on the intake of food proteins to explain the mechanism as the main issue as discussed in this chapter. Our combined use of RCO-HSS was the first attempt in order to accomplish our objectives. Apparently, after the most important tripeptide or tripeptides were selected in Table 14.1, which is digestion-resistant according to BIOPEP (Figure 14.2), the HSS was applied to sequences of unknown peptide/protein to evaluate the functionality of interest. The assumption made here was that the potential functional activity would be proportional to the sequence similarity. If this assumption was proved acceptable, this approach may have a great use in elucidating the functional mechanism of biopeptides in the future. No doubt the capacity of tripeptides in the function of the entire molecule must be most critical.

Meta-analysis of a large collection of analytical results obtained from individual studies has been integrated into the major findings (Pripp, 2008). However, use of meta-analysis was criticized due to the fact that a complex intervention with a complicated outcome should not be combined (Cappuccio et al., 2004). Originally, the meta-analysis of randomized testing trials intended to draw the same conclusion as those of randomized control trials conducted on a large scale of patients. Thompson and Higgins (2002) discussed limitations and pitfalls of the metaregression analysis. Unfortunately, any potentially consistent effect might have been underestimated. One reason may be because too many risk factors of already relatively known causes are included in the computation, such as aging. As a result, the normal distribution of data that is the essential requirement of modern statistics is unwittingly interrupted by logical expectation. Apparently, the two measurement effects between hypertensive activity expressed by using ACE and antioxidative activity are not in good concordance. Bias was already contained in meta-analysis in comparison with large randomized controlled trials according to Egger et al. (1997).

The general consensus is that without experiencing time-consuming interruption during observation of the medical outcomes of high BP, there is enhanced interest in knowing how individual diet, medication, and lifestyle can easily control hypertension in the subject and his or her family. Although many studies utilize large populations for monitoring BP, grouping based on old age should be useful because one important factor at least can be eliminated from the entire statistical larger scale survey when the valuable statistical summary is able to draw from a small sample group. Caution is, however, required to distinguish the effect of BP on cardiovascular diseases and aging, simultaneously or separately.

According to the survey of more than 4,000 patients at Veteran's Health Administration Clinics, among adults of age 80 and older, "High Normal" systolic BP of 130–139 mmHg or even higher was safer for survival than those with lower pressures, probably because of loss in artery flexibility (Braith and Stewart, 2006; Oates et al., 2007). Reconsideration may be required because the antioxidant effects are more closely related with aging rather than with ACE inhibitory activity. This fact may justify recomputation of the BP controlling effects as discussed previously. A great portion of the population has reported dissatisfaction through the Internet with the side effects caused by medication. Immense variability in physiological reactions of the human body is apparent.

14.7 Conclusion

Computer-aided optimization of peptide sequences has selected rice bran and barley hordeins to be the most active antihypertensive proteins. However, there was difficulty in the oldest-old patients, which was discordant with other age cohorts. This discrepancy of the oldest-old would impose a new question: what is the lower bound risk BP value for this age group, 200 mmHg or lower? Most recently, however, these reported paradoxical aging effects were denied according to the large-scale survey result (Beckett et al., 2008). The antihypertensive treatments were beneficial including death rates of patients 80 years of age or older. Barley hordeins and rice bran both appeared to be effective in BP reduction but hardly replaced medication. At any rate, larger numbers of the oldest-old cohort should be used separately to obtain more reliable statistically supported conclusions in the future. Employing HSS after optimization of tripeptides using RCG may be a useful efficient approach to achieve our objectives in the peptide QSAR study.

14.8 Supplements (Prospective Uncertainty and Experimental Approach)

14.8.1 Home Blood Pressure Monitoring

During mild exercise for 15 min (on a Schwinn 203 Recumbent Exercise Bike at pulse 12–13 with simultaneous weight lifting of 5-lb dumbbell 40 times every 3 min) in addition to the belly-muscle training (while laying on the floor, bend legs and place heels on the floor. Lift arms up and push them toward ankles, while tightening abdominal muscles. Ten times per set, two sets), several readings were taken at 1–2 min intervals (Williams et al., 2004). When the readings decreased (>10 mmHg) each time (Williams et al., 2004), they continued to be taken until they leveled off (four times <10 mmHg differences), and then the average of the leveled-off values were expressed as the patient's reading (Blood Pressure Association, 2004). Blood pressure of elderly patients (male age 80, body mass index 23.3, and female age 76, body mass index 23.6) was monitored in the afternoon before 4 p.m. of the same day (Bobrie et al., 2004).

We have found that exercises requiring brain action are more effective in BP reduction, such as heavy-load carrying, gardening, snowplowing, lawn mowing, raking fallen leaves, or driving long distances, than monotonic exercises. The importance of brain function in preventing hyper-BP has already been recognized (Waki et al., 2007). However, those exercises alone could not effectively replace medication for reducing BP <150 mmHg (level C). A recent encouraging finding is that introduction of the above intensified exercise in addition to medication has reduced the BP level to A or B almost constantly (120–140 mmHg). Considering the BP <140 mmHg (level B), which may be the appropriate value to the high-normal patients, this approach could be valuable for preventing fatal diseases.

The meaning of this "enforced exercise" could be of extreme importance because this may immediately link to lifespan extension. What other exercises can replace driving long distances, raking leaves, or snowplowing? Climbing up 60 steps or three stories is added to the above machine pedaling exercise. Advantages of this enforced exercise over walking or running depend on weather; it is difficult to perform in a rainy season or area.

14.8.2 Computer-Aided Programs Developed for Peptide Sequence Analysis

The random-centroid optimization (RCO) computer software was first developed in our laboratory to find food processing conditions to obtain the best, either maximum or minimum, functionality of products. This RCO program was then modified to apply to bioactive compounds such as peptides (RCG; i.e., RCO for genetics). The property of amino acid residues in the peptides to be used in optimization computation as an independent variable was selected from a list of chemical/physical properties of the residues (Nakai et al., 1998b).

In addition to the optimization, a pattern similarity search was carried out for the peptide sequence analysis: homology similarity analysis (HSA). Using the best tripeptide thus selected as a reference, the classification of tripeptides was attempted first in HSA, then followed by HSS.

Hordeins were separated from barley flour purchased from a local market by washing with 0.05 M NaCl to remove albumin and globulin first, then were extracted using 0.03 N KOH. In the case of rice bran, "Pure Rice Bran" manufactured by Ener-G Foods, Inc. (Seattle, WA) was used in this study.

14.8.3 Downloading of Computer Programs

The address ftp://ftp.landfood.ubc.ca/foodsci/ contains RCO as well as RCG and both SPCS (Vodovotz et al., 1993) and HSS (included in the HSA program). To download a program in PC computers, it is recommended to use the executable version, such as SPCS.exe.

14.8.4 Antiaging Effects of Antioxidative Oligopeptide Segments

Upon completion of the work on antihypertensive activity of peptides consisting of 3–7 amino acid residues that affect BP, tripeptides with low IC_{50} values selected from 151 ACE inhibitory tripeptides (Table 14.1) were used for reducing BP. Homology similarity search and the average of residue indices (for instance, hydrophobicity) were used for characterization of the tripeptides such as GPL, GPH, GPP, and GPM (Wu et al., 2006). Oat globulin among other food proteins could be one of the most effective BP-reducing enhancers by preparing peptide concentrates using a membrane process (Nakai et al., 2008).

To investigate "antiaging" activity, the most appropriate biomarkers, for example, lipid oxidation, scavenging free radicals, and the like were selected (Dröge, 2002). Superiority of C hordein in barley was revealed by demonstrating the highest reducing activity compared with B and D hordeins (Chanput, Theerakulkait, and Nakai, 2008). However, interpretation of the results obtained based on biomarker effects, thereby reflecting human responses such as declining memory and physical strength, or enhancement of specific activity of enzymes and hormones would invite additional complexity in the search. Antioxidative activity of peptides is a critical factor of aging (Dröge, 2002). Effects of antioxidative tripeptides on aging were excellently reviewed by Saito et al. (2003). In their study, an attempt was made to apply the approach used for antihypertensive activity (Nakai et. al., 2008) to antiaging based on restrained variability of the antiaging tripeptides; however, the mechanism of antiaging is still not accurately defined to utilize in the investigation of the background mechanism of animal aging.

As well, the RCG optimization of synthetic tripeptides has been initiated in our laboratory to find the most potent tripeptides in terms of scavenging free radicals. Meanwhile, the ability of HSS that has been validated recently to predict antioxidative activity of tripeptides could be a new approach in elucidating the mechanism of antioxidative activity to exert antiaging activity of peptides. Availability of C hordein extracted from acid-solubilized barley flour resulted in eliminating needless organic solvent for extracting this gliadin-like protein. Therefore, it is possible to develop a low-cost extraction of antiaging peptides in comparison to rather expensive use of dipeptide, that is, carnosin (Hipkiss, 2005).

14.8.4.1 *Acid-Solubilization of Barley Flour*

RCO software was applied to validate the concentrations of flour (5–20%) and HCl (0.1–0.3 N), autoclaving duration (1–30 min), and pH of maximum precipitation (Fung et al., 1977). Effective isoelectric precipitation of intact proteins cannot be expected under the acid hydrolysis conditions used, therefore full recovery of intact water-soluble albumin and salt-soluble globulin in the obtained precipitates cannot be expected. The response values used in this RCO were the overall antioxidative activity no matter what protein fractions were recovered.

14.8.4.2 *Homology Similarity Search (HSS) of Tripeptides*

RCO software was applied to find the highest antioxidizability of tripeptides. Within 13 vertices in cycle 1 (Table 14.4), the best regression line was obtained using exponential fitting after eliminating vertices with negative similarity constants, thereby demonstrating the highest antioxidative value of WPN (vertex 11). Using WPN as a reference, HSS was carried out for barley C hordein after digestion with pepsin and trypsin using BIOPEP (Figure 14.2; Iwaniak, Dziba, and Niklewicz, 2005). Potential tripeptides in C hordein detected were 24QPQ, 17QPF, 5XPQ, and 3XQQ along with some other single tripeptides, 317 residues in total. Acid solubilization causing Q→E did not significantly alter the antioxidative activity of the above tripeptides similar to N→D change as already stated earlier.

Although B hordein content is greater than C hordein (70–80% compared to 10–20% of total protein in the flour; Slafer et al., 2002), potential tripeptides in B hordein were 6QPQ, 5QPF, 4QQX, and 3XQQ, 285 residues in total. This explains why C hordein plays a more important role than those of other hordeins in antioxidative effect. D hordein contains 42 effective tripepetides, 101 residues in total; there were about 27 tripeptides after adjusting due to difference in the molecular sizes. Furthermore, because of the much smaller concentration of 2–4% among total barley hordeins (Slafer et al., 2002), the contribution of D hordein to aging may be minor.

14.9 Prospective Remarks

At present, the conditions for BP reduction acceptable to the oldest-old patients are medications such as metoprolol and amilodipine, in addition to indoor exercise such as staircase climbing and cycling on a stationary bicycle. Metoprolol and amilodipine appear to relieve potential side effects and have reduced BP from 180–200 mmHg to 120–150 mmHg. However, the simplicity of home blood pressure monitoring along with regular exercise is definitely advantageous as there is no need to take blood samples. More detailed information about preventing a heart attack or stroke, if applied in practice, may provide very valuable evidence of this.

14.10 Conclusion

Hypertension is generally considered the major cause of cardiovascular diseases. Blood pressure is one of the most important factors for controlling a variety of symptoms. Food protein can moderate high blood pressures. However, this hypotensive effect is generally less powerful than that of medication.

A number of studies have indicated that food-derived peptides have antioxidative activity to prevent chronic- and age-degenerative disease. Therefore, in addition to lowering blood pressure, antioxidative properties of food proteins could be helpful in slowing aging. To investigate the antioxidative activity, the novel peptide QSAR strategy was proposed to optimize protein or peptide functionality using a unique combination of computer programs: RCO/RCG and HSA/HSS. This strategy could be a new approach in elucidating the mechanism of antioxidative activity to exert antiaging activity of peptides.

References

Ardiansyah, Shirakawa, H., Koseki, T., Ohinata, K., Hashizume, K., and Komai, M. 2006. Rice bran fractions improve blood pressure, lipit profile, and glucose metabolism in storke-prone spontaneousely hypertensive rats. *J. Agric. Food Chem.* 54: 1914–1920.

American Heart Association, Learn and Live. Retrieved 2008 from http://www.heart.org/HEARTORG/

Beckett, N.S., Peters, R., Fletcher, A.E., Staessen, J.A., Liu, L., Dumitrascu, D., Stoyanovsky, F., Antikainen, R.L., Nikitin, Y., Anderson, C., Belhani, A., Foette, F., Rajkumar, C., Tihijis, L., Benya, W., and Bulpitt, C.J. 2008. Treatments of hypertension in patients 80 years or age or older. *New Engl. J. Med.* 358: 1887–1898.

Benzie, I.F.F. and Strain, J.J. 1996. The ferric reducing ability of plasma (FRAP) as a measure of "antioxidant power": The FRAP assay. *Anal. Biochem.* 239: 70–76.

Blood Pressure Association. 2004. *Guidelines for Measuring Blood Pressure at Home.* London: BPA.

Bobrie, G., Delonca, J., Mooulin, C., Giacomino, A., Postel-Vinay, N., and Asmar, R. 2005. A home blood pressure monitoring study comparing the anhypertensive efficacy of two antiotensin II receptor antagonist fixed combinations. *Amer. J. Hypertens.* 18: 1482–1488.

Braith, R.W. and Stewart, K.J. 2006. Resistance exercise training: Its role in the prevention of cardiovascular disease. *Circulation* 113: 2642–2650.

Bulpitt, C.J., Bechett, N.S., Cooke, J., Dumitrscu, D.L., Gil-Extremera, B., Naches, C., Nunes, M., Peterss, R., Staessen, J.A., and Thijis, L. 2003. Result of pilot study for the hypertension in the very elderly trial. *J. Hypertension* 21: 2409–2427.

Burke, D.F. and Deane, C.M. 2001. Improved protein loop prediction from sequence alone. *Protein Eng.* 14: 473–478.

Cappuccio, F.P., Kerry, S.M., Forbes, L., and Donald, A. 2004. Blood pressure control by home monitoring: Meta-analysis of randomized trials. *Biomed. J.* 329: 145–160. doi: 10.1136/bmj.38121.684410.AE.

Chanput, W., Theerakulkait, C., and Nakai, S. 2009. Antioxidative properties of partially purified barley hordein, rice bran protein fractions and their hydrolysates. *J. Cereal Sci.* 49: 422–428.

Dathe, M., Schümann, M., Wieprechy, T., Winkler, A., Beyermann, M., Krause, E., Matsuzaki, K., Murase, O., and Biernert, M. 1996. Peptide helicity and membrane surface charge modulate the balance of electrostatic and hydrophobic interactions with lipid bilayers and biological membranes. *Biochemistry* 35: 12612–12622.

Dröge, W. 2002. Free radicals in the physiological contro of cell function. *Physiol. Rev.* 82: 47–95.

Dziuba, D. and Iwaniak, A. 2006. Database of protein and bioactive peptide sequences. In *Neutraceutical Proteins and Peptides in Helath and Disease* (Y. Mine, and F. Shahidi, Eds.), Boca Raton, FL: CRC Press, Taylor & Francis, pp. 543–563.

Egger, M., Simith, G.D., Schneider, M., and Minder, C. 1997. Bias in meta-analysis detected by a simple, graphical test. *BMJ* 315: 629–634.

Fung, C.P., Fong, A.S., Lam, A.S. M., Lee, G.L., and Nakai, S. 1977. Preparation and properties of acid-solubilized wheat flour. *J. Food Sci.* 42: 1594–1599.

Han, L.-K., Takaku, T., Li, J., Kimura, Y., and Okuda, H. 1999. Anti-obesity action of oolong tea. *Int. J. Obesity* 23: 98–105.

Hipkiss, A.R. 2005. Glycation, ageing and carnosine: Are carnivorous diets beneficial? *Mech. Age. Dev.* 126: 1034–1039.

Iwaniak, A., Dziuba, J., and Nikleiwics, M. 2005. The BIOPEP database: A tool for the in *silico* method of classification of food proteins as the source of peptides with antihypertensive activity. *Acta Alimentalia* 34: 417–425.

Jensen, M.K., Koh-Banerjee, P., Hu, F.B., Franz, M., and Sampson, L., Grønbaek, M. and Rimm, E.B. 2004. Intake of whole grains, bran, and germ and the resk of coronary heart disease in man. *Am. J. Clin. Nutr.* 80: 1492–1499.

Korhonen, H. and Pihlanto, A. 2006. Bioactive peptides: Production and functionality. *Int. Dairy J.* 16: 945–960.

Langer, R.D., Ganiats, T.G., and Barrett-Connor, E. 1989. Paradoxial survival of elderly men with high blood pressure. *Brit. Med. J.* 298: 1356–1357.

Li, F.-H., Le, G.-W., Sji, Y.-H., and Shrestha, S. 2004. Angiotensin I-converting enzyme inhibitory peptides derived from food proteins and their physiological and pharmacological effects. *Nutr. Res.* 24: 469–486.

Mattila, K., Haavisto, M., Rajala, S., and Herikinheino, R. 1988. Blood pressure and five year survival in the very old. *BMJ* 296: 887–889.

Mitsuda, H., Yasumoto, K., and Iwami, K. 1966.Antioxidative action of indol compounds during the antioxidation of linoleic acid. *Eiyo to Shokuryo* 19: 210–214.

Nakai, S. 2004. Modelling protein behaviour. In *Protein in Food Processing* (R.Y. Yada, Ed.), Cambridge, UK: Woodhead, pp. 245–269.

Nakai, S. and Alizadeh-Pasdar, N. 2005b. Rational designing of bioactive peptides. In *Neutraceutical Proteins and Peptides in Health and Diseases* (Y. Mine and F. Shahidi, Eds.), Boca Raton, FL: Taylor & Francis, CRC Press, pp. 565–582.

Nakai, S., Alzadeh-Pasdar, N., Dou, J., Butimore, R., Rousseau, D., and Paulson, A. 2004. Pattern similarity analysis of amino acid sequences for peptide emulsification. *J. Agric. Food Chem.* 52: 927–934.

Nakai, S., Chan, L.C.K., Li-Chan, E.C., Dou, J., and Ogawa, M. 2003. Homology similarity analysis of sequences of lactoferrin and its derivatives. *J. Agric. Food Chem.* 51: 1215–1223.

Nakai, S., Dou, J., Lo, K.V., and Scaman, C. 1998a. Optimization of site-directed mutagenesis. 1. New random-centroid optimization program for Windows useful in research and development. *J. Agric. Food Chem.* 46: 1642–1654.

Nakai, S., Horimoto, Y., Dou, J., and Verdini, R.A. 2008. Random-centroid optimization. In *Optimization in Food Engineering* (F. Ergogdu, Ed.), Boca Raton, FL: CRC Press.

Nakai, S., Li-Chan, E.C.Y., and Dou, J. 2005a. Pattern similarity study of functional sites in protein sequencing: lysozymes and cystatins. *BMC Biochem.* 6: 9.

Nakai, S., Li-Chan, E.C.Y., and Dou, J. 2007. Experimental design and response-surface methodology. In *Handbook of Food and Bioprocess Modeling Techniques* (S.F. Sablami, M.S. Rahman, A.K. Data, and A.S. Mujumdar, Eds.), Boca Raton: FL: CRC Press, Taylor & Francis, pp. 293–322.

Nakai, S., Nakamura, S., and Scaman, C. 1998b. Optimization of site-specific mutagenesis. 2. Application of RCO to one-site mutation of *B. stearothermophilus* neutral protease to improve thermostability. *J. Agric. Food Chem.* 46: 1655–1661.

Natesh, R., Schwager, S.L.U., Sturrock, E.D., and Acharya, K.R. 2003. Crystal structure of the human ACE-lisinopril complex. *Nature* 421: 551–554.

Oates, D., Berlowitz, D.R., Glickman, M.E., Silman, R.A., and Borzecki, A.M. 2007. Blood pressure and survival in the oldest old. *J. Amer. Geriat. Soc.* 55: 383–388.

O'Rourke, F., Dean, N., Akhtar, N., and Shuaib, A. 2004. Current and future concepts in stroke prevention. *Can. Med. Assoc. J.* 170: 1123–1133.

Pates, D.J., Berlowitz, D.R., Glickman, M.E., Silliman, R.A., and Borzeccki, A.M. 2007. Blood pressure and survival amongst the oldest old. *J. Am. Geriat. Soc.* 55: 383–388.

Polacco, G.J. and Babbitt, P.C. 2006. Automated discovery of 3D motifs for protein function annotation. *Bioinformatics* 22: 723–730.

Pripp, A.H. 2008. Effect of peptides derived from food proteins on blood pressure: A meta-analysis of randomized controlled trials. *Food & Nutr. Res.* 52.

Saito, K., Jin, D.-H., Ogawa, T., Muramoto, K., Hatakeyama, E., Yasuhara, T., and Nokihara, K. 2003. Antioxidant properties of tripeptide libraries prepared by the combinatorial chemistry. *J. Agric. Food Chem.* 51: 3668–3674.

Satish, S., Freeman, D.H., Ray, L., and Goodwin, J.S. 2001.The relationship between blood pressure and mortality in the oldest old. *J. Am. Geriat. Soc.* 49: 367–374.

Seppo, L., Jauhiainen, T., Pousa, T., and Korpela, R. 2003. A fermented milk high in bioactive peptides has a BP-lowering effect in hypertensive subjects. *Am. J. Clin. Nutr.* 77: 326–330.

Slafer, G.A., Molina-Cano, J.L., Savin, R., Araus, J.L., and Romagosa, I. 2002. *Barley Science: Recent Advances from Molecular Biology to Agronomy of Yield and Quality.* Binghamton, NY: Haworth Press.

Thompson, S.G. and Higgins, J.P. 2002. How should meta-regression analysis be undertaken and interpreted? *Stat. Med.* 21: 1559–1573.

Vermeirssen, V., Van Bent, A., Van Camp. J., van Amerongen, A., and Verstraete, W.A. 2004. A quantitative in *silico* analysis calculates the angiotensin I-coverting enzyme (ACE) inhibitory activity in pea and whey protein digests. *Biochimie* 86: 231–239.

Vodovotz, Y., Arteaga, G.E., and Nakai, S. 1993. Principal component similarity analysis for classification and its application of GC data of mango. *Food Res. Int.* 26: 355–363.

Waki, H., Liu, B., Miyake, M., Karahira, K., Murphy, D., Kasparov, S., and Paton, J.F.R. 2007. Junctional adhesion molecule-1 is upregulated in spontaneous hypertensive rats. Evidence for a prohypertensive role within the brain stem. *Hypertension* 49: 1321–1327.

Williams, B., Poulter, N.R., Brown, M., Davis, McInnes, G., Potter, J.F., Sever, P., and McG Thom, S. 2004. *British Hypertension Society Guidelines for Hypertension Management* 2004 (BHS-IV) summary. BMJ 328: 634–640.

Wu, J., Aluko, R.E., and Nakai, S. 2006. Structural requirements of angiotensin I-converting enzyme inhibitory peptides: RCG study of di- and tripeptides. *J. Agric. Food Chem.* 54: 732–738.

Index

Printed and bound by CPI Group (UK) Ltd, Croydon, CR0 4YY

21/10/2024

01777107-0012